IET TRANSPORTATION SERIES 1

Clean Mobility and Intelligent Transport Systems

Clean Mobility and Intelligent Transport Systems

Edited by Michele Fiorini
and Jia-Chin Lin

The Institution of Engineering and Technology

Published by The Institution of Engineering and Technology, London, United Kingdom

The Institution of Engineering and Technology is registered as a Charity in England & Wales (no. 211014) and Scotland (no. SC038698).

First published 2015

The Institution of Engineering and Technology
Michael Faraday House
Six Hills Way, Stevenage
Herts, SG1 2AY, United Kingdom

www.theiet.org

British Library Cataloguing in Publication Data
A catalogue record for this product is available from the British Library

ISBN 978-1-84919-895-0 (hardback)
ISBN 978-1-84919-896-7 (PDF)

Typeset in India by MPS Limited

Contents

Preface

In the era of perpetual connectivity, digital data have become a catalyst for existing business models. Intelligent Transport Systems (ITSs) are the integration of information and communication technologies with transport infrastructure and are considered to be the framework for the digitalization of transportation. This is no longer considered to be an isolated phenomenon; rather, it goes beyond the boundaries of the sector, generating synergies with other sectors such as energy, tourism, safety, and environment. The need to satisfy the demand for 24/7 seamless connectivity has increased in both human and machine domains, thus resulting in various sophisticated business models, especially when users are "on the move" and expect to be able to communicate seamlessly with one another.

This book aims to serve as a timely reminder regarding the scale of the issues comprising the full spectrum of the transportation model—from land to sea and aircraft domains—including research from economics and social sciences without pretending to be exhaustive in all domains. This book is designed for system architects, system engineers, professors, and students who are interested in ITS, clear mobility, and smart systems, including software design, testing, performance analysis, and validation. This book is also attractive for administration professionals and businessmen involved in mobility and transportation systems who are willing to oversee various transportation modes.

This book consists of 18 chapters. Each chapter was reviewed by several independent reviewers. The Editors would like to express their gratitude to the distinguished authors and reviewers for their great contributions to the publication. The Editors also congratulate the authors for their excellent work. The 55 authors contributed from 16 countries (Australia, Canada, China, France, Germany, Italy, Kingdom of Saudi Arabia, Korea, Poland, Portugal, Senegal, Sweden, Switzerland, Taiwan, United Kingdom, and United States of America). The book is a globally cooperative achievement from different experts who share an engineering—in the Latin, *ingenium*—interest.

This book is organized as follows:

Chapter 1, "*Towards Contributive Development of Services*," was authored by Michel Leonard and Anastasiya Yurchyshyna. This chapter presents the service science in ITS.

Chapter 2, "*The Importance of ITS in Urban Movement*," was authored by Steve Kearns. This chapter focuses on the function of ITS in urban traffic management for modern towns and cities.

Chapter 3, "*ICT for Intelligent Public Transport Systems, State of Knowledge and Future Trends*," was authored by Marion Berbineau, Mohamed Kassab, Christophe Gransart, Martine Wahl, Juliette Marais, and Divitha Seetharamdoo. This chapter presents knowledge and future trends in ICT for intelligent public transport systems.

Chapter 4, "*ITS and Freight Transport: Stakes and Perspectives,*" was authored by Corinne Blanquart and Thomas Zéroual. This chapter provides innovative examples of ICT implemented in freight transport before examining its suitability for the needs of shippers and the abilities of transportation undertakings.

Chapter 5, "*Energy-Efficient and Real-Time Databases Management Techniques for Wireless Sensor Networks*," was authored by Ousmane Diallo, Joel J. P. C. Rodrigues, and Mbaye Sene. This chapter describes research and development activities on energy-efficient and real-time database management techniques for wireless sensor networks.

Chapter 6, "*Proactive Safety – Cooperative Collision Warning for Vehicles*," was authored by Chung-Ming Huang, Shih-Yang Lin, Alexey Vinel, and Magnus Jonsson. This chapter presents a proactive safety technique called cooperative collision warning to detect potential collisions between vehicles in a cooperative manner.

Chapter 7, "*Electronic Toll Collection Systems in Europe*," was authored by Mike Hayward. This chapter provides an overview of the toll systems used on motorways, tunnels, and bridges and the corresponding regulations in Europe.

Chapter 8, "*Business Model and Solutions for User-Centric ITSs*," was authored by Chi-Sheng Lin and Jia-Chin Lin. This chapter exhibits telematics solutions driven by integration and collaboration of information communication technology in the automotive industry.

Chapter 9, "*Digitalization for Sustainable Sea Transports*," was authored by Mikael Lind, Anders Brödje, Sandra Haraldson, Mikael Hägg, and Richard Watson. This chapter presents reconfigurations and changes toward digitalization in a sustainable sea transport and logistics chain.

Chapter 10, "*Team Management and Soft Skills: Case Studies and Cross-Connections on the Creation of an Effective Working Team*," was authored by Cristian De Mitri, Raffaele Loscialpo and Michele Fiorini. This chapter provides a novel soft skill platform that can facilitate team building and management by exploiting two cross-connected models.

Chapter 11, "*Integrated Visual Information for Maritime Surveillance*," was authored by Domenico D. Bloisi, Luca Iocchi, Daniele Nardi, and Michele Fiorini. This chapter introduces a framework for integrating visual information from user-controlled cameras in existing maritime Vessel Traffic Services and coastal surveillance systems; it is also suitable in airports to control aircraft movement on land.

Chapter 12, "*AIS Signal Radiolocation, Tracking and Verification*," was authored by Pier Francesco Sammartino, Michele Vespe, Dario Tarchi, Franco Oliveri, Francesco Papi, Francesco Borghese, Giuseppe Aulicino, and Antonio Vollero. This chapter provides an analysis of the Automatic Identification System (AIS) for vessels as an intelligent transportation system.

Chapter 13, "*The Impact of Satellite AIS to the Environmental Challenges of Modern Shipping,*" was authored by Nicole Schill and Margaret Browning. This chapter describes vessel tracking by space-based sensors, satellite AIS (S-AIS), and its use for environmental protection.

Chapter 14, "*How 'Green' is e-Navigation?*", was authored by Jan-Hendrik Oltmann. This chapter discusses the maritime e-Navigation definition regarding its claimed protection of the marine environment.

Chapter 15, "*Optimal Ship Operation: Monitoring Technology of Ship Overall Heat Balance,*" was authored by Yung-Ho, Yu. This chapter also analyzes protection of the marine environment and compares ship engine efficiency and wasted heat.

Chapter 16, "*Regulation of Ship-Source Pollution through International Convention Regimes,*" was authored by Proshanto K. Mukherjee and Jingjing Xu. This chapter studies regulatory conventions in the field of ship-source marine pollution.

Chapter 17, "*Foresight Application for Transport Sector,*" was authored by Joanna Ejdys, Joanicjusz Nazarko, Łukasz Nazarko and Katarzyna Halicka. This chapter presents good practices and potential benefits of using foresight studies in the process of creating the future of the transportation sector.

Chapter 18, "*Aeronautical Air-Ground Data Communications: Current and Future Trends,*" was authored by Fabien Garcia, Alain Pirovano, Michaël Royer and Quentin Vey. This chapter reports the current state of aeronautical air-ground communication (between airlines and their crew or between pilots and air traffic controllers) as well as some new solutions proposed recently.

The broad spectrum of topics and the quality of the contents of the book will undoubtedly stimulate further research and development to improve human knowledge of the transportation sector and inspire sustainable and effective mobility.

Michele Fiorini
Jia-Chin Lin
Scientific Editors

June 2015

Acknowledgements

The Scientific Editors, Michele Fiorini and Jia-Chin Lin, are gratefully indebted to many people for having inspired, supported—with positive and negative comments—and encouraged them on a number of occasions. Technical discussions and international cooperation form the foundation of this book. Opportunities to consult with experts and exchange opinions and information have been invaluable for enriching the contents of this book.

Our appreciation also goes to all authors of the chapters and their families for agreeing to contribute to such an ambitious project that required many weekends and overnight hours while sacrificing family life during the preparation of the manuscript. The Scientific Editors are fully confident that this book is a noteworthy exhibit of the chapter authors' professions.

The Editors would also like to acknowledge the professional skills of the IET Editorial Team, especially the Senior Book Commissioning Editor, Paul Deards, and Assistant Editor, Jennifer Grace. We thank all of them for their excellent work in maintaining the high quality of this book.

Foreword

This book is a testament to the international nature of scholarship, to its broad base, and wide-ranging interconnections. Such meeting of minds was epitomized in the gathering of some thousand participants in the historic city of Xian, China, a place famed as the starting point of the ancient Silk Road, from which the treasured exports of the East were brought to Europe on centuries old tracks through difficult terrain. Nowadays, conferences such as the Euro-Asia Economic Forum, meeting in Xian and hosted by the People's Republic of China, enables those at the forefront of new thinking to hear of the latest advances and discuss their import.

The topics are comprehensive, from developments in World Finance, Economics, Tourism and Education to a range of science-related areas pertaining to the overarching problems that the world needs to consider: energy, economic growth, culture, fostering educational talent, science, international ecological safety, Euro-Asia continental bridge logistics, eco-agricultural products and food safety, and investment for environmental promotion and project matchmaking.

The unusual mix of speakers from both scientific and social science backgrounds, a feature of Chinese international conferences, enabled the idea of this book to take root. The editors are themselves drawn from West and East, from industry and academia. Michele Fiorini chaired a session of the Low Carbon Earth Summit on the topic of Clean Mobility and Intelligent Transport Systems. The vibrancy of the session was such that those involved felt that the range of topics discussed warranted an international audience and a book of general interest to all readers worldwide might be welcomed. The topics are touched on daily in our newspapers and are of urgent interest.

This book aims to cover the full spectrum of transport models, on land, sea, and in flight. It draws on a rich background of current knowledge, encompassing research from the economic and human sciences. Those who have contributed chapters are principally drawn from the speakers at the Euro-Asia Economic Forum 2013, with others who were keynote speakers at the Intelligent Transport Systems and Telecommunications Conference 2012. Professor Jia-Chin Lin, the General Chairman of that conference, has kindly agreed to act as co-editor.

The contributors are at the forefront of new knowledge on Intelligent Transport Systems, and represent a pool of experts drawn from a huge range of countries.

As one who has a general interest in these topics, and found the conference in Xian a stimulating environment for the cross-fertilization of new ideas and approaches, I would hope that this volume will prove valuable to all those who have a lively interest in Systems Engineering and its related issues.

Heather Eggins, London 2015

Chapter 1

Towards contributive development of services

Michel Leonard and Anastasiya Yurchyshyna**

1.1 Introduction

The human world is undergoing a startling transformation; rapidly evolving as the discovery of new cognitive environments give rise to innovative models essential for creative coexistence to flourish in every facet of society; in all public, private and international sectors; and in regional, national and international contexts. These cognitive environments will transcend those which we have known. In particular, they will offer new answers to crucial challenges, not least in the domain of Intelligent Transportation Systems (ITS). In the early days of transport design, when the spellbinding perfection of nature's own mass movement systems seemed an unattainable ideal, the creation of technological equivalents was an impossible dream. Today, as a European space probe, the size of a refrigerator, lands upon a moving comet six billion kilometres from earth and a blind man 'drives' an automated car through American city traffic, we stand upon the very threshold of that dream; its realisation has already begun.

Defined in EU Directive 2010/40/EU [1], ITS are viewed as systems in which information and communication technologies (ICT) are applied in the field of road transport, including infrastructure, vehicles and users, and in traffic management and mobility management, as well as for interfaces with other modes of transport. We see in the world of nature that millions of years of evolution have produced complex organic systems to facilitate the coordination of navigation by a shoal of fish, a flock of birds or an army of ants. ITS are largely reliant on technologies in different spheres of transport-related activities: for example, traffic signal control systems, speed cameras, road signage and messaging, in-vehicle navigation, container management systems, automatic number plate recognition, security CCTV systems, electronic toll collection, rear-end collision avoidance systems, dynamic message signs, parking guidance, weather information, road works information systems. Such technologies include, but are not limited to, computational technologies and wireless communication, video vehicle detection and Bluetooth-enabled detection, as well as Big Data related to transport circulation, all of which have been created in the human world, within a few generations.

*Institute of Services Science, University of Geneva, Geneva, Switzerland

To improve transportation safety and mobility, as well as to increase productivity through the integration of advanced communications technologies into the transportation infrastructure and in vehicles, multiple research and practical programmes throughout the world have recently been conducted. Among such fruitful research projects are the outstanding works of Michele Fiorini in Maritime e-Navigation [2], Jean-Pierre Birat on the Green Car initiative [3] and Scott Sedlik on the Internet of Automobiles [4]; the research of Steffen Preissler identifying the engine for Open Innovation through a gap between publicly funded R&D organisations and private companies [5]; the approaches towards sustainable transport in London, argued by Steve Kearns [6], and driverless cars; and a set of initiatives developed in the scope of the ITS programme of the U.S. Department of Transportation's Research and Innovative Technology Administration [7], to name but a few. Practical implementation of this research through ITS applications may now focus on infrastructure and vehicles and the integrated applications between them, as well as dissemination of transport-related technology and skills, in the creation of intelligent movement strategies as effective as some of those wonders we may observe in the natural world.

The important acquired results of these and other works in the ITS domain imply a complex approach for their systematisation, precipitating the application of this knowledge and its implementation in future systems, services and information technologies supporting ITS.

Such a systematisation approach is the formal Transport Maturity Model, proposed by IBM [8]. The Transport Maturity Model is based on the level of system, data and service integration across multiple transport modes, and its objective is to enable authorities to view their businesses and the ITS perspectives holistically and capture their purposes for future development. The analyses of recent trends in ITS development, however, run the risk of focusing upon the technology-oriented development, at the expense of consideration for the human-oriented development. In other words, by giving the priority exclusively to technological achievements, one might lose the consistency of such development: to be able to answer the situations in which people find themselves. This need attracts the interest of leading scientists to follow human-oriented approaches with ITS technologies to achieve innovative competitiveness in the domain: for example, the programme towards Humanity Oriented Development (HOD) of Transportation in Taiwan presented by Tien-Pen Hsu [9].

In this scope, it is particularly interesting to explore: how innovations are encountered in the domain of ITS, which dynamic changes drive our society, how one can encourage and manage such innovative changes, how one can support the corresponding design, which aspects should be automated, and which remain under human control and supervision, which new situations appear in the domain of transport that have not been recently addressed by ITS and, finally, what is the added value of these innovations.

This multifaceted, multidisciplinary challenge can be met through a Service Science approach, by taking into account deep transformations in current environments, integrating the results of all relevant research projects to consider a complex approach for their analysis, systematisation and knowledge capitalisation, followed by its further implementation in future systems, services and information technologies supporting all spheres of the economy. More precisely, it becomes

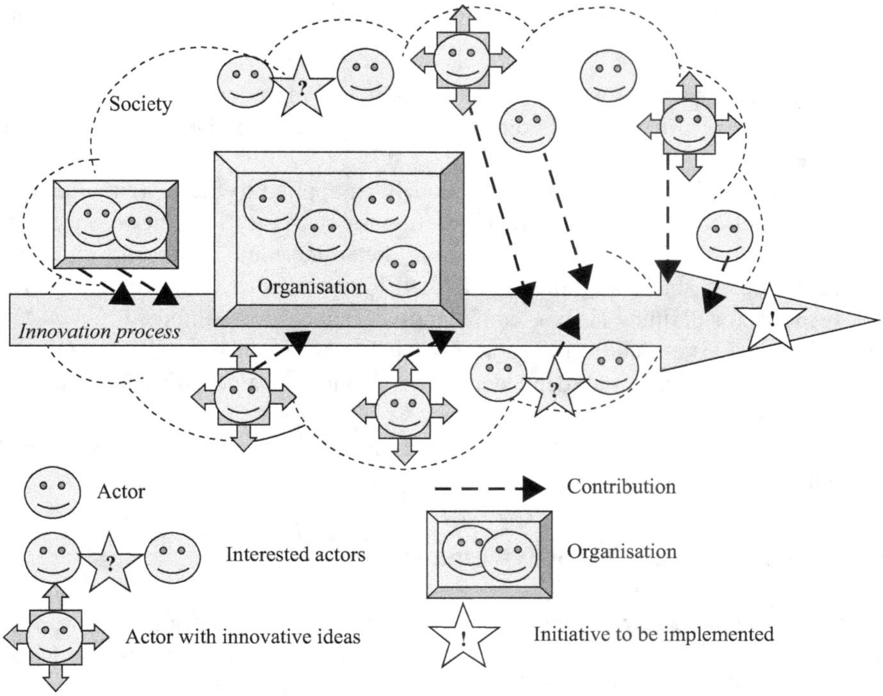

Figure 1.1 Society, organisations, actors and innovations

possible by creating and progressing design which is oriented not only towards various innovative aspects of ITS but also to ensuring sustainability of ITS, by taking into account dynamic changes in the environment and active contributory participation of the involved actors from different domains (Figure 1.1).

Indeed, the innovation process in our society benefits from contributions from organisations and actors, some of whom have well-defined innovative ideas and are active contributors in the innovation process, some of whom may be interested people who link their efforts around new situations and have fruitful discussions yet without concrete suggestions, while other actors are just interested people, who would like to explore the situations and contribute in the best way they can. By virtue of its inclusion and evaluation of all these contributions, the innovation process leads to clarifying the ideas of actors and develops an initiative, which should be implemented as a service.

Such exploration might not be fruitful without a deep understanding of the following fundamental aspects:

- the characteristics of the creative and innovative nature of our society, in general, and ITS, in particular;
- the democratisation of initiatives and involvement of different actors into public decisions tend to make impacts on the development of innovative technologies and systems;
- the motivation of actors in multidisciplinary innovation and development.

Facing these aspects, such exploration will enter the following crucial questions:

- Why these contributions are becoming more and more multiple and effective? How to construct the economy of contribution with the help of Service Science?
- Can any of the existing economic models be applied to the current situation, and which fundamental differences define the necessity for a new conceptual approach? How to ensure that these heterogeneous approaches are not contradictory, but complimentary in their joint constructive movement? What is the factor defining such a 'confederative' balanced development?
- Which values are addressed today? What is the role of human value in such development and its interdependence with economic value? What is the role of services in creating added value?
- Is it possible to create collaborative co-creative environments for service-centred innovations? How?

1.2 Exploration with Service Science

Today, when our society becomes more complex and heterogeneous, it has to face the challenges of new situations typical for new domains, like for example Clean Mobility and ITS, very different from those to which its members were previously accustomed, and in order to prevail, they must transcend all their previously acquired experience and training. Societies are inevitably servitised in all their domains: this servitisation has a boundless potential to address their mounting complexities and to reveal fundamentally simple antidotes within organisational forms and collective human activities. In order to meet these challenges, it is now crucial to give a chance for real explorers to discover, promote and perform trans-disciplinary/organisational[1] services, allowing overcoming the new complexities.

The plan of Service Science is to develop such knowledge which actually helps to facilitate the work of these explorers in ITS and to place it in the context of the development of our society. This also includes offering them environments of creative freedom to achieve such a goal. Isolated explorers have always existed; but it is now for the first time, we have the opportunity to develop the means to integrate their discoveries into practical, operational realisations, which can be integrated into any aspect of the world, in general, and ITS, in particular.

Indeed, ITS knowledge must take into account the fact that explorers face multidisciplinary and multiorganisational complexities, which are the result of their multiple responsibilities. They are to make a profound analysis of the involved activities, recompose and rebuild them along the exploration process. They create new professional situations and new economic values. They have to take into

[1]A service is trans-disciplinary and trans-organisational, as a service is created by a multidisciplinary approach with the involvement of multiple organisations, but it does not belong to any discipline or any organisation.

account profound impacts on the activities of organisations and their management. They are required to disclose the *raison d'être* of their own responsibilities within their organisation and even *raison d'être* of their own organisation. Within their organisation, their position is rather paradoxical, for example in relation to project managers or investment: at the beginning of an exploratory phase of a service, they do not even know what exactly they are aiming to implement! Indeed, a service is not discovered as a whole finished product (as it is in traditional or design methods), but it is built in an evolutive enriching way.

The situations related to ITS explorers are, therefore, encouraging the researchers into exploratory situations, where they can use the knowledge of their own scientific discipline for consecutive exploration and discovery of services. In other words, we argue that the knowledge for ITS development is in fact of the same exploratory and actionable nature, which is inherent for Service Science, and that Service Science and its approaches might answer – in a more intelligible and creative way – those challenges, which the ITS domain faces today.

Furthermore, by addressing a wide range of classical disciplines (management, transport, climatology, logistics, etc.), which are now susceptible to be re-constructed, in order to integrate the impacts of ITS-related innovations into these domains, this creates new challenges not only for explorers outside the domains, but also for the domain scientists, whose research is focused on enriching their domains with service-based approaches.

These advantages seem to be crucial to take into account continuously emerging situations, with higher effectiveness, by decreasing uncertainty and by clarifying processes to be triggered to face complex situations. They appear to be fruitful as a basis to develop new approaches for systematisation of ITS activities.

1.3 Service Science approach

In this context, by supporting the joint effort on developing a balanced approach, we suggest the approach of Service Science for the ITS developments and their systematisation.

Let us analyse the possible advantages of this approach.

- Service orientation. By underlining the role of service orientation at multiple levels of technology and business, it is essential to clarify its meaning. It is stated that everything (i.e., good or activity) is seen as a service, and is analysed by an interdisciplinary approach that brings together study, design and implementation of services in which specific arrangements of people and technologies take actions that provide value for others [10].
- Possibility to address new professional situations, in absence of a 'usual' approach, only driven by the 'common sense'.
- Directed exploration and design towards a constellation of interwoven and interoperable activities.
- Situation-orientation, in contrast to problem-orientation.

- Knowledge discovery and capitalisation through initiatives as a key element of direction-oriented development.
- Rigorous cognitive results.
- Trans-disciplinary, trans-organisational, transnational: it does not depend on a specific country or domain, but is applicable for all of them.
- As the result, the Service Science approach is worldwide, confederal and cognitive.
- Co-creation of value for ITS by the multitude of actors involved in the development process, and aspects of such contributory development.
- Formalisation of ITS activities, which can consequently be implemented as a service and/or constellation of services.

These advantages seem to be crucial because of continuous emerging situations, with higher effectiveness and performance characteristics, new approaches in usages, which decrease uncertainty, the simplification of complicated situations and, consequently, the reorganisation of all the involved activities. They appear to be fruitful as a basis to develop new approaches for systematisation of ITS activities.

What is the original position of Service Science in this perspective? Service Science aims to address the inherent complexities of human activities from the vantage point of emerging services, in order to discover and develop simple activities-oriented forms/patterns, which are masked by initial complexities. It is interested in the role of services in these complexities, in the processes of their emergence and sustainability. Consequently, Service Science is focused on discovering how individuals, businesses or governments can actually live together in an ever-changing environment.

It is crucial, however, to note that such an original position is – at the same time – fundamental. Indeed, Service Science is not centred on the 'exterior' as exact sciences are, since one of its major explicit concerns is the development of the society. From a different side, in contrast to some researches in human sciences, it is not enough for research in Service Science to focus on objects of study, in order to standardise or formalise them with the help of quantitative or empirical models, as it seeks to pierce the complexities evolving in the process of the creation of services. It cannot be built only upon theoretical thinking, because it directly faces the uncertainty of the development of society. It is not only intended to build solutions to problems through the application of technologies such as IT technologies, in the way it is done in ITS, for example since its domain, the domain supporting innovation in transport systems, is much larger than the transport domain alone.

Thus, by respecting the original domain of ITS, Service Science also produces its own knowledge and research methods, including those methods that provide knowledge and the capacity to create it by explorers. As an example, it creates the knowledge to construct languages, allowing explorers of different responsibilities, disciplines and organisations, not only to understand each other, but also to work together in exploring services.

Service Science offers a scientific frame to tackle new challenges of society, thanks to its methods of transforming ideas into concrete services. Service Science requires account to be taken of social and economic contexts at the design level, thanks to Service-Dominant Logics [11] and its focusing on knowledge and skills.

According to Service-Dominant Logics, a service is defined as the applications of competences (knowledge and skills) for the benefit of a party [12]. The following main principles are respected:

- Service is the fundamental basis of exchange.
- Indirect exchange masks the fundamental basis of exchange.
- Goods are distribution mechanisms for service provision.
- Operant resources are the fundamental source of competitive advantage.
- All economies are service economies.
- The customer is always a co-creator of value.
- The enterprise cannot deliver value, but only offer value propositions.
- A service-centred view is inherently customer oriented and relational.
- All economic and social actors are resource integrators.
- Value is always uniquely and phenomenologically determined by the beneficiary.

Thereafter, Service Science allows society to leave the passive 'static' position, where society only passively contemplates and applauds ICT successes, and to take the active 'dynamic' position, in which society has to actively re-design its organisational parts. It is a great challenge of innovation to open information through services and create new economic values, by respecting the principles of cognitive social responsibility [10] to strengthen the sense of inclusion and to contribute to social stability.

1.4 Services as engines for innovation

Despite the importance of information in design and innovations, on its basic level it may seem somewhat tedious and such aridity can inhibit innovations. Thus, the concept of informational service appears as much more appropriated. Even if an informational service is defined upon the concept of information, it provides a deeper conceptual semantics, by describing how information contributes to the development of society by the means of smart access and smart treatment.

Generally, an information service is seen as a part of an information system that serves data/knowledge/information to customers and collects it from its contributors, to manage and store it by optionally using administrators.

Services are characterised by four main factors [13,14]. They are as follows:

- information is the core element of the design, production and management of services, so services are information-driven;
- customers are co-producers of services; they may require the adaptation or the customisation of services, so services are customer-centric;

- digital orientation of services is explained by the achievements in information and communication technologies, the (semi)automation of main services-oriented activities and the creation of new domains: for example, e-commerce, e-business, e-collaboration, e-government, e-environment; and
- services are driven by their performance criteria and as such are productivity-focused.

We argue that these phenomena related to services can be analysed in the wider context of a service-oriented society in general and, as such, are seen as characteristics of this service society.

Guided by the current experiments in developing economies and the evolution of the underlying technologies, our society is based on information and knowledge that become the main sources for value creation. New, emerging situations lead to new challenges we should now face; and new ideas addressing these challenges require to be identified, analysed and – more importantly – concretised. This is achieved, thanks to services [15]. Generally, the phenomenon of service innovation mirrors the requirements of the knowledge society where knowledge is both the primary production resource and the tool of value (co-)creation and in which, information lies in the centre of service creation and functioning. We point to the practical importance of this phenomenon that is implemented by informatics: it not only supplies the tool for IT development but also guarantees the consistency of the sustainable co-creation of its fundamental concepts [16].

1.5 Triple push

Continuous emerging situations are due to the 'technological push', with higher effectiveness and performance characteristics, new approaches in usages, which decrease uncertainty, the simplification of complicated situations and, consequently, the reorganisation of all the involved activities. They appear to be fruitful as a basis to develop new approaches for systematisation of innovative activities, in general, and innovative ITS, in particular.

When new situations appear and current existing situations change, that is, are simplified, dissolved or, in contrast, turn to be more complicated, it becomes clear that old approaches are no longer applicable to process these situations (Figure 1.2). Indeed, previous 'problem-solving' approaches cannot be applied to these situations, as they represent a different reality, different challenges and different environments. From a different point of view, there is no generalised 'usual' approach on how to process these new situations. There are no techniques universally applicable for each and every new situation, and the general 'if-then' analysis cannot help, simply one never knows where to start! These new situations are transforming competencies, know-how, ways of thinking, so a multitude of academic, entrepreneurial, public, private, worldwide efforts are needed to face, study, master and process these situations leading towards discovering and capitalising newly created knowledge and, consequently, new human activities appear and become more 'condensed' in time and speed. However, we are now equipped

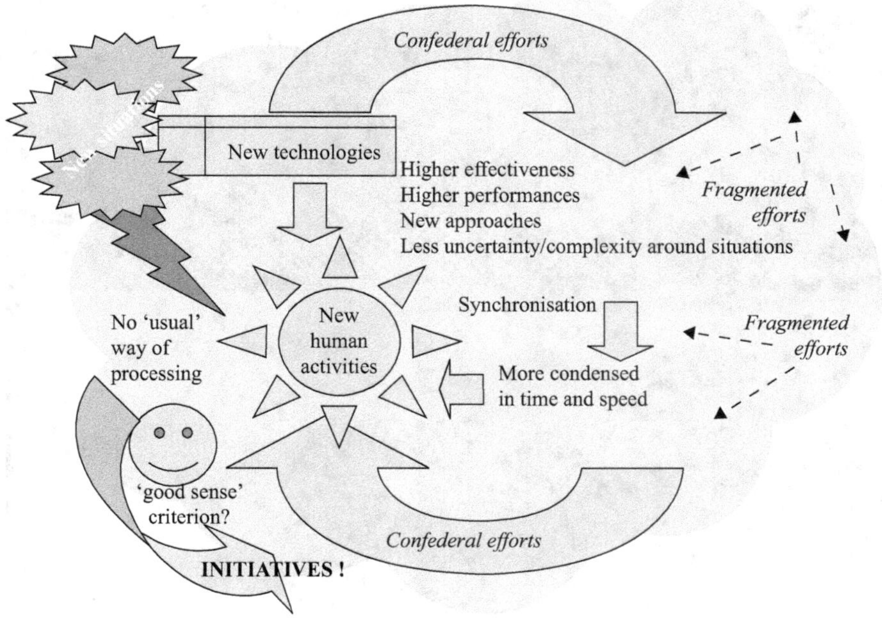

Figure 1.2 Towards triple push

with new technologies, which decrease the uncertainty/complexity around situations, and it is thus possible to achieve higher effectiveness and performances. In this context, will it be enough to simply apply 'good sense' criterion, in order to develop new ways of addressing these situations, as well as unify initiatives of different actors, in their confederal efforts?

Here, a technique of triple push is applicable.

The primary technological push is related to the 'obvious' cognitive value of the discovery. As an example, we take an innovative camera developed by FOXEL [17] allowing 3D capture and digital processing of the data (Figure 1.3).

Once such a breakthrough innovation is invented, a second push of cognitive assimilation of this invention is required. It addresses the analysis of technological possibilities of such an invention in comparison to the situations, which might be addressed, the implementation through technological parameterisation, the particularities of its implementation and execution, as well as the technological and organisational steering of this innovation.

Finally, a third technological push is oriented towards development of a constellation[2] of interconnected and interoperable services allowing further exploration of this innovation. This concerns not only the known situations, in which this innovation is applicable, but also all the multitude of yet unknown ones, where it might be applicable, that is, the ones to discover.

[2]Constellation of services is a myriad of interconnected and interoperable services.

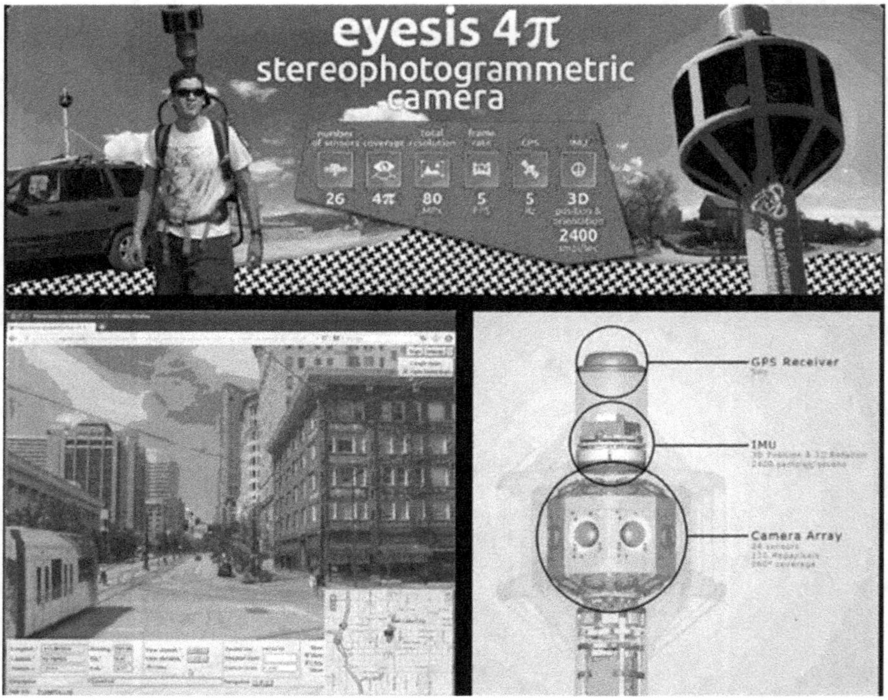

Figure 1.3 Triple push

As a result, it should lead to the development of a service with a cognitive unity around the services aiming to address larger (more complex) situations in future.

1.6 Collaborative development supporting innovation

An idea of joint efforts aiming to develop our society or its different domains is not new. Indeed, a large number of collaborative groups, public–private partnerships (PPP) and other forms of organisations have already proved to be a successful answer in innovative projects [18]. It is also noticeable that these recent works are not exclusively focused on the institutional principles of PPP functioning [19], but particularly address the core elements of different types of partnerships, which guarantee their success: skills and knowledge from both private and public sectors [20].

Traditional PPP are beneficial for complex projects involving the government and the private sector. In this context, what are the challenges and advantages brought by industrialisation and globalisation, and by open and big data, that become more and more important for the development of enterprises and society?

To analyse this phenomenon of collaborative development, we investigate the three levels of classical PPP. From one side, PPP represent a tool to construct, structure and maintain PPP projects. From the other side, PPP are working environments allowing, defining and examining different processes leading to creation of projects-oriented PPP. Third, PPP are defined by their concrete aim: achieving a defined result. By its dynamics, PPP attempt to make our society 'stable', in the meaning of guiding and controlling its sustainable development, according to the current working environments.

We also underline that, in the case of classical collaborative partnerships, they are mainly focused on infrastructure projects, when the return on investments is not expected at the moment of signing the PPP contracts or even finalising the projects. It is throughout a certain number of years after the projects are terminated that the private partners have an exclusive right to exploit and/or maintain the results of the project and, as such, receive the return on their initial investments. (Indeed, a private company financing the construction of the public road that does not have its investments returned during and before the project is finished will be benefiting from exclusive rights on its maintenance and/or receipt of pay tolls during its functioning.)

The principles of collaborative development within classical PPP create a solid background for extending this vision for exploring the process of service creation, the results of which are intangible and often more complex to be valorised.

To do this, we first refer to the principles of development of services, or more precisely, sustainable services. In Reference 16, a sustainable service is defined as a service that is capable to adapt to its environment, to dynamically integrate the ever-changing conditions of this environment and as such, to be sustainably coherent with its evolving challenges. It is important to underline that, according to this approach, sustainable services are dynamically built thanks to all the relevant semantics: that is, the information kernel that relies on domain, legal and knowledge ontologies and defines the semantics of services. This semantics is verified and enriched through situation-based validation and existing by practices of usage, and as such, a new semantics will be identified and integrated into the information kernel.

This implies that service creation relies on its sustainable development, enabled by co-creation of its members; the process is similar to the one traditional to PPP. As such, those PPP, which are oriented towards creation of services (PPPS), have naturally become sources of sustainable creation through a service-capacitating process for all the involved actors and stakeholders. Indeed, such PPPS not only benefit from the advantages of traditional PPP, but also enable contributory service empowerment process for all their partners (the government, private partners, individual experts, etc.).

By their composition and through their functioning, PPPS create an innovation-oriented environment and, as such, are themselves both the source and the result of innovation. Moreover, the results of contributory development within PPPS are increased, thanks to contributions of actors interested in developing new services.

Furthermore, the resulting new services are themselves likely to be sustainable, as they are developed on the basis of sustainable services within PPPS.

In other words, PPPS extend the advantages of traditional PPP: when the government provides direction, the private sector provides the drive to ensure success and people (actors) provide initiatives and skills to concretise ideas. As a result, PPPS allow concretisation of initiatives through services.

Following the increasing impact of contributory development, it is important to underline that it is mainly due to the creativity and motivation of actors (individual persons, formal working groups, as well as informal groups of civil society) in PPPS. Furthermore, its innovation-oriented processes are not restricted by the 'desired' predefined results. Besides, given the freedom of exploration, such contributory development concerns the common interest and then turns to be more sustainable within the developing society.

In this context, the services developed through PPPS are, in fact, common goods (i.e., rivalrous and non-excludable goods shared by and beneficial for all or most members of a community), or more precisely, the myriad of common goods, which serve the common interest and are free.

1.7 Towards contributory development of services

The explorers of a service shall gather around a table of cross-fertilisation where they will understand each other and work in an effective way together, despite the fact that they all have different jobs and different responsibilities and come from various organisations and have widely differing origins. To make such understanding rigorous, they will share an effective language, which is independent from all disciplinary and organisational languages. This language will allow them to drive the implementation of the service at all levels – organisational, informational, budget-related or staff-oriented – and to cherish its perennial features while yet facilitating their evolution and transformation. To our knowledge, the only language with all these properties is the language of information knowledge, in the meaning specialists in information systems have developed over time. Each explorer will thus be the path towards the knowledge of this informational language and towards understanding a successful marriage of human activities and their informational support, which is inherent to any created service.

A team of explorers will maintain its unity, internal cohesion and identity with respect to related organisations and – in a more general sense – their environment. The difficulty in this case arises, in particular, within the multitudinous facets of a service described with the help of models – those of human values, economic values, activities, ethical, informational, informational, ontological and regulatory. Every service will consistently maintain an informational unity, which is clear for all explorers.

In the context of the traditional economy, guided by added value and the copyright principles of protecting the rights for the goods developed by businesses, it becomes unclear why they should be encouraged to contribute 'free of charge';

why the actors need to share their own knowledge, skills and make efforts to develop something, the results of which do not belong to them.

Indeed, in the world of commercialised services, the notion of a service as common good seems to be lacking sense. The hot questions from the traditional businesses are focused on the following:

- Who is the 'consumer' of free services?
- Who 'pays' for service creation?
- What outcome can the businesses have from them?
- Which are human and economic values of services?
- Why are enterprises interested in having co-creators?
- Why should enterprises agree to pay for creating anything, from which they cannot profit?

By implementing the vision of the traditional economy, no positive answers to these questions can be given. Instead, this problematic should be addressed from a different viewpoint.

According to the exploratory approach of Service Science, the accents of this discussion are changed from seeking the answers to these questions within the traditional economic vision and focus on the analysis of how to address the existing situations.

Indeed, with the growing impact of services in identifying and enabling our society, one cannot ignore the importance of creating services. However, the life cycle of services is very different from that of tangible goods, it is impossible to have clear criteria on the material value of services, neither to have classical return on investments from their creation. Consequently, it is impossible for the private sector to accurately evaluate the risk of participating in projects of service creation, or to estimate the expected income as the result of creating a free common good service.

It becomes obvious that this, in fact, is not needed. In the context of knowledge-based and service-enabled society, the main risk is not the one of not returning one's investments, but the risk of 'no innovation'; the risk of being outside the revolutionary tendencies identifying the dynamics of society and participating in innovations arising around them, the risk of losing the knowledge and skills allowing sustainable leadership in each domain.

The economy of contribution is described in References 21 and 22 as the one having the following characteristics:

- no longer is there separation between economic actors as producers and consumers; all actors are now seen as contributors,
- the value produced by contributors is not totally monetisable, it is a 'positive externality',
- it is as much an economy of existence (as the production of '*savoir vivre*') as it is an economy of subsistence.

It is first important to underline that the economy of contribution does not aim at putting into question the traditional market economy but to broaden its vision by taking into account the new phenomena, such as technological progress, networking and social media, allowing more possibilities to create and share value.

Contributory development appears to be much more efficient to highlight the intangible part of the technology that is essential to its social assimilation.

Economy of contribution does not aim to create a universe of services and products, free of charge, but a world, in which knowledge is available, free of charge, to promote its spread, its application and its exploration, in favour of the development of innovation. At the same time, a free study – or method, or research or technology – becomes free of charge only if its development was fully funded.

It is remarkable that the economy of contribution does not exclude alternative means of production and exchange, but rather combines with them, whilst its contributors participate in chosen activities, creation of social value and have an interest in selflessness (there is an interesting discussion on the phenomenon of selflessness in References 21 and 22).

The main challenge here, in comparison to the traditional approach, is to accept the vision that a service is not a product, that PPP-oriented products cannot be applied and that this classical vision should be extended.

In other words, each actor (private, public or individual) is not any more seen as just a consumer or creator of a service, but has become a co-creator: 'paying' by their commitment and efforts in service development and by being 'paid' through an exclusive right to define the dynamics of the development of the domain, access to the most recent technologies and methods, as well as initiatives of various actors, and finally by forming the cognitive unity in service creation. This also means that, in its complexity, the contributory approach in service development through PPPS allows the creation of a new market for initially planned and completely new services, which is not dependant on the initial estimation of the utility of a service but coordinates itself sustainably, according to new situations and guided by democratically developed initiatives.

In this context, Internet and new technologies offer collaborative multicultural environments whose actors are put in the position of active participants, supported by free- and open-source software. It thus represents the source of collective creative practice and new value-creating mechanisms, whilst digital technologies have intensified the exchange of information and knowledge.

Creating a new market for services as common goods has a crucial effect for commercial activities of the involved actors. Indeed, 'free' services allow the development of a new market, which offers an environment for further development of a vast amount of commercialised services with economically defined added value, based on and disseminated by virtue of the initial common goods services.

An example of contributory participation for supporting a free service can be illustrated by CamSam, the speed and red light camera warning system [23]. CamSam is an application for a smart phone that turns it into a real-life speed camera detector. It also informs about any other obstacles on the road (Figure 1.4). The application has a database of about fifty thousand cameras and uses GPS to warn a driver when a vehicle approaches one. Warnings are issued regarding

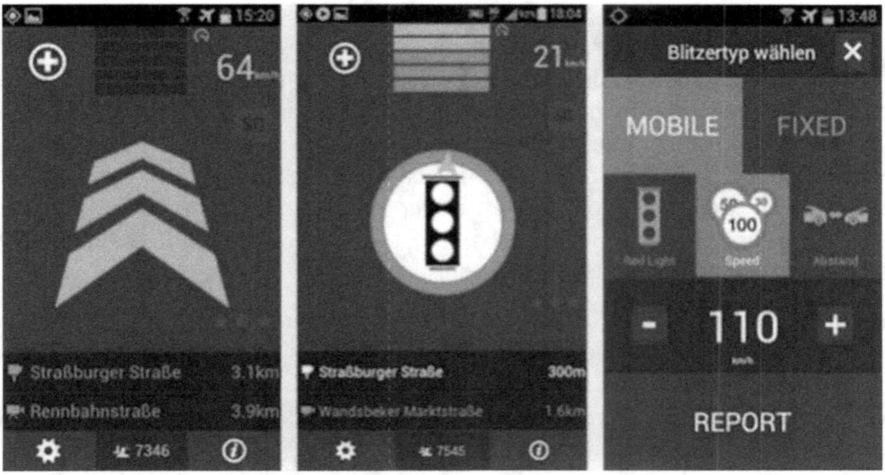

Figure 1.4 CamSam

cameras in proximity (displayed as a detailed list, including their geographical location and the distance towards them), as well as when a new obstacle is approaching (displayed as symbol of radar, or traffic lights, or any other type of obstacles related to traffic). Distances to cameras are displayed, giving users enough time to react accordingly and adjust their speed if necessary. Both options can be displayed separately or together. Moreover, the application is interactive: drivers regularly update this database to keep track of recently installed cameras or any other obstacles of the road, by adding the information dynamically.

1.8 Tiers-Lieu of general interest for creating services as common goods

To face new challenging situations and create a service with a cognitive unity, the concept of Tiers-Lieu provides an exploratory environment for service-centred innovations, in which collaborative work is concretised by the development of a constellation of services. Tiers-Lieu represents an instrument of the contributory economy for the development facing People-Public-Private-Partnerships situations based on initiatives, that is, inter-organisational, interdisciplinary, inter-domain breakthrough propositions, targeting the creation of human-oriented and/or economic values and concretising the semantics of the innovation.

With the help of Tiers-Lieu, which encourages co-creation and exploration of challenging situations in the ITS domain, it becomes possible to offer a formal systemic approach for exploring these situations, concretising the innovation-based initiatives oriented to improve ITS activities, as well as enable their realisation through a constellation of the corresponding services.

Creating services of the 'common good' nature can be possible, thanks to a collaborative environment which offers the possibility to identify the initiatives-oriented common goods, and detail them and develop the corresponding services for commercial projects, research or business-oriented methodologies.

The first results were discussed in Reference 15, which introduces the concept of Tiers-Lieu and argues the characteristics of Tiers-Lieu, which reveal its added value in comparison to other collaborative environments.

According to this approach, Tiers-Lieu are envisaged as an open environment that motivates collaboration, intellectual creativity and surpasses the limits of traditional disciplines-defined collaborative spaces, by allowing to defining new services. All initiatives of Tiers-Lieu must:

- be inter-organisational (by representing a general interest),
- be interdisciplinary (not related to the only domain, but aiming to address interdisciplinary situations),
- take into account international/intercultural aspects (lead to value creation in human, social and economic spheres) and, finally,
- be concretised in the form of one or several trans-organisational and trans-disciplinary services.

The participants of Tiers-Lieu share 'think contribution' vision, supporting creating services of the 'common good' nature, which are also characterised by human-added value.

This also includes understanding by all the participants that these services are created under Creative Common License [24], and no exclusive ownership rights can be demanded by any actor of such service innovation environments.

First, it is important to underline that Tiers-Lieu are developed as collaborative environments, which aim to overcome the generic problems potential actors may face. This includes the so-called resistance to change when actors are invited to change their way of work, communication, usage of systems, in order to find a 'better' way of doing (even if they do not particularly want to change them). Instead of this, in Tiers-Lieu actors receive a possibility to take part in co-creation as active designers in changes. Moreover, within Tiers-Lieu, absence of immediate tangible results is not seen as blockage, since all actors contribute to elaborate shared thinking, with added human value. It is only from this shared thinking that it is possible to formalise initiatives and then to concretise them as services.

The activities that are supported by Tiers-Lieu are naturally discussed during the meetings of cooperative nature concerning strategic questions typical for complex competition-based environments.

Despite certain contradictions between the value intentions of each actor, Tiers-Lieu are characterised by a high level of collaboration in achieving common objectives. For example, while introducing new forms of medical services, IT standards or compliance norms, the necessity of coherent collaboration between direct competitors (e.g., leading companies in the sector), standardisation organisations, policy-making bodies and other interested parties has become a crucial factor.

The forms of meetings within Tiers-Lieu are multiple: they can be face-to-face, diffused by Internet, supported in real-time or asynchronous, or a mixture of different forms. The participation is open for other contributors: actively interested people, and is highly beneficial if various interesting – and multidomain – ideas are exchanged.

In this context, it is important to underline that the participations in Tiers-Lieu are based on the acceptance of its members of the main principles of team creativity, their desire and ability to create collectively, to share the expert knowledge and the acquired results, to avoid innovation resistance [25], to ensure participative safety and to improve the quantity and quality of attempts to introduce or develop new ideas.

One of the most important conventions within Tiers-Lieu is the mutual agreements between all the involved actors that innovative ideas are represented through initiatives which can dynamically change, according to the discussions. Each actor should share the interest in selflessness [21,22]: before being selected and approved by all actors, initiatives can be modified, reorganised, abandoned, etc.

In order to guarantee smooth functioning of Tiers-Lieu in creation and implication of initiatives for service creation, it is necessary to establish its infrastructure, allowing tracking the history and dynamics of ideas exchange, some principles for regulating roles and access of actors.

A few crucial aspects should be noted.

First, Tiers-Lieu are motivated by the spirit of the PPPS approach: they are supported by private (P), public (P), partnerships (P) and are oriented to service (S) creation.

Second, to ensure the balanced organisation of discussions and the effectiveness of taken decisions, there should be a system of roles within Tiers-Lieu.

The following six roles are identified:

- Initiators: actors, who come with a new innovative idea, define an initiative and invite other actors to discussions. Initiators are those who take the final decision, once the initiative is discussed and developed by others. For example, there is an initiative to develop a special watch for aged people, which would evaluate their blood pressure, as well as speed and amplitude of their movements, analyse the results and trigger the other related services, if necessary.
- Participants: actors who actively contribute in discussions and help to develop the proposed initiative.
- Moderators: actors supporting the process of Tiers-Lieu functioning, that is, 'neutral' actors who guide the discussion towards concretisation of initiatives.
- Observers: actors who assist at discussions and follow them, but are not actively participating in them, that is, the ones who do not have a word. Observers might have educational purposes (e.g., students in medicine) or just share the general interest for the discussed subject, without offering any concrete solutions (e.g., public who would like to participate in creating services for aged people).

- Historians (or secretaries): actors who play a supporting role: helping to register and track discussions and contributions of participants, introducing required information, keeping in order the agreed planning, etc.
- Developers: actors whose aim is to develop a service, once the initiative has been defined and validated. These can be the same actors as initiators and participants, after having accepted the validated initiative and acting within the defined framework for service creation.

The role of the initiator is characterised by a high level of responsibility and is crucial for functioning of Tiers-Lieu. Indeed, it is the initiator who not only introduces a new initiative as a subject of innovation, but also defines the scope of participation within the scope of Tiers-Lieu. The initiator is also the one who evaluates the expressed ideas and has the final word on accepting or refusing them.

To facilitate the discussion procedure and to minimise the uncertainty in discussions, the initiator has a set of measures to keep the discussions fruitful, by attributing the participants a set of warnings: (warning about the semantic inconsistency or non-respect of the ethics of Tiers-Lieu) or a red card (serious breach of the rules or consistent contradiction with the main idea of the current initiative; this leads to the exclusion of the participant from this Tiers-Lieu). Analogically to football rules, two yellow cards in the same meeting constitute a red card.

A participant with a red card (or in fact any participant at any time) may leave this initiative and eventually launch an alternative initiative and a different Tiers-Lieu, which might have the same participants of the initial Tiers-Lieu. All initiatives are launched under the Creative Commons License [24].

It is remarkable that there are no limitations for the participants to contribute for multiple initiatives, as well as to leave them at any time.

It is important to develop a balanced system of ethics principles concerning the supported activity, and, consequently, the ethics principles defining the developed service.

Despite the self-motivation of the actors to participate in Tiers-Lieu, a balanced approach should be developed for supporting their interest in sharing and increasing their knowledge about complex situations that require common effort, even under the risk of competition. Tiers-Lieu are thus becoming a good choice for a neutral environment, which can put together various actors for their 'winning-winning' collaboration.

1.9 Conclusion

This exploratory paper addresses the challenge of facing new, emerging situations and supporting innovations in the domain of ITS, by studying the nature of these situations like more traffic and more complex transport infrastructure, higher requirements for security, ecological challenges, ever-increasing amounts of information being extracted from big data and new technologies supporting different aspects of the ITS domain. It is particularly interesting to witness the role of multidisciplinary actors in contributing to the support of these changes.

It is by virtue of the Service Science approach that it becomes possible to ensure contributive development of innovation-oriented services and facilitate the corresponding design. The role of PPPS, as engines impelling such innovative development, and the involvement of actors are of vital importance to this 'confederative' balanced development. Whereupon, following a concept of the economy of contribution and analysis of the environments-oriented service creation, a path forged towards positive outcomes by the proposed development of Tiers-Lieu (collaborative environments for service creation) describes the conceptual aspects of organisation of Tiers-Lieu.

We may thus address the multiplicity of situations facing our society and develop situational method components, offering to all the actors the possibility of collaborative and effective co-creation of services, in particular services of 'common good' nature for the ITS domain.

For millennia, human kind has stood in wonder before the spectacle as a thousand brilliant birds soar and turn, flashing in the sunshine, in perfect agreement upon an instant, without collision. The early human observers unable to conceive with what means these creatures are able to achieve such harmony could barely dream that their own kind may one day participate in such movement. How could so many be so immediately able to conceive, agree, plan and execute such coordinated manoeuvres? Today, we have Services Science to provide the potential. With child-like fascination, we have beheld in the society of ants, the unimaginable communications of vast, alien races, marching unstoppable across the mighty forest. Today, Services Science can outpace their communications in speed and complexity. We have perceived the patient work of eons in the fabulous designs of nature all about us and now, for the first time, it is possible to create and develop comparable potentialities, on every scale, within our own lifetime. Our microscopic ants may be nano technological agents and our soaring flocks may be space-capable flights, but whatever forms they manifest, we have today already embarked upon the realisation of these possibilities in the ITS that will bring much-needed efficiency, safety and sustainability to our own personal and public journeys on earth.

Our ongoing and future works include the further conceptualisation of Tiers-Lieu within the economy of contribution and the analysis of the impacts of the proposed approach for every area of societal development. In an integrated, intelligent, system of everything, Tiers-Lieu will become a sustainable environment for exploring the phenomena of service society and all the situations we face, by offering all the actors the possibility to collaboratively create services of 'common good' as we progress within the human and the natural world, as one, harmonious whole.

References

[1] Directive 2010/40/EU of the European Parliament and of the Council of 7 July 2010 [online]. Available from http://eur-lex.europa.eu/LexUriServ/LexUriServ.do?uri=OJ:L:2010:207:0001:0013:EN:PDF [Accessed 14 Oct 2014]

[2] Fiorini M. 'From vessel traffic services (VTS) to e-Navigation service portfolios'. *Proceedings of the 10th MAST Conference*; 4–6 June 2013, Amber Expo, Gdańsk, Poland

[3] Birat J.-P., Hoffmann, O. 'Steel and the green cars initiative in Europe'. *Revue de Métallurgie*. 2013;**110**(6):415–27

[4] Sedlik S. *Cloud, crowds and traffic: the Internet of the automobile* [online]. Available from http://www.automotiveworld.com/analysis/cloud-crowds-traffic-internet-automobile/ [Accessed 14 Oct 2014]

[5] Preissler S. *Technology transfer in Slovakia and abroad: models of technology transfer* [online]. Available from http://nptt.cvtisr.sk/buxus/docs//podujatia%201/prezentacie/5_Technology_Transfer_-_Models_and_Examples_S.Preissler_MOEZ_Fraunhofer_Lipsko.pdf [Accessed 14 Oct 2014]

[6] Kearns S. *Transport for London: the move towards sustainable transport* [online]. Available from http://sustainabledevelopment.un.org/content/documents/3742kearns.pdf [Accessed 14 Oct 2014]

[7] ITS program of the U.S. Department of Transportation's Research and Innovative Technology Administration [online]. Available from http://www.its.dot.gov [Accessed 14 Oct 2014]

[8] IBM publication. *Delivering intelligent transport systems: driving integration and innovation* [online]. Available from http://www-935.ibm.com/services/us/igs/pdf/transport-systems-white-paper.pdf [Accessed 14 Oct 2014]

[9] Hsu T.-P. 'Program towards humanity oriented development (HOD) of transportation in Taiwan'. *Proceedings of the Low Carbon Earth Summit-2013*; Sep 26–28, 2013. Xi'an, China

[10] Demirkan H., Kauffman R., Vayghan J., Fill H., Karagiannis D., Maglio P. 'Service-oriented technology and management: perspectives on research and practice for the coming decade'. *Electronic Commerce Research and Applications*. 2008;**7**(4):356–76

[11] Vargo S.L. 'Service-dominant logic reframes (service) innovation'. *Highlights in service research. VTT*, 2013

[12] Lusch R.F., Vargo S.L., Wessels G. 'Toward a conceptual foundation for service science: contributions from service-dominant logic'. *IBM Systems Journal*. Jan–Mar 2008;**47**:5–14

[13] Le-Dinh T., Léonard M. 'A conceptual framework for modelling service value creation networks'. *Proceedings of the 1st International Workshop on Information Technology for Innovative Services (ITIS-2009), IEEE 12th International Conference on Network-Based Information Systems (NBiS-2009)*; Indianapolis, IN, USA, Aug 2009

[14] Tien J.M., Berg O. 'A case for service systems engineering'. *Journal of Systems Science and Systems Engineering*. 2003;**12**(1):13–38

[15] Yurchyshyna A., Opprecht W., Leonard M. 'Collaborative decision constructing supported by Cross-Pollination Space'. *Proceedings International Conference on Advanced Collaborative Networks, Systems and Applications*, COLLA'11; Luxembourg, 2011

[16] Opprecht W., Yurchyshyna A., Khadraoui A., Léonard M. 'Governance of initiatives for e-government services innovation'. *Proceedings Electronic Government and Electronic Participation: Joint Proceedings of Ongoing Research and Projects of IFIP EGOV and ePart*; 2010, pp. 203–10

[17] FOXEL [online]. Available from http://www.foxel.ch [Accessed 14 Oct 2014]

[18] UNCITRAL United Nations Commission on International Trade Law, *International Colloquium on PPPs Discussion Paper* [online]. Available from http://daccess-dds-ny.un.org/doc/UNDOC/GEN/V13/821/00/PDF/V1382100.pdf?OpenElement [Accessed 20 Jun 2014]

[19] UNECE Guidebook on Promoting Good Governance in Public-Private Partnerships. New York and Geneva: UN; 2008

[20] UNECE PPP ICoE, *International PPP Centre of Excellence* [online]. Available from http://www.unece.org/ceci/ppp.html [Accessed 20 Jun 2014]

[21] Ars Industrialis. Available from http://www.arsindustrialis.org/ [Accessed 20 Oct 2014]

[22] Stiegler B. *We are entering an era of contributory work* [online]. Available from http://www.samkinsley.com/2013/02/06/bernard-stiegler-we-are-entering-an-era-of-contributory-work/comment-page-1/ [Accessed 30 Sep 2014]

[23] CamSam [online]. Available from http://www.camsam.com/ [Accessed 14 Oct 2014]

[24] Creative Commons Licenses. Available from http://creativecommons.org/licenses/ [Accessed 20 Sep 2014]

[25] Sheth J.N. 'Psychology of innovation resistance: the less developed concept (LDC) in diffusion research' in Sheth J.N. (ed.). *Research in Marketing*. Greenwich, CT, USA: Jai Press Inc.; 1981. pp. 273–82

Chapter 2

The importance of ITS in urban movement

*Steve Kearns**

Towns and cities across the world are growing at a rapid pace as the process of global urbanisation continues, seemingly at an unremitting pace.

This process is seeing the expansion of cities in both developing and developed societies; the phenomenon differs as many cities in developing countries have experienced and continue to experience largely linear growth patterns, for example Latin American, Asian and African cities, whilst large cities in Europe and North America, which previously experienced a decline of population following the growth of suburbanisation, are now seeing that trend reversing with populations in central and inner urban areas once again starting to increase.

The urbanisation process has resulted in an ever-increasing demand for movement in urban areas, often using modes of transport that are subject to high levels of congestion, overcrowding and lack of capacity.

Intelligent Transport System (ITS) has a key role to play in facilitating urban-based journeys for those using both private and public transport. It can aid journey planning; provide higher quality information, often directed at the individual traveller; provide the basis of vehicle-to-infrastructure (V2I) and vehicle-to-vehicle (V2V) communications; and provide a quick and efficient public transport ticketing system.

This chapter will focus on the function of ITS in urban traffic management. It will draw significantly on work undertaken by the European Commission's ITS Urban Expert Group between 2010 and 2012 and will aim to highlight the wide range of applications that ITS can offer in aiding individuals' movements in urban areas. The second half of the chapter will use Congestion Charging and Olympic Legacy in London as case studies to exemplify the action that is being taken to utilise ITS in traffic management.

Urban mobility can be seen as a mixture of needs and constraints alongside a series of triggers that can determine the nature and character of any ITS intervention.

Urban mobility has traditionally been closely correlated with economic development. As a city, urban area or conurbation grows, so the demand for

*Transport for London, London, UK

movement increases. Certain initiatives have recently been introduced to try to break this link, for example home working, and certain technological advances also indicate that there does not have to be an irrevocable link between economic development and the need/desire to travel, for example online shopping.

Figure 2.1 sums up the interrelationship of the needs, constraints and triggers, each of which acts as an input or potential input into the concept of urban mobility.

The categorisation above highlights a number of inputs and potential inputs that define the concept of urban mobility. Needs are balanced by constraints and respond to certain identified triggers.

Figure 2.1 importantly shows the ever-growing importance of environmental issues in an urban context. Public realm is being prioritised as a key issue by many urban authorities across the world and it is vital that ITS applications conform to this priority, that is, they should be as unobtrusive as possible and should in no way have a detrimental effect on the urban realm. Often promising ITS-based traffic management projects do not come to fruition because they are environmentally unacceptable to local or regional authorities or the population in a given city area. As technology evolves this issue should, hopefully, recede in importance as sensors, cameras and other devices decline in size and become more discrete.

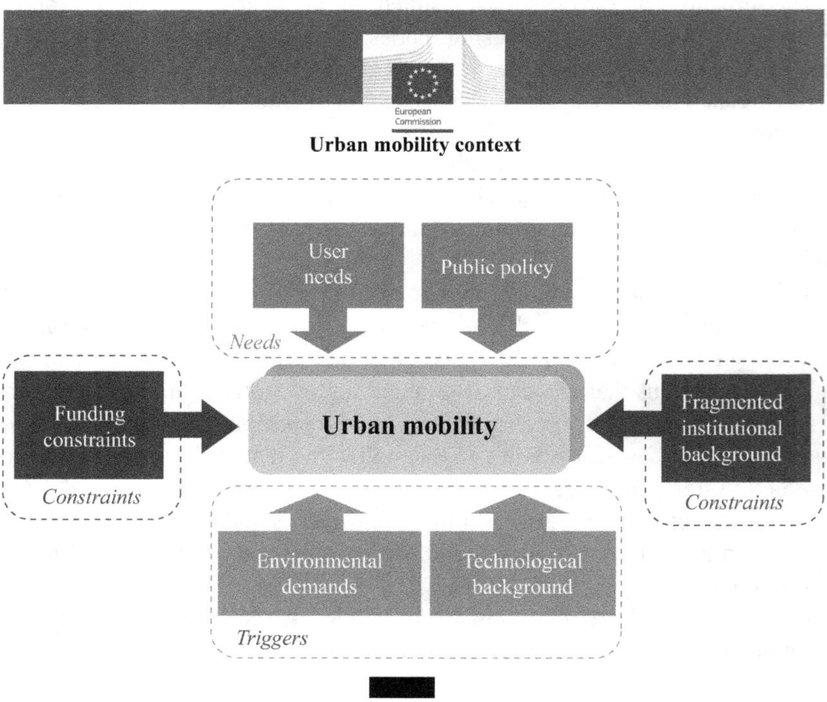

Figure 2.1 Urban mobility context – diagram of interrelationships

2.1 Traffic management and urban logistics

It is important to establish a framework of policy goals in relation to traffic management, which would provide the basis and rationale for deployment of ITS projects. A typical set of urban transport policy goals is shown in Table 2.1.

Table 2.1 Typical urban transport policies

(i)	Reduce congestion
(ii)	Reduce energy consumption and traffic emissions
(iii)	Improve quality of life in city centres
(iv)	Increase market share of clean vehicles in private and public fleets
(v)	Increase efficiency of transport systems
(vi)	Increase attractiveness of public transport and encourage modal shift
(vii)	Facilitate freight delivery and services
(viii)	Enhance road safety
(ix)	Decrease parking pressure

ITS has a central role to play in providing operational systems aimed at achieving each of the policy goals mentioned in Table 2.1.

A transport system could be defined as smart if it could deal efficiently with new situations (e.g. congestion, modal integration, safety) by linking data in order to produce valuable information for transport users and operators. Being able to create or develop a traffic and transport network to behave in a manner that acts intelligently is a key challenge and key application for ITS.

2.2 Existing technologies

2.2.1 Traffic signals

There are various examples of long-established usage of ITS in urban traffic management. These include *traffic signal control.* Originally traffic signals did not require any intelligence; the advancement in computer technologies junction control has become more sophisticated since the 1980s. Rather than just being a tool that typically separates conflicting movements at an individual junction, microprocessor modules enable modal-based, artificially intelligent, ubiquitous control using data from inductive loops and other detector systems. This has enabled the development of features such as prioritisation of public transport (e.g. through systems that recognise buses, trams) and optimisation of traffic throughput at individual arms of junctions.

2.2.2 Satellite tracking

Satellite tracking of vehicles has existed for some time but mostly used by freight transport and public transport operators for their own fleet management purposes. A big challenge for traffic and transport authorities is to use the vast amount of data that is generated from vehicle tracking systems and data emitted by mobile phones and other electronic devices in a coherent and intelligent way.

A source of floating vehicle data that may be more accessible to the highway authority is through public transport systems such as Real Time Passenger Information (RTPI) and systems designed to give private motorists information based on satellite tracking of their route and location. RTPI systems are fundamentally designed to inform users when their next bus/tram/train is due to arrive. RTPI and systems aimed at the private motorist typically use some form of tracking either through Global Positioning System, GPS (satellite), or dedicated short range communications (DSRC), Bluetooth or GSM (V2I) to identify positioning that can be used to measure delays and congestion in areas of the road network not otherwise monitored.

2.2.3 Probe vehicles

Access to floating vehicle data and cellular data is a relatively new concept. Where this information is available to the local authority, it enhances the view of the urban road network performance and can serve as a source for switching signal programmes to keep traffic flowing in an optimum fashion. In this case, travel time data can be derived from *probe vehicles* with the data being used to refine traffic signal programmes and could have the added benefit of reducing the need for roadside infrastructure such as local sensors.

Often local traffic and highway authorities have access to large quantities of valuable vehicle data through the bus fleets that operate in their areas. A more valuable source probe vehicle data can often be taxis as they have the flexibility to deviate off route when they experience a problem rather than wait in traffic congestion on a proscribed route.

2.2.4 Cellular data

Cellular data (the tracking of mobile phones and other electronic devices) can vary greatly from country to country with some countries allowing it whilst others have legislative restrictions. Cellular data has previously required the cooperation of mobile phone companies and where it is made available no additional roadside equipment is required. However, the raw data often needs to be converted so that it can be used by the highway authority. Bluetooth readers can easily be converted for roadside use and the data is collected in such a way that it can be used by the urban road authority. This technology is evolving rapidly with the advent of 3G and 4G allowing faster and more precise identification of location of individuals based on their electronic devices.

2.2.5 Automatic number plate recognition

Automatic number plate recognition (ANPR) systems for journey time monitoring are widely used in some European urban area whilst again local legislation in some area does not allow such systems. ANPR is beneficial in identifying traffic location and movement based on vehicle registration plates and has been used productively across Europe and North America. It does, however, tend to be quite labour intensive with significant manual intervention required to verify 'doubtful' vehicle reads.

2.2.6 CCTV

CCTV has existed for many years and it is typical to find cameras monitoring the highway network, particularly at critical junctions, linked back to the highway authority traffic control centre. Historically, CCTV has enabled the highway authority to view what is happening and intervene through the Urban Traffic Control (UTC) system based on what is visible. It is often the case that whilst junction-specific intervention, based on images received from CCTV cameras, will help, it may not be that easy to spread the benefits to the wider road network when incidents and accidents occur. It is preferable to gather a wider range of data sources (detectors, ANPR, CCTV, floating vehicle data, including buses, trams and freight vehicles) and feed these into the UTC system to allow area-wide network management to be maintained. With a wide range of data sources, central control and management systems and strategies, highway authorities are finding it much easier to deal with incidents as they occur. Such developments have also pushed the boundaries on access and control possibilities where congestion and air quality issues continue to persist.

2.2.7 Vehicle detection systems

Vehicle detection systems, as already highlighted, can be used for a variety of purposes from car park management to dedicated parking for just in-time freight deliveries at the kerbside. The potential for vehicle detection systems, often using infrared or comparable technology, to facilitate more efficient traffic movement in urban areas should not be underestimated. A prime example of such functionality is on-street and off-street parking in urban areas. Increasingly evidence is emerging that circulating traffic seeking a parking space is a significant contributor to congestion in urban areas. ITS can have a key future role in reducing the negative impact of such traffic by informing drivers of the location of the nearest unoccupied parking space.

2.2.8 Pollution monitors

Pollution monitors assist the highway authority in understanding the impact of vehicle pollution within the urban environment. Such data fed in real time into the UTC system can be used to develop strategies that can be implemented to specifically change traffic patterns through signal timings in response to poor air quality.

2.2.9 Variable message signs

Variable Message Signs (VMS) are now common place across larger towns and cities in many parts of the world. Whilst these may be considered to be a tool for delivering travel information to the roadside, they are also a valuable traffic management tool. VMS can be used to inform and direct drivers of problems and around incidents when they occur, reducing their impact on the highway network. VMS can also be used to inform on specific reduction in capacity in the network:

for example when an incident makes it necessary to give priority to public transport or when a major event (sports, exhibition or other) requires implementation of a special parking and traffic management strategy. A further application is to inform the general public on pollution levels caused by traffic.

2.2.10 Road user charging

Road User Charging can be challenging to introduce politically but has proved to be very effective in its primary objective in cities such as Singapore, London, Stockholm and Milan of reducing traffic volumes and levels of congestion.

The ITS element of Road User Charging schemes include payment, enforcement and back office functions.

Payment systems should ideally be as user-friendly as possible. In order to minimise the possibility of incurring a penalty charge, either through deliberate non-payment or through forgetfulness, it is desirable from an operational and customer relations perspective to offer account-based systems involving customers registering and the service provider offering a post-payment facility.

Enforcement systems for road charging schemes vary across the world. ANPR systems record a photographic image of vehicle registration plates and associate this information with owner details of the vehicle. DSRC, often more commonly referred to as 'tag and beacon', utilises wireless radio communications between roadside beacons and electronic tags in the windscreen of vehicles. The beacons read the tags in the windscreens as they pass a given point. They can use DSRC technology that generally offers greater potential for a more sophisticated charging system such as one which can set a charging level based on type of vehicle, direction of travel, time of day, distance travelled, etc. The disadvantage of DSRC is the amount of infrastructure that such a system requires; often roadside beacons have to be supplemented by overhead gantries in order to get an accurate reading from a vehicle at a given location.

A further enforcement technology used for distance-based charging is Global Navigation Satellite System (GNSS), often in Western Europe, and North America, using GPS. Usage of GPS has the advantage of not requiring roadside infrastructure and has been used successfully for national road charging systems. In dense urban areas for this technology to be successfully applied, it is necessary for there to be a good degree of visibility between the satellite system and the vehicle in question. The more satellites that are directly visible from a vehicle, the more accurate the locational reading relating to the vehicle is.

Low Emission Zones (LEZs) that generally aim to improve air quality within a given area by targeting the worst polluters have proved easier to implement. As LEZs directly affect freight operators, their requirements need to be fully taken into account, for example through exemptions and alternative loading arrangements. Revenue raised from access control schemes can also be productively used to further traffic and transportation goals – for example in Oslo, revenue from the toll ring systems provides a valuable contribution towards investment in new infrastructure, both road and rail/Metro and running costs for public transport.

2.3 Evolving technologies

ITS, by its very nature, is constantly evolving. In the context of urban traffic management, development is currently being undertaken on *vehicle-to-infrastructure (V2I)* systems. Pilot projects such as the German simTD, with a fleet of 120 vehicles, test and demonstrate the suitability of various V2I systems, amongst them road works information systems, local traffic-adapted signal controls, traffic sign assistants and traffic light assistants. Some V2I data is also collected through extended floating vehicle data to provide updates on weather conditions or the availability of parking spaces. Other examples of pilot projects research work undertaken by German vehicle manufacturers leading to on-street trials in London, giving drivers information on traffic signal status; the 'Spitsmeiden' project in the Netherlands and the Co-Operative Vehicle Infrastructure Systems (CVIS) and related projects that have taken place throughout Europe where communication between vehicle and infrastructure has been a key feature in drivers receiving information about parking spaces, traffic disruption, public transport services, etc.

Vehicle-to-vehicle (V2V) technology may still be at the research stage but clearly shows potential to play a significant role in urban traffic management. The capacity for vehicles to communicate with one another, thereby acting as probes or mobile sensors is an exciting concept. There is considerable scope for V2V and V2I to play an increasingly prominent role in urban traffic management by providing real-time information on location and speed of individual vehicles, with this information being communicated to other vehicles, to traffic signals, traffic control centres and being incorporated in traffic management plans. The information generated can be used in daily decision making which helps to keep traffic flowing. The re-routing of traffic in the case of incidents is important in this respect. The possibility of direct messaging to the navigation systems of freight vehicles will also help to ease the flow of commercial transport.

It is essential that stakeholders involved in initiatives such as V2V and V2I establish and develop productive working relationships with other relevant actors, such as vehicle manufacturers and academic institutions, who are at the forefront of this research. This will be necessary to ensure that traffic management applications flowing from these technologies are fully realised.

Local road authorities often are not in close contact with academic institutions and, particularly, vehicle manufacturers. However, the collaboration of the German regional road authority Hessen Mobil with the automotive industry and with various academic institutions can be cited as one practical example of such cooperation in the field of V2V technology. It is highly desirable that the links between local road authorities, industry and academia are improved. One potential channel may well be the involvement of road authorities in regional and national ITS organisations, who have direct links with vehicle manufacturing sector. Local authority representation on ITS bodies is important in creating and defining an agenda.

Local road authorities can also influence the development of V2V and V2I in other ways. The roads they control can be an ideal, real-life testing ground for new technologies. Offering to host trials can influence how these technologies evolve. Laboratory and off-road testing are important but, at some stage, all traffic-related technologies need to be tested in a real-life situation and urban road networks generally offer the most challenging environment. By hosting technology trials urban authorities maximise their opportunity to influence their development.

Road and telecommunications infrastructure form a key part of V2I technology. Its location and manner of usage should generally be agreed by the local authority, which, in turn, gives that authority leverage over the development process. However, the infrastructure may also be provided by communication suppliers, thereby reducing local authorities' financial input – there is not necessarily a need for the entire infrastructure to be provided by the road authority themselves.

Local road authorities need to be able to gain access to information emanating from V2I and V2V in order to ensure that data arising from these technologies are fully available to the local urban traffic management control system and consequently ensure that these technologies are used to their optimum effect.

2.4 Key factors for success

2.4.1 Cooperation, partnership and interoperability

These attributes may appear self-evident when designing any ITS project but they are especially applicable to ITS in an urban environment.

Local authorities often have limited capacity to specify and procure the ITS facilities that they are seeking to optimise traffic performance in their areas. This is especially important at urban/suburban/regional interface where the jurisdiction of roads changes. A typical urban car journey could often involve the driver using roads that are managed by two or more authorities.

It is essential that these authorities work in conjunction with one another to achieve maximum equalities of scale and optimise the level of service that is provided to the individual motorist. Joint procurement and contract management will help in this respect.

2.4.2 Targeting individuals – optimising network performance

As communications technology and ITS develop, it is becoming increasingly clear that individual travellers benefit particularly from individually targeted information.

With respect to traffic management in urban areas the prevalence of a road network that is at or near capacity for long periods of the day presents particular challenges. Roadside cameras, satellite tracking devices and information from electronic devices can give valuable information about the position of an individual and their relative position in relation to any perturbation on the highway, for example broken down vehicle, site of road traffic accident, etc.

The challenge is to use this information as intelligently as possible, for example in a situation where there is an obstruction on the highway should all vehicles be advised or instructed to take a specific diversionary route, this could well result in the diversionary route itself becoming congested as there is often only limited marginal capacity on this route. ITS can be applied in an intelligent way by advising certain drivers to take one diversionary route, others to take a different diversionary route and others to remain where they are currently located and continue their journey on their original route.

Individually targeted information is not only more efficient for the person at whom it is directed, but it also has a beneficial effect for the network as a whole as it results in more efficient network operation.

2.4.3 Maximising ITS potential/minimising human intervention at operational level

A successful ITS traffic management policy or project should aim to limit the amount of human intervention at basic operational level as this tends to be costly and often time consuming. The role for humans should be focused at a strategic level determining policy and managing processes.

For example, camera coverage of traffic and highway networks in urban areas is often comprehensive with many city authorities having a proliferation of roadside cameras that are used to facilitate management of the road network. The challenge for camera operators and traffic managers is to identify which of the often numerous cameras they should view and what action they should take in order to optimise the performance of the network.

ITS techniques such as video analytics are helpful in detecting unusual patterns of movement captured by roadside cameras, sensors, loops, etc. These can be used to draw the attention of the camera operator to specific locations where intervention would be beneficial in expediting traffic movement. Currently, there is still an element of human intervention inherent within this process but ITS has the potential to offer a wholly automated operational system, which, for example, would introduce a given traffic plan when particular patterns of vehicular movement are detected.

Future development of V2I and V2V systems should also result in the introduction of automated processes, many of which will be carried out in real time or near real time. Their advent will increase the role and relevance of ITS at operational level in urban traffic management.

Standards and harmonisation are important in delivering effective ITS traffic management projects. They serve to improve the quality of a project and enhance the concept of interoperability. However it is important that standards are used in an appropriate manner in an urban project, as being overly proscriptive could limit technical innovation and undermine the EU subsidiarity principle, which is important in an urban context.

Specifically in the area of traffic management the rationale to establish and use standards are quite different.

- Rationale for regional 'standards': *Reducing procurement costs for field devices and fostering innovation*

  Some manufacturers have established a de facto monopoly situation by providing silo systems for traffic management. Due to the absence of standardised interfaces of central systems, field devices have to be procured from the same vendor in order to be connected to the centre ('vendor lock-in'). The resulting absence of competition and lack of exposure to market forces not only leads to high price levels, but also hinders innovation since new, innovative companies are denied access to the market.

- Rationale for supra-regional 'standards': *Networking centres and enabling data exchange*

  Networking existing traffic management systems and creating new traffic management services require *interoperability* and data exchange from centre to centre. On supra-regional and international level, centre-to-centre data exchange is a mandatory prerequisite for continuity of services. The major benefit of centre-to-centre communication standards therefore lies in the networking capabilities and interoperability of subsystems from different operators and manufacturers. Reducing cost is less important here since the cost of central facilities such as accommodation costs, computing costs, monitoring facilities, software are usually only a fraction of the overall cost of a traffic management system, which would consist of traffic management and Traffic Control Systems (e.g. loops, signal controllers, cabling, VMS signs).

Creating interoperability does not nullify the importance of specification or even standardisation of components of ITS services. This is crucial since each modification (enlargement, modification, new procurement, etc.) of IT infrastructure should be based on existing (local, national) or market-available products (protection of investment). Furthermore, an approach that stipulates adherence to a rigid standards and harmonisation protocol could well be expensive and time consuming. A key element of a productive approach to standards and harmonisation is the creation of an open platform for ITS services. This allows the advantages of technological interoperability to be captured and applied, whilst allowing individual authorities to maintain autonomy in deciding the precise platform(s), form of technology and services that they wish to use, thereby avoiding potential vendor lock-in.

Initiatives such as Universal traffic management Control (UTMC) have created open standards and specifications for system to system data exchange. This has led to the creation of an integrated network strategy tool enabling better highway network management and incident detection and response. Open Traffic Systems City Association (OCA) has created an open specification for traffic signals and detector data exchange with the UTC system which offers similar advantages to UTMC. A key feature of UTMC, which broadly operates in cities in English-speaking countries and OCA, which broadly operates in German-speaking cities, is that they both provide an open platform for progressive traffic signal control. Not only can they each provide a platform for UTC in their own right but urban road and traffic authorities also have the opportunity of interworking the two systems by selecting features of each to meet their own requirements, whilst retaining a common open operating platform.

2.5 The importance of freight in urban areas

A feature of traffic management in urban areas is the necessity of providing adequate facilities for freight vehicles to service residential, commercial and retail premises.

This is not such a challenging task where there are adequate loading/unloading facilities provided off-street. However properties in the central areas of many towns do not enjoy the luxury of off-street servicing facilities. In such cases the local traffic/highway authority has to prioritise the scarce resource of kerbside capacity. It is generally not sufficient to merely allocate sufficient roadside space for freight vehicles; it is usually the responsibility of the local authority to ensure that other vehicles do not use the space in question. Methods of enforcement of parking regulations, therefore, have to operate as efficiently as possible. ITS has a role to play in optimising the usage of on-street parking spaces. Examples include pre-booking of on-street parking spaces by freight companies. This can be done relatively simply through creation of a website to allow advance on-street booking and an enforcement system that recognises the registration plate of a specifically booked vehicle. More refined systems could also allow usage of such a freight facility by general vehicles at times when it has not been pre-booked by a freight vehicle.

2.6 ITS traffic management links with sustainable transport modes

This chapter has focused on ITS functions relating to road-based, specifically private transport in urban areas. Transport policies adopted by cities are usually designed to promote sustainable modes, leaving the scarce resource of highway capacity to those vehicles who are priority users, for example public transport modes (buses, trams, etc.) and delivery vehicles.

ITS information needs to be targeted at individual motorists but also needs to make them aware of other modes of transport, thereby helping to encourage them to use more sustainable modes.

ITS plays a key role in providing enhanced public transport services, not solely in relation to vehicle technology, but also core services such as smart ticketing and journey planning.

A comprehensive ITS service aimed at facilitating the journey experience of individual motorists should also incorporate features designed to encourage modal shift. Examples include notification of the operational status of public transport services and identification of unoccupied parking spaces at station car parks.

2.7 London – Case studies for its traffic management

2.7.1 Background

London is a world city with a current population of 8.6 million residents and approximately 330,000 businesses, which directly employ about 4.6 million people.

It is a centre of commerce and finance and major tourist destination, attracting more than 30 million visitors per year.

One critical facet of the population trend that has had a direct impact on transport planning is the fact the numbers of people living in the city declined steadily from the end of World War II to the mid-1980s when a large-scale flight took place to the suburbs and 'Green Belt' area including the new towns that were built some 20–30 miles outside the city. These new towns were built to house those Londoners who were no longer able to live in their homes due to the bomb damage sustained during the war and those who wished to live in a more suburban, less dense environment.

From the 1940s to 1980s it was generally anticipated that this depopulation of London would be on-going phenomenon and the city would continue to decrease in size. This perception relating to population trend was reinforced by the ever-increasing rise in car ownership from the 1950s onwards, which was not only seen as heralding a new era of personal, independent travel, it also redefined the relationship between town planning and transport planning.

It became possible to live some distance outside London and to use a car to drive into the city each day to work. In many ways a car became a prerequisite for those people in areas outside London as there was not the level of public transport provision in the suburban areas and new towns as there was in inner and central London.

In the mid-1980s the population of London started to stabilise and from the 1990s the earlier trend started to reverse with the population once again starting to increase. Initially the reason behind the reversal of this population trend was primarily assigned to rise in immigration, particularly from Commonwealth countries. This trend has not only continued since the 1990s but has become greater since the turn of the twenty-first century. Forecasts now indicate that London's population will grow to over 10 million by 2030 with high forecasts estimating a population of 13.4 million by 2050 (Figure 2.2). The main reason for this sustained growth is no longer immigration, though many additional people from the European Community accession countries have become resident in London over the past 10 years. The major factor influencing population growth is now the relatively high birth rate among the resident population and high level of internal migration.

The reversal of the population trend from the 1990s onwards has presented many challenges for transport planners. The downward trend in public transport usage was reversed and congestion became an ever more pressing problem as car ownership continued to grow and unfettered usage of cars became a dominant feature of transport movement across the city.

Transport for London (TfL), created in 2000 by the Greater London Authority Act 1999, which also created the position of the first directly elected Mayor of London, has responsibility for a range of transport modes across the city. TfL is answerable directly to the Mayor of London for transport provision in the capital. In addition to 8,500 buses, 12 London Underground lines, the London Overground suburban rail service, river transport, Docklands Light Railway, Tramlink in the

Figure 2.2 London population growth

Croydon area, TfL is responsible for licensing 22,000 black taxis and over 40,000 minicabs. With respect to road transport, TfL has responsibility for 6,000 sets of traffic signals and is the highway authority for 580 km of the major road network, whilst this represents only 5% of the total road network – the other 95% of the network is under the control of the 32 borough councils and the City Corporation – this section of the road network contains all the major roads in London and carries over 33% of the total traffic. TfL also has direct responsibility for promoting walking and cycling facilities across London.

TfL is taking significant steps to improve capacity to meet ever-growing demand and, at the same time, introducing measures to mitigate against the negative environmental impacts of public transport – introducing buses with lower CO_2 and NO_x emission levels and some of the zero emission vehicles; designing a new Ultra Low Emission Zone, which is scheduled to be opened in 2020; constructing new rail lines and line extensions to meet the ever-increasing demand for services, particularly the new east-west Crossrail line, scheduled to open in 2018, which will increase total rail network capacity by 10%, linking Heathrow Airport, central London and the Docklands regeneration area; a comprehensive programme to upgrade rolling stock, signalling and increase station capacity on the underground network; updated RTPI using a variety of channels including texts and twitter and many other initiatives to enhance the experience for people using TfL's services.

2.8 Case study 1 – Central London Congestion Charging scheme

ITS in traffic management became an immediate issue for TfL on its creation as one of the candidates at the first mayoral elections in 2000, Ken Livingstone stated that road-based congestion had become such an overriding problem that, if elected, he would introduce a Congestion Charging scheme in the central area. He was elected with just under 70% of the vote and immediately instructed TfL to introduce a Congestion Charging scheme by early 2003.

Congestion Charging represented a significant political risk for the Mayor. It had many opponents among certain residents and particularly among businesses. However, TfL had to approach the task set by the mayor and introduce a scheme as efficiently and expeditiously as possible. It was important to identify a technology that was proven to be successful and given the strict time constraints, it was decided to opt for ANPR technology for payment and enforcement purposes.

ANPR was proven technology, even in the early 2000s and a decision was made to use an ANPR roadside camera network to capture vehicle registration number plates as they passed the boundary point. In order to achieve the requisite accuracy, the cameras had to be located at sites where they could capture the image of the number plate whatever its position on the carriageway, which meant deployment of more than one camera on larger roads with two or three lanes (Figure 2.3). These cameras had to be sited so they could capture a close up image of the registration plate, a wider angle image of the make, model and colour of the vehicle for verification purposes but also sited so they could not capture an image of the driver of the vehicle, which could have led to legal complications.

Care was taken by TfL staff in formulating the contravention of being in Congestion Charging zone without having made a valid payment. The person responsible for paying the ensuing fine would be the owner of the vehicle, not the driver ensuring that the owner could not maintain that he/she was not driving the vehicle at the time of the contravention.

On the inception of the scheme particular attention was paid to the capture rate of the cameras, which was in excess of 90%. It was decided to supplement the boundary cameras with additional ones inside the zone as the charge itself related directly to being in the zone rather than entering the zone, that is crossing the zone boundary. This increased the capture rate to the high 90%'s.

Various ploys were tried by drivers who wanted to avoid paying the charge. These included resiting their number plates away from the centre of the vehicle, which were dealt with by police officers being alerted to this practice, intercepting offending vehicles and issuing penalty charge notices. To improve capture rate of the cameras in hours of darkness and bad weather conditions, infrared illuminators were incorporated into the second generation of cameras when the cameras were refreshed some years into the scheme. Foreign number plates also posed a particular challenge for the cameras – they are able to read all Latin script but cannot yet identify Arabic script. Enforcement of violation by foreign vehicles is dealt with a bailiff company, which buys the debt from TfL and pursues the drivers to obtain

Figure 2.3 Boundary point of central London Congestion Charging scheme

payment of the fine. This now works well with registration plates of violating vehicles in most Western European countries but is still not comprehensive in capturing all Eastern European violating vehicles.

The Congestion Charge in London is a flat fee (£5 when introduced in 2003 and since risen incrementally to its current level of £11.50). The flat fee is payable as a daily charge irrespective of the amount of time a vehicle spends in the zone on any given day, so the £11.50 is payable if a vehicle spends 11 hours in the zone or merely enters the zone for a matter of minutes or even seconds.

A number of technology trials took place in the mid-2000s to address this blunt aspect of the scheme including replacing ANPR with a GPS system or DSRC tag and beacon scheme. The GPS trial when carried out did not achieve the required level of accuracy due to the relatively narrow streets in the city coupled with tall buildings blocking sight of the four satellites, which is ideally required to get a locational fix within 1 m, which was the parameter set for the test. The accuracy of GPS should improve as GALILEO comes on stream.

The DSRC trial using tags and roadside beacons showed that achieving accuracy in identifying a vehicle's location within 1 m could be undertaken successfully in a dense urban environment. The main issue relating to the DSRC trial was not a technical one; it related to the urban realm. From the outset of the trial, local residents, businesses, community groups and other interested stakeholders made it clear

Figure 2.4 DSRC charge point – Singapore

that installation of overhead gantries, similar to ones used in road pricing schemes in Singapore and Stockholm, would not be acceptable in London (Figures 2.4 and 2.5). Pressure to enhance the urban realm is increasing in large cities across the world and London is no exception. TfL, during the DSRC trial, used roadside columns rather than overhead gantries, modifying the length of the arm emanating from the column to match the width of the carriageway (Figures 2.6 and 2.7).

The DSRC trial concluded that it was possible to utilise this technology for an area-based road pricing scheme in a city such as London. The cost of introducing the required infrastructure to convert from ANPR to DSRC was estimated to be in the region of £60 million, which was considered to be prohibitively expensive and the scheme has continues to operate using ANPR enforcement and payment technology to this day.

In 2011 Congestion Charging Auto Pay was introduced and has rapidly become the most common means of payment for those people wishing to pay the charge. Auto Pay involves an owner registering one or more vehicles and giving credit card or debit card details to TfL with payment being taken automatically on the occasions when the registered vehicle(s) is identified inside the zone. The great advantage of Auto Pay is that the vehicle owner will never again incur a Penalty Charge Notice (PCN) and this brought a consequent benefit to TfL as the number of PCNs reduced from a high of about 8,000 per day down to a current level of about 3,000 per day. TfL is able to make considerable savings through processing

Figure 2.5 DSRC infrastructure – Stockholm

Figure 2.6 DSRC roadside column – London road pricing technology trials

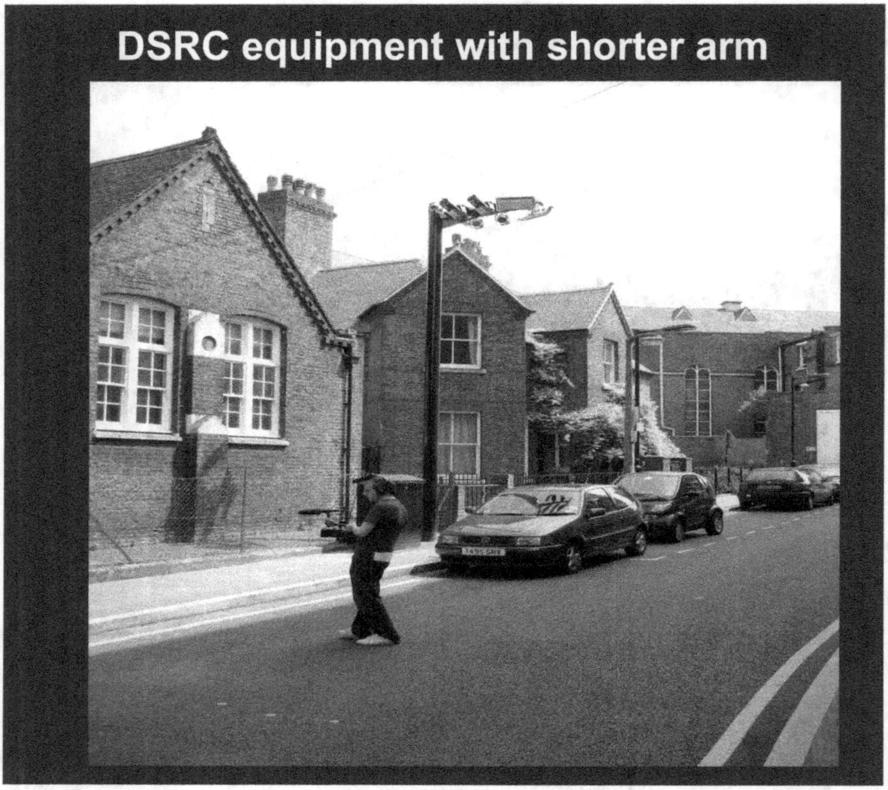

Figure 2.7 DSRC 'Short Arm' column – London road pricing technology trials

fewer PCNs as each contravention has to be checked manually to ensure that the details are correct before the details are sent to the driver.

Although entering the Congestion Charging zone without having made a valid payment is a civil offence, dealt with by TfL rather than the police, an interesting by-product of the Congestion Charging scheme is the correlation that has emerged between non-payment of the charge and involvement in more serious criminal activities. Since the introduction of the scheme TfL has been able to provide the police with information on extremely serious criminal offences based on information derived from Congestion Charging non-paying vehicles.

Congestion Charging remains TfL's flagship project and has resulted in a decrease of 28% of general traffic in the central area since its introduction in 2003. Although perhaps not at cutting edge in terms of ITS usage, it has established a robust and effective operating regime based on proven technology and has been refined and modified over the past 12 years to ensure that it continues to make a real difference to the level of traffic and quality of life for people in central London.

2.9 Case study 2 – 2012 Olympic legacy

The staging of 2012 Olympic and Paralympic Games in London presented TfL with some very serious challenges in terms of managing the numbers of people attending the Games events.

2.9.1 Managing large groups of public transport users

The need to manage large groups of people travelling on public transport was integral to transport planning for the London Olympics in 2012. TfL is used to managing large numbers of people – over 6.5 million use London's buses each day and 4.5 million use the tube in addition to those people attending sporting events and one-off occasions such as Royal Weddings, Tour de France Grand Depart, London Marathon and New Year's celebrations. The challenge during the Games was to cope with patterns of movement of large numbers of Olympic spectators overlaid on regular patterns of movement a number of times each day. This was because events at the Games' venues were scheduled for three separate time slots, therefore the Olympic Stadium in Stratford hosted three separate sessions – morning, afternoon and evening – each of which were attended by 80,000 people. Similarly on some days Wembley Stadium hosted football matches at three sessions, many of which also attracted the capacity crowds of 80,000. Added to this were equestrian events at Greenwich which attracted 50,000 people to each session, tennis at Wimbledon (30,000 per session) and a variety of smaller venues hosting sports such as basketball, swimming, beach volleyball (in Horsguard's Parade in the heart of the city) and many other sports.

Public transport was the key means of getting people to and from the events. Given the capacity problems being faced by London's public transport on a daily basis, it was clear that a strategy of managing people movement through the transport system was required. Passenger numbers using all central London tube stations were modelled based on overlaying predicted flows generated by Games events on existing travel patterns. Data regarding predicted journeys of Olympic visitors was based on post-code information derived from ticket sales.

Pinch points in the tube network became apparent through this exercise and regular travellers, who TfL could track through information derived from their Oyster card (Public Transport Smartcard) usage, were discouraged from travelling when congestion on the network was forecast to be at its greatest. When registering for an Oyster card, a traveller is encouraged to give their e-mail address and mobile phone number (useful for ongoing communication between TfL and the individual about service disruptions, fares revision, etc.). E-mailing and texting relevant individual travellers with bespoke messages such as 'Avoid London Bridge station between 3.00pm and 6.00pm tomorrow afternoon. It will be exceptionally busy' were extremely well received during the Games by regular travellers. This service has continued post-Olympics with individual passengers being pre-warned about disruptions via e-mail and text. TfL now also makes significant use of twitter and Facebook for real-time communications with its customers.

2.9.2 *Managing road-based traffic*

Whilst TfL prioritised movement by public transport during the Games, unusual patterns of road-based traffic were also forecast and indeed did materialise. The Games family were based in hotels in central London and TfL made a commitment to the International Olympic Committee that the journey time between Park Lane, where most of the hotels used by the Games' family were located, and the Olympic Stadium in Stratford in East London would not exceed 21 minutes. This involved a number of strategies being employed in setting traffic signals to give priority to certain routes and use make became known as Active traffic management (ATM). The roads linking central London and the main Games venues were known as Olympic Route Network.

An essential precursor of ATM modelling was the construction of a series of interrelated traffic models ranging from a pan-London model through to an area model, detailed models of strategic corridors down to models of individual key junctions on the road network (Figure 2.8). This series of models was supported by an interventionist programme of ATM based on camera observation by traffic controllers of all the key junctions and corridors on the Olympic Route Network.

Traffic coordination for the Games was carried out at a newly formed London Surface Transport Control Centre (LSTCC), which brought together controllers of road network, bus controllers and the Metropolitan Police's traffic controllers. The advantages of having these key agencies working together under one roof cannot be overstated. Critical decisions could be taken at speed in a highly coordinated manner

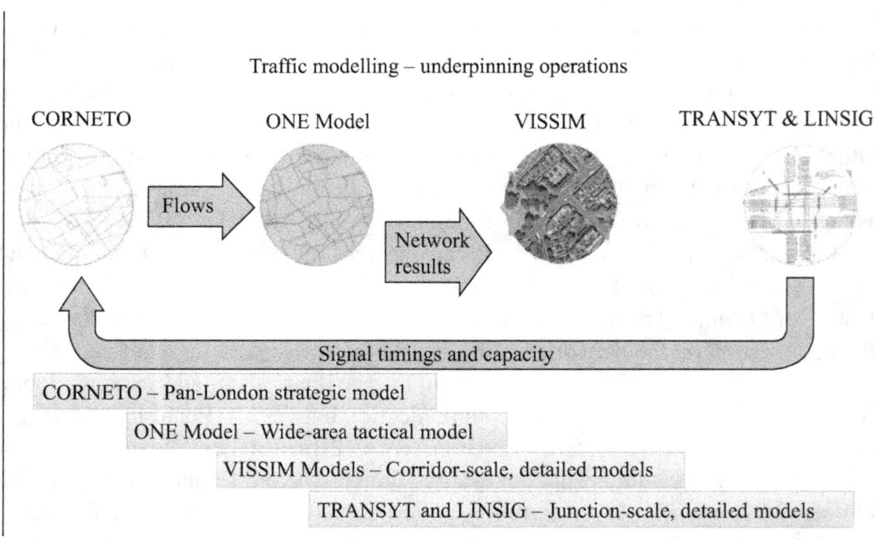

Figure 2.8 2012 Olympic Games – traffic modelling

Figure 2.9 Image recognition and incident detection

involving all relevant agencies. The LSTCC is now a permanent feature of traffic management in London but it was the Olympics that were the catalyst for its creation.

Comprehensive camera coverage on the road network was a vital component of ATM. The challenge for traffic controllers during the Games (and since then) has been to make optimum usage of the camera system. The number of cameras on the highway network in London now runs to several thousands. Not all of these are used for traffic management – some are police or borough cameras that are used for safety and security purposes. TfL engineers have devised a facility using video analytics called Image Recognition and Incident Detection (IRID), which identifies a camera image, where an unusual pattern of movement is detected (Figure 2.9). For example, if the usual traffic speed on a particular road at 8.00 a.m. on weekday morning is 15 mph, the camera is programmed to recognise a pattern of movement that deviates sharply from that speed. In this case the camera can be set to report a pattern of movement of less than 8 or 10 mph and once the average traffic speed has reduced to this figure, the camera will flag up to the operator with a visual and/or audible warning.

IRID obviates the need for the camera operator to scroll randomly through large numbers of cameras until he/she finds a situation where ATM intervention would be appropriate and allows the operator to make an early intervention to alleviate potential problematic traffic situations. This facility has proved a most efficient Olympic legacy project.

2.9.3 Managing pedestrian flows

The Games also saw large numbers of pedestrians in London, many of whom were not accustomed to the road layout or geography of the city. London has traditionally been a challenging city to negotiate on foot. It does not have a neat grid pattern of streets like New York. Even native Londoners often use the tube for relatively

Figure 2.10 Legible London

short journeys, as the London Underground map is so well known and is often relied on by local people and tourists alike. The reality is that walking distances between points are often shorter than they appear on the tube map, for example travelling between Bayswater (District and Circle Line) and Queensway (Central Line) would involve a change at Notting Hill Gate, with a journey time of up to 15 minutes, whereas both stations are, in fact, located in the same street, less than 5 minutes walk away from one another.

To encourage more people to have the confidence to walk in the central area, TfL introduced a programme called 'Legible London', which has resulted in the proliferation of new information roadside boards, maps and signage with a characteristic yellow and black colour scheme. Legible London also makes innovative use of 'heads up mapping' (maps orientated from the directional perspective that the viewer has when reading the map rather than always having North located at the top of the page) and 3D landmark building representation. Legible London is supported by a range of in-house and external mobile phone apps (Figure 2.10).

Another feature of large pedestrian numbers that became apparent during the Games was the lack of flexibility of traditional traffic signal programmes in recognising the quantity of pedestrians wishing to cross the road at any given signalised crossing. Since the Games TfL has devised a system, again using video analytics, to identify precise numbers of pedestrians waiting to cross the road at a signalised crossing.

Figure 2.11 Pedestrian SCOOT at traffic signals

The introduction of Pedestrian Split Cycle Offset Optimisation Technique, or 'pedestrian SCOOT', is the first of its kind in the world and uses state-of-the-art video camera technology to automatically detect how many pedestrians are waiting at crossings. The camera image is divided into a number of discrete sections and the formation of the pixels in each section indicates whether it is empty, partially occupied or fully occupied (Figure 2.11). From this information, video analytics are used to estimate the number of people waiting to cross the road.

The system enables the adjustment of traffic signal timings automatically to extend the green pedestrian invitation to cross phase when large numbers of people are waiting, allowing more people to cross the road.

In addition, TfL is developing a 'call cancel' technology, which can detect when a pedestrian who has pushed the crossing button has either crossed before the signal goes green or walks away, and therefore cancels the pedestrian crossing.

2.9.4 Managing freight

A feature of freight servicing during the Games was the success of the campaign launched by TfL to persuade freight operators to re-time their deliveries to avoid not only peak times but also pre-peak, inter-peak and post-peak times. The central thrust of the campaign was to try to persuade freight operators to service premises in non-peak times.

Out of hours servicing has traditionally been quite difficult to achieve as freight companies and their customers are not structured to carry out and receive freight deliveries at this time. During the three weeks of Olympic Games and two weeks of the Paralympics freight companies made considerable efforts to radically alter their delivery schedules. Obviously, this had to be done sensitively in residential areas and measures were taken to keep noise associated with deliveries to a minimum, for example by using mobile-carpeted walkways when delivering or collecting materials such as metal beer barrels.

Out of hours deliveries are easier for large companies such as department stores and supermarkets, where there is sometimes security staff on the premises during the night. The same is generally not the case for smaller businesses. One positive outcome of freight operation during the Games was the channel of dialogue that was opened between larger businesses and smaller, neighbouring businesses. If the only impediment to a smaller business receiving a delivery or collection during night hours was the absence of staff, TfL encouraged larger businesses in the vicinity to accept the delivery or collection on behalf of the smaller business and arrange for the goods to be transferred between the businesses the following morning. This proved successful in some cases and the plan is to look at its longer term potential.

The Games offered TfL a golden opportunity to reach large numbers of businesses and freight operators through its communications strategy, some of whom had no previous history of direct contact with TfL. This contact has been sustained through:

(a) The establishment of a Freight Forum representing all interested stakeholders, which was established to improve communication with the industry and help coordinate planning and preparations for the 2012 Games.

(b) The strengthening of the Freight Operator Recognition Scheme (FORS), through which TfL recognises freight operators who behave responsibly on London's roads – the number of FORS registered vehicles now stands at almost 167,000.

(c) Establishment of London Out-of-Hours Consortium, consisting of key boroughs, retailers, London Councils, the Freight Transport Association and Road Haulage Association to begin a comprehensive review of re-timing activity and take the lead in delivering a sustainable increase in out-of-hours operation. This will include trial deliveries, production of a briefing document for London boroughs and driver training programme for quieter deliveries.

(d) Provision of more than 300 training courses to promote and share effective practice on safety (which remains the primary focus for freight operation in London), air quality and compliance issues.

(e) Training more than 2,000 drivers in safe, urban driving, which includes practical cycle awareness from both the driver's and cyclist's perspective.

(f) Continuing to examine potential for increasing efficiency through creation of consolidation (break bulk) centres outside central London with emphasis being placed on sustainable delivery modes for the 'last mile' of the journey chain.

(g) Issuing a weekly e-bulletin to more than 7,500 operators to advise of planned events and traffic disruption, and help improve planning and delivery efficiency.

(h) Future production of guidance for Delivery and Servicing Plans for new developments (April 2015).
(i) Ensuring that freight and servicing movement, space for parking and loading are reflected in street-type criteria for TfL roads (Summer 2015) and included in design and management of all TfL-funded schemes and promoted in all highway and traffic schemes across London (December 2015).

Essentially, the Games time communications message to freight operators and businesses was re-time, reduce (the number of deliveries) and re-route and, where possible, remode. This is the ongoing message that TfL wishes to promote across the business community in London and those organisations that serve them.

Bibliography

Guidelines for ITS deployment in urban areas – traffic management – DG MOVE 2013

Best practices in urban ITS – collection of projects – DG MOVE 2013

Potential needs for standardisation of urban ITS – DG MOVE 2013

Chapter 3

ICT for intelligent public transport systems, state of knowledge and future trends

Marion Berbineau, Mohamed Kassab*, Christophe Gransart*, Martine Wahl*, Juliette Marais* and Divitha Seetharamdoo**

3.1 General introduction

The European Rail Research Advisory Council (ERRAC) has targeted for the year 2020 to double the passenger traffic by rail. Such a goal should be achieved by reducing costs, enhancing environmental sustainability, and offering new services to passengers. These objectives will be reached, thanks to a huge increase of system-wide information exchanges between stakeholders. As a consequence, many wireless communication devices operating at different frequencies are widely deployed inside the trains and along railway lines including metro and tramway lines.

Two main application families can be considered for railway wireless communications. Safety applications, generally related to command and control, have strong requirements in terms of robustness, availability, low latency, and packet loss. Nevertheless, these applications have generally low data rate. Non-safety applications, such as passenger information, Internet access, remote diagnosis, maintenance, monitoring, have strong requirements in terms of throughput, quality of service (QoS), etc. Today, the complexity of public transport systems and the necessity to optimize traffic management require satisfying three interconnected vital functions: communication, localization, and surveillance. The information provided by these functions is shared between the different stakeholders: infrastructure managers, train managers, maintenance managers, time table managers, passenger managers, etc. These functions constitute the pillars of optimized traffic management.

This chapter will be organized in three main sections.

The first section will be devoted to wireless communications for public transport applications. We will first recall some challenges for sustainable mobility and the difficulties encountered to implement on-the-shelf wireless systems for public transport. Then, we will give a brief overview and some examples of the

*IFSTTAR, COSYS Department, France

wireless technologies deployed today: beacons, bifilar communications, radio systems, etc. After this, we will present the future trends in the domain of wireless systems for public transport applications and the ongoing research in the railway domain as well as the metro and tramway domains. Particularly, we will emphasis the railway applications of Internet of Things.

The second section of the chapter will be focused on the localization function for public transport applications. The position is the core of every intelligent transportation system, for non-safe as well as safe systems. After a brief description of the principles and the advantages and drawbacks of the systems, we will describe the main existing applications and the one foreseen in the future. We will then extend to the case where satellites are not available, particularly in the case of tunnels. We will describe more solutions deployed or planned for the future.

In the last section, we will highlight the problems set up by the specific context of electromagnetic (EM) and propagation environments for railways. We will show the necessity to develop EM modeling of some environments, to set up *ad hoc* channel models as well as noise distribution. It is also important to cope with non-metallic material of the trains and to find solution for the implementation of multiple antennas at various locations and at very different frequencies (from several kHz to several GHz).

Finally we will conclude this chapter.

3.2 Wireless systems for public transport applications

3.2.1 Introduction

The continuous development of electronics, informatics, and automatics has drastically multiplied the needs for transmissions in public transports to increase travel safety, to optimize the use of infrastructures by increasing the frequency of vehicles (bus, metros, tramways, trains), to reduce operational and maintenance costs, and to reduce the impact of public transport on environments. Needs for information and services related to crew and customers and also information related to security (monitoring of the inside of the vehicles and infrastructures) are added to transmission needs related to operation and maintenance of vehicles and infrastructures. All these services help public transport to become more user-friendly and attractive, to increase multimodal behaviors and security feeling of customers [1,2].

3.2.2 Communication applications in public transport

The first transmission systems in the field of public transport have been designed to retransmit lateral signaling onboard the driver cabin. (In France, one can mention the *brush-crocodile* system that appeared since 1872 and the radio links between regulation points and engines deployed by the Société Nationale des Chemins de fer Français (SNCF) at the beginning of the 1960s.)

Since several years, the needs for information exchanges have drastically increased in all the transportation fields but particularly in the public transport domain. This trend is also known as Intelligent Transport Systems (ITS) to translate the introduction of electronics and informatics that contribute to the development of automation. The systems are "informed" and the different vehicles are "connected." The objective is to increase mobility of persons and goods. This trend is also called "intelligent mobility." The information exchanges in the transportation field are classified in three categories that generally answer different needs: vehicle-to-infrastructure (V2I) communications, vehicle-to-vehicle (V2V) communications, and intra-vehicle communications. Among these, one can distinguish three application families:

- *Vital applications or safety applications* (command and control of train movements and automatisms, remote control of trains, obstacle detection, driving assistance, etc.);
- *Non-vital applications* (information to drivers and passengers, remote diagnostic and remote maintenance of equipment, rolling stock and infrastructures monitoring and surveillance, potential dangerous situations detection inside the vehicles or along the infrastructures, etc.);
- *Non-vital applications related to comfort* (entertainment applications such as television and Internet onboard, which request ever-increasing data rate and QoS).

The vital applications or safety applications[1], related to command and control of equipment require generally low transmission rate (around 10 kbps) but a high availability rate (at least 99.99% of the time) and high robustness and reliability level (typically a packet error rate equals to 10^{-3} for packet length equals to 200 kb). Key performance indicators should be verified whatever the mobility conditions are: time needed for radio access point change (handover) lower than 10 ms when the terminal is moving, time to establish a communication, etc. These applications dedicated to control and command are also called Communication Based Train Control (CBTC). They cover train localization applications, ground to train transmissions, and all the communications between the computers onboard the trains and those on the ground. An IEEE standard describes some of the CBTC requirements [3] but from a general point of view, it is the customer (transport operator) who defines the requirements of the applications he wants to deploy. In the context of high-speed trains, works coordinated at European level allowed to define a unified system: the European Train Control System (ETCS)[2]. The communication systems are resilient to errors, thanks to a safety layer often proprietary in the case of metros and tramways but standardized, known as Euroradio [4], in the case of high-speed trains in the framework of ETCS.

The non-vital *applications are not related to system safety*. They require high throughputs (tens of Mbps in both transmission direction – uplink and downlink)

[1]http://www.uic.org/spip.php?rubrique850

[2]The concept of security uses the concept of probability of error or failure of the function. For example, a SIL 4 application (application integrity level security) requires the highest level of safety integrity, that is, a probability of failure on demand (PFD – Probability of Failure on Demand) between 10^{-5} and 10^{-4} (standard EN 50126 and 50128).

with lower robustness constraints (packet error rate around 10^{-2} for packet length around 1,000 kb). All these applications are called in the public transport field as Closed Circuit TeleVision (CCTV). These applications can share the same transmission medium.

The non-vital applications for comfort have really high requirements in terms on throughputs and QoSs particularly for the downlink.

Specific classes of services and traffic should be available for the three families of applications that can also share the same medium of transmission. They must operate in various contexts of mobility even at very high speed. They have to take into account requirements concerning interoperability between systems and applications and, finally, they should operate in very complex radio environments (obstacles, masking effects, urban canyons, high density of vehicles, various EM noises, tunnels, etc.).

Numerous communication solutions are deployed today to meet the needs of public transport systems. The integration and interoperability of all these telecommunication systems constitutes today a major technical challenge in order to answer the requirements related to spectral and energy efficiency and to reduce the deployment and maintenance costs. Tomorrow, wireless communication systems should be able to adapt and reconfigure themselves in real time. The applications and protocols should be interoperable, several physical layers should cohabit, and more and more functions will be implemented in reconfigurable intelligent components. A global system approach is necessary for the design of these systems and this requires the collaboration of several scientific communities (software design, information and signal processing, hardware, electronics).

3.2.3 The main deployed technologies

3.2.3.1 Introduction

Existing communication systems developed for buses, metros, tramways, and trains applications mainly rely today on technologies such as spot communications with beacons, continuous communications with crossed and non-crossed bifilar lines, proprietary radio technologies such as the Professional Mobile Radio and TETRA which offer mobile telephony services for "transport operators" or radio technologies derived from existing standards such as GSM-R [5] for European high-speed line and WLAN (Wireless Local Area Network) systems such as IEEE802.11x [6].

3.2.3.2 Short range communications or beacons

These systems allow the transmission of information generally between two relatively close points. These communications are generally short range and not necessarily bidirectional. They allow the transmission of information valid only locally, for example in the area of the beacon transmission. In the coverage area of another tag, the information previously received can be modified. These short range communication systems operate in a frequency range between several kHz and several GHz. The coverage area of these systems must necessarily be limited. Like all communications, they must not cause interference to other communications systems (public or not), especially on systems of the same family.

These systems are used, for example, for vehicle identification, localization, and transmission of lateral signals to the driver cabin, speed control, integrity control, and the transmission of any kind of information that allows driving assistance, the control of onboard automatisms, and sometimes diagnostic and maintenance. In France, for example, we can mention [7]:

- The various SACEM beacons (initialization, re-localization and end of SACEM) and also DAM (*Dispositif d'Aide à la Maintenance*) beacons at 9.9 GHz that collect information related to maintenance of equipped trains.
- KVB beacons (uplink 27 MHz – downlink 4.5 MHz) used at SNCF for speed control on some lines as well as on some Paris metro operator (RATP) lines.
- TVM 430 beacons (*Transmission Voie-Machine*) used for transmissions of signaling between tracks and trains on high-speed lines (e.g., Lille-Paris). The principle of these balises relies on the use of local inductive communications between track circuit and the train. Similar systems are used on regional trains in the Paris suburbs.
- Eurobalises [8] are used on European high-speed lines. They operate in the 27 MHz range and should be compatible with the KVB system.

Track circuits can be classified in this category. These are circuits obtained by isolating a section of track called block. They are used to detect the presence of a train and to transmit track-side signals on trains as well as speed limit instructions, the next speed limit, and distance to this limitation. They still constitute on many lines the fundamental element to control the movement of trains. The principle of communication by track circuits is detailed in References 7 and 9.

It is important to mention here the DSRC (*Dedicated Short Range Communication*) system developed for road pricing in the 5.8 GHz frequency range. A frequency band has been allocated at European and international level. The IEEE 802.11p technology, also known as "Wave," is recommended for mobile vehicular application in urban areas. It is the basis of the ITS-G5 standard developed by European Telecommunications Standards Institute (ETSI) for ITS.

In France, the RATP and SNCF have requested to ETSI frequency allocations in this 5.8 GHz frequency band [10]. It is also important to mention a new concept of tags that relies on Ultra Wide Band (UWB) technology, which is under development [11].

3.2.3.3 Continuous communication system using magnetic coupling

These systems are still widely used in the public transport field. The basic technology consists in a simple wireline. The electric current circulating in the metallic conductors induces a magnetic field that allows the transmission of the information detected by a magnetic antenna onboard the train. In their most common form, they offer the advantage of being economical. However, as low frequency systems, they are subject to limitations related to the useful bandwidth and poor signal to noise ratio. The wireline allows continuous transmission of information. The wires can be crossed on a regular basis or not. In the last case, the system is named automatic piloting carpet.

Crossings of the lines are used to encode a speed program or information for train re-localization. These systems are used for driverless systems such as VAL (Automatic Light Vehicle) (Lille, Toulouse, Rennes, Roissy, Orly).

For automated metro in Lyon (MAGALY) and Paris (METEOR), manufacturers are reusing a regularly crossed wireline called "ribbon" because the implementation of nonregular crossings of the lines for encoding information greatly complicates maintenance and industrialization but also inspection. The positioning is then operated with specific tags.

3.2.3.4 Radio communication systems – V2I

Radio technologies are the more suited solutions for medium and long distance communications between the ground and the vehicles in all types of public transport. So far, the most used frequency bands for public transport context are the VHF (Very High Frequency) in the 70–88 MHz band or in the 155–220 MHz band, and the UHF (Ultra High Frequency) in the 420–470 MHz band. However, the VHF–UHF spectrum is very busy today. The trend is to raise the frequency spectrum to increase the number of available transmission channels and their width.

The use of radio has grown considerably, particularly in France with the development of ground to train radio equipment at 450 MHz (radio networks UIC – International Union of Railways) for voice services, and with the development of a data transmission system for remote control and remote maintenance applications (from 1988 on the Paris-Rennes line and 1989 for the TGV Atlantique). The deployment of different radio systems has increased with the arrival of the PMR (Professional Mobile Radio) systems and particularly the TETRA system. Table 3.1 gives an example of frequency occupancy at SNCF in Paris region for various communication systems.

The levels 2 and 3 of the ETCS system, mentioned previously, rely on the use of a digital radio system called EIRENE (European Integrated Railway Radio Enhanced Network) in the 900 MHz frequency band derived from the GSM phase 2+ cellular technology [5,12].

Today, the development of wireless technologies, such as IEEE 802.11a/b/g/n for general public applications, led the public transport operators to rely on the potential of these inexpensive systems to meet their growing needs in telecommunications. As an example, driverless subway systems used transmission

Table 3.1 Example of frequency allocations for SNCF in Paris region

31–32 MHz	Voice
50–65 MHz	TV semi-embedded
70–151 MHz	Remote control
152–180 MHz	Operation, construction, analog 3RP
414–429 MHz	Intelligent radio solutions - safety radio
876.2–879.8 MHz	Voice and data
921.2–924.8 MHz	GSM-R, railway control systems
2.4 GHz and 5.8 GHz	WiFi access

systems based on WiFi-like modems on which specific protocol layers as well as specific radio deployment engineering have been developed. This is the case of New York metro, line 1 of RATP in Paris, the metros of Malaga, Marmaray, Beijing, Shanghai, and very soon the line 1 of the metro of Lille.

One can mention also that RATP has developed specific proprietary radio technologies to enable communications between rolling stock and some fixed points in stations. RATP received the permission to operate a digital system in the 5.915–5.935 GHz band with four channels of 5 MHz each [10]. RATP operates this technology for train to ground transmissions in stations. The IEEE 802.11p standard also known as "Wave" has been developed in a 30 MHz band between 5.875 GHz and 5.905 GHz. The main applications of WAVE concern buses and vehicles but this technology can answer the telecommunication needs of public transport systems. Radio systems can also carry the Internet traffic for passengers. For example, in the buses of Madrid (EMT operator), Internet service via WiFi is offered, thanks to the local telecommunication operator Gowex. In France, SNCF has developed projects for TV broadcasting and Internet access [13,14].

Mobility management is a major key challenge for wireless systems in public transports. The mobility may lead to two types of connectivity change: in the same radio access network between two access points (horizontal handover) and between two different networks ("roaming" or "vertical handover"). Mobility management in a wireless network requires the establishment of specific mechanisms and protocols to ensure continuity of communications and the transparency for the users. Unlike cellular technologies (e.g., GSM, UMTS, LTE), where mobility management is taken into account by design, many technologies, particularly the WiFi, are designed without management of mobility user. Furthermore, mobility between various networks remains an unsolved problem in a lot of deployed systems. Numerous solutions to avoid this difficulty are proposed at different levels of the OSI (Open System Interconnection) model [15]: application layer with the use of middleware (cf. Section 3.2.5), transport layer (e.g., mobile Stream Control Transmission Protocol (mSCTP)), network layer (e.g., Proxy Mobile IP, Hierarchical Mobile IP, and Fast Mobile IP), and link layer [16,17]. The mobility management has to take into account key performance indicators, especially not only those required by safety applications such as CBTC and ETCS but also non-safety application such as CCTV. Among these key indicators we can mention maximum handover time, maximum packet error rates, and maximum packet loss rate (cf. [18,19]).

The issue of mobility management is also compounded by the question regarding the integration and interoperability of several heterogeneous wireless technologies on networks. This constitutes a key technological challenge in order to enhance the global efficiency of public transport systems by reducing drastically the deployment of dedicated radio systems along infrastructures and onboard vehicles as well as associated antennas. The concept of flexible and cognitive radio, in development by the wireless communication community, could change that in a near future [20]. Cognitive radio pave the way toward a new type of wireless communications in which mobile terminals interact with the radio environment in order to optimize the use of frequencies and available channels. This concept meets all the requirements while guaranteeing communications with better quality, better reliability, safer and potentially resilient to traffic overload and attacks.

In the framework of the EU FP6 project, InteGRail [21], an intelligent telecommunication architecture, ICOM, has been developed in order to integrate all the communication systems (vital applications are excluded) in the railway domain, thanks to the use of middleware (cf. Section 3.2.5). This solution is a first step toward interoperability and integration of the communication systems in the public transport world. The French research projects CORRIDOR [22] and SYSTUF [23] treat these questions.

3.2.4 Intra-vehicle communications and between interconnected vehicles

3.2.4.1 Onboard communication networks

Always more systems and devices are deployed aboard public transportations (trains, buses, tramways, etc.) for the automation of vehicle functions. Onboard communication networks were installed aboard trains since the end of the 1980s to reduce the cable beams used to transfer information between different devices. Multiplexing digital information technics over a serial cable replaces the classical point-to-point analogical copper lines. Wired communication networks were standardized for onboard railway applications in the end of the 1990s (IEEE-1473-1999 and IEC 61375 – First edition, 1999-09 standard).

These wired networks allow remote information sharing and the control-command of the distributed applications embedded devices. For example, they play a role in the applications of preventive maintenance by forwarding diagnosis data toward the wireless interface of the vehicles.

Railway embedded network topology enables "open trains." They consist of a set of vehicles or train-sets, in which arrangements can change during the day: vehicles are added or removed from the train. Therefore, train network consists of a hierarchical two-level topology. A first "open" network supports the coupling and the decoupling of vehicles. Detecting any change in the number of connected nodes, it triggers its configuration. Other networks are closed. They interconnect the various devices aboard vehicles [24,25]. The nodes of the open network are gateway, with bridging and routing functions. Also interfaced to closed networks, they can forward the information toward the whole distributed part of any application function (Figure 3.1). Sensor busses or point-to-point links connect sensors and actuators to the input–output ports of the equipment [24,25].

The first railway embedded networks used were railway dedicated technologies such as Tornad token ring solutions that have been embedded into several materials in France such as the MAGGALY driverless subway of Lyon (Line D), MP 89 vehicles of the Paris metro, or the "TGV Atlantique" of year 1988 [24]. Standard technologies have then been deployed, coming from market technologies (WorldFIP, CANOpen, LonWorks, Profibus, . . .) or dedicated to railway (Train Communication Network, TCN). For instance, a WorldFIP technology was deployed in Citadis

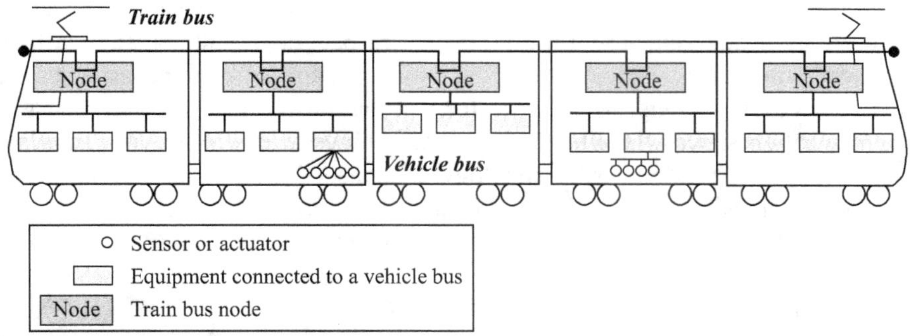

Figure 3.1 Communication busses in trains (IEEE-1473-1999 and IEC 61375 – First edition, 1999-09 standard) [24]

tramways of Barcelona, Dublin, Melbourne, Montpellier, Orleans, and in Singapore metro [24]; TCN technologies were deployed in Citadis tramways of Rotterdam and Strasbourg, and in the German high-speed train ICE3 [26].

All the tested technologies had different medium-access protocols. However, a common scheme of implementation has been chosen in Europe to guarantee determinism [25]: synchronous and asynchronous traffic is defined and cyclic scheduled exchanges are specified. During the synchronous traffic, time-triggered exchanges take place for the refreshment of the application periodic data and these data are broadcast on the bus. During asynchronous traffic, information is exchanged following a remote event. It can be a point-to-point exchange or a point-to-multipoint one. The traffic is organized under control by a centralized network controller which redundantly is planned for in order to replace a faulty controller.

In the 2005s, manufacturers considered the Real-Time Ethernet (RTE) technologies. In addition to the control-command functions offered by the classical fieldbuses technologies, RTE provides Internet Protocol (IP) traffic. The relevance of Ethernet technologies stemmed for the need to transfer high amounts of data for passenger information, to embed IP in public transport to help the integration of new applications to make data available on long distance via Internet, and to reduce costs by reusing and sharing the network. In 2012, the new release of IEC 61375 TCN proposes new networks. Part 2 is then concerned with train open networks. It adds two new technologies, Ethernet Train Backbone (ETB, IEC 61375-2-5) and *Wireless Train Backbone* (WLTB, IEC 61375-2-7), to the one standardized in 1999, *Wire train Bus* (WTB, IEC 61375-2-1). Part 3 is concerned with vehicle closed networks.

At this stage, these networks are not fault-tolerant or safety compliant, this means that they are not concerned with the train safety. Dedicated TDMA (time division multi-access) protocols have been studied for aeronautics and automotive areas. Solutions such as TTCAN, TTP-C, and FlexRay address X-by-Wire applications in which fault-tolerant electrical commands replaced mechanical commands.

In the recent years, PLC technology (Power Line Communication) for communications inside vehicles in the field of aerospace and automotive industries experience important developments. This is also true for public transport vehicles. This technology exploits the network for distribution of electric power to carry communication signals. One of the major difficulties to be solved is to translate this well-developed "indoor" technology to transportation fields because high frequency signals should coexist on the same medium, with the DC voltage and current spikes at the start or stop of electrical power systems [27,28].

3.2.4.2 V2V communications

Communication between vehicles in the public transport sector covers several applications. The first one, often referred to as the concept of *carrier pigeon*, consists in providing information on the fly between two vehicles. The use case is often a disabled vehicle, out of range of a communication network, that will transmit information to another vehicle passing nearby [29–31]. Another very present-day use case, still in the research labs, is virtual coupling of two vehicles (car trains, subways, trams) this means no mechanical coupling but wireless coupling. Indeed, in the case of mechanical coupling of two vehicles (subways, trams, trains, TGV), it is also necessary to interconnect high-speed networks embedded in both vehicles. This is now carried out mechanically using specific connectors that deteriorate quickly under hard railway operation conditions. The goal today is to replace these mechanical connectors by a safe wireless high-speed broadband connection [32]. One of the candidate technologies is the very high-speed technology like UWB as the IEEE 802.15x standard. The UWB links are deemed more robust to frequency-selective fading [33]. Other technologies in the range of 60 GHz like the IEEE 802.11ad and communication systems like machine-to-machine being defined in 4G and 5G may also be considered.

3.2.5 Middleware for communication system in the transportation field

3.2.5.1 What is a middleware

To solve the heterogeneity problem implied by the use of subsystems and to homogenize the interfaces in order to simplify the integration and to build new applications, the rolling stock manufacturers and the stake holders use more and more solutions based on middleware. A communication middleware offers a well-known set of interfaces that can be used by the applications to be able to communicate together. The middleware uses the low-level functionalities offered by the computer and its operating system. It uses also the communication stack (classically a TCP/IP stack).

The middleware goals are to hide the hardware and operating system heterogeneity, to hide the distribution of the data on the network and it offers a standard set of Application Programming Interfaces (API) as depicted in Figure 3.2.

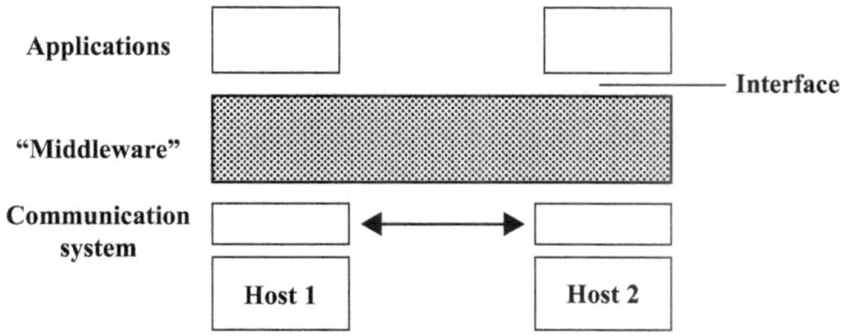

Figure 3.2 Generic view of a middleware

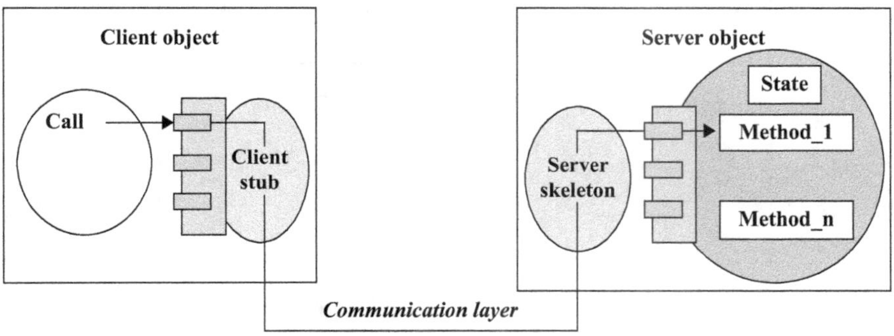

Figure 3.3 Object-oriented invocation

According to the kind of exchanges required by the applications and also according to the communication model, the middleware can be classified in different classes. For synchronous exchanges (see Figure 3.3), like Remote Procedure Call (RPC) or Remote Method Invocation (RMI), they can be implemented with CORBA (Common Object Request Broker Architecture) or with web services based on the HTTP (Hyper Text Transfer Protocol). For asynchronous messaging using message queues like in Figure 3.5, the implementation can use messaging system like JMS (Java Messaging Service) as in Figure 3.4.

3.2.5.2 Asynchronous messaging

Asynchronous messaging uses queues or mailboxes as in Figure 3.5 to exchange messages between transmitter and receiver. The emitter and the receiver do not need to be online at the same time to send and receive messages. Figures 3.6 and 3.7 present the two solutions to exchange messages: point-to-point to exchange a message from one emitter to a particular receiver and Publish/Subscribe to send a message to a pool of receivers.

For periodic messages requiring real-time capabilities, the implementation can use the DDS (Data Distribution Service) standard. Figure 3.8 presents a schema of

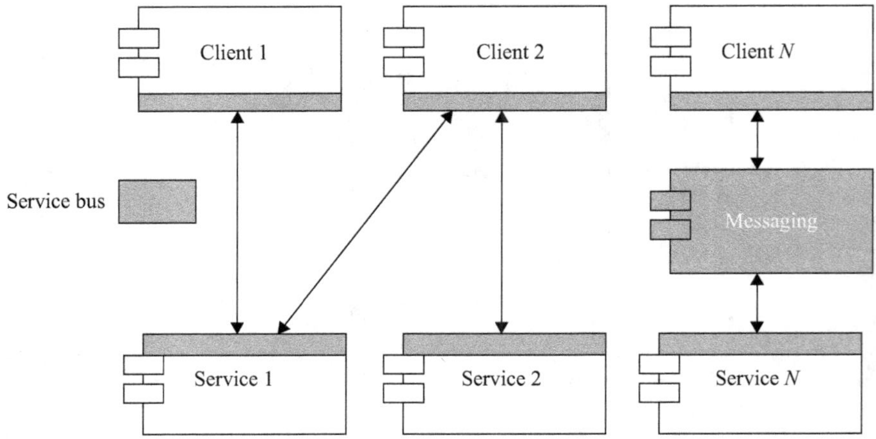

Figure 3.4 SOA service bus © IONA Technologies 2005

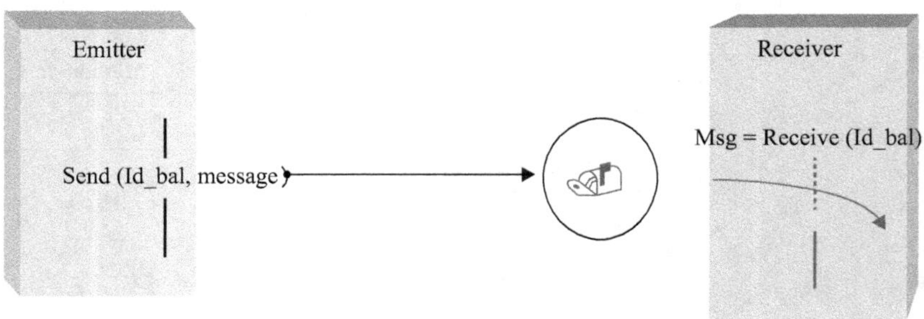

Figure 3.5 Asynchronous communication with message

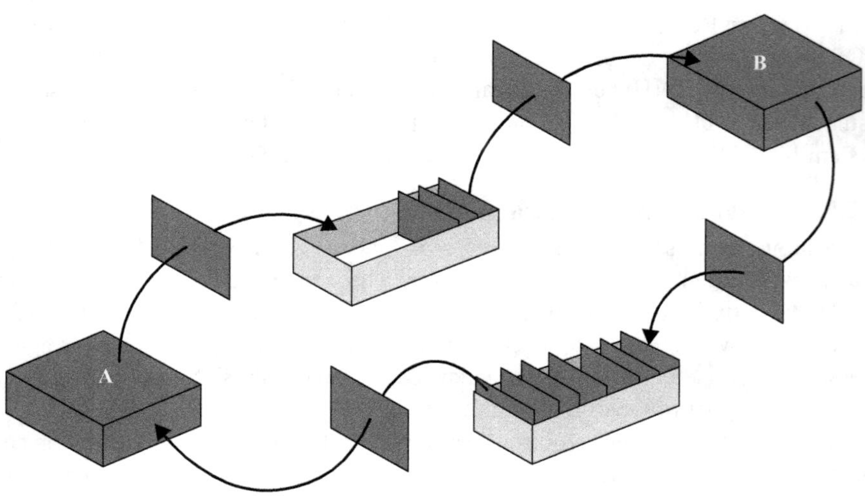

Figure 3.6 Point-to-point communication

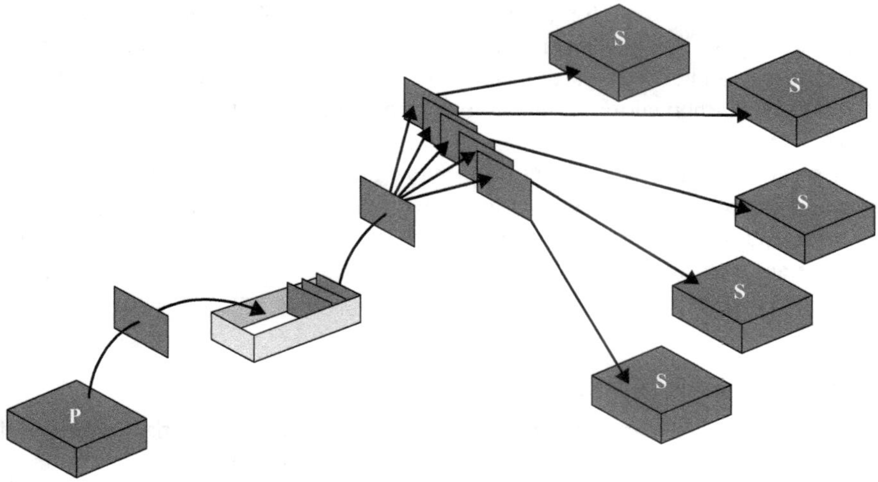

Figure 3.7 Publish/subscribe communication point communication

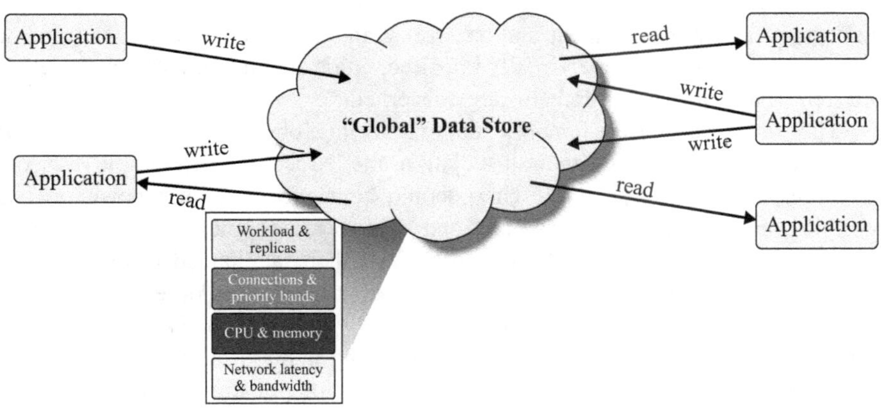

Figure 3.8 DDS global data store (from [38])

this architecture with several applications acting as producers and several applications acting as consumers. According to the use cases, some characteristics are required for the middleware: real-time, QoS, traffic prioritization, reliability, and rugged exchanges.

A transport system is mobile and its environment constantly changes. The software (and the middleware) must adapt itself to this environment. Different works have taken into account this context [34–36]. In the InteGRail EU project [21], a communication architecture named ICOM [37] has been defined that involved several kinds of middlewares compliant with railway constraints: for embedded systems they proposed real-time and fault-tolerant middleware; for train to ground communication they proposed a service-oriented architecture (SOA).

Building middleware for the new transportation systems is a big challenge. Future works on middleware must take into account safety requirements, limited resources for embedded systems, and the increase of geolocated applications. Model-Driven Engineering and automatic code generation is also a promising research area.

However, even using middleware technologies, there is still a problem of interoperability between information systems build by different operators (e.g., booking, ticketing, and route planning). A global interoperability between all the applications could be achieved through the use of Web Semantic and ontologies to build translation and communication rules between the various information systems.

3.2.6 Internet of Things and wireless sensor networks

The development and miniaturization of wireless sensor networks (Internet of Things) offer very large potential for public transport operators and infrastructure managers. Indeed, these systems provide information feedback at large scales for everything concerning rolling stock, infrastructure, diagnostic equipment, etc. Data collection concerns weather information, behavior and aging of infrastructure and vehicles, energy consumption, maintenance and diagnostic (doors, motors, switches, etc.), the property management and art works, and customer service management (caddies in stations, assist the visually impaired, guide and indoor localization, etc.). Different stakeholders and domains are concerned.

The widespread use of mobile communicating objects and the Internet of Things involves rapid and simple installation, the least intrusive possible. Sensors are increasingly often wireless. They should consume very little power and if possible they should be autonomous and self-powered. They must be able to communicate indicators (not the raw data) and then facilitate decision making. Generally, data are routed to a centralized processing center, in order to extract the information that will permit the decision making by the concerned actors. Thus, the entrance of Internet of Things and Big data in the public transport mode [39] is promoted. These technologies open the door to new applications and communication means for the users, the infrastructure owners and managers and the public transport vehicle. These stakeholders will benefit from the richness and relevance of content, thanks to the diversity of information collected. The interaction will also be improved through the availability of multiple sources of information that can be merged and contextualized in real time. Much progress will be implemented through the effective use of sensor networks to improve innovative wireless user access to services based on availability and updates. New means of communication and dissemination of information emerge through innovative applications.

Numerous research and development works are ongoing on these topics in the public transport world. We can mention projects for SNCF commuter trains, SNCF [40,41], works on routing protocols taking into account QoS to increase efficiency of information transmission [42] and also works on low energy consumption and reconfigurable sensors.

3.3 Localization for public transport systems

3.3.1 Introduction

Localization plays a major role in many functions in the public transport field. The position is the heart of a large panel of applications: navigation of course but also eco-driving, variable speed limits, bus priority, etc.

Everybody can now follow the movement of an urban bus thanks to messages broadcasted at the bus stations or see the next stop of its regional train or bus, thanks to announcement inside the vehicle. Even the trains can now be tracked, as French ones on the SNCF website[7] (Figure 3.9). This information is visible to passengers for information service. Less visible but also important, localization is a synchronization element in an information system and is a fundamental input for traffic management or maintenance operation to locate accurately the track or rail defaults.

In railways (urban, regional, or high speed), this information is used in the signaling system devoted to control, command, and protection of train exploitation [43]. Furthermore, train detection activates the closing gates of level crossings. Nowadays, the safety function that consists in "forbidding the presence of two

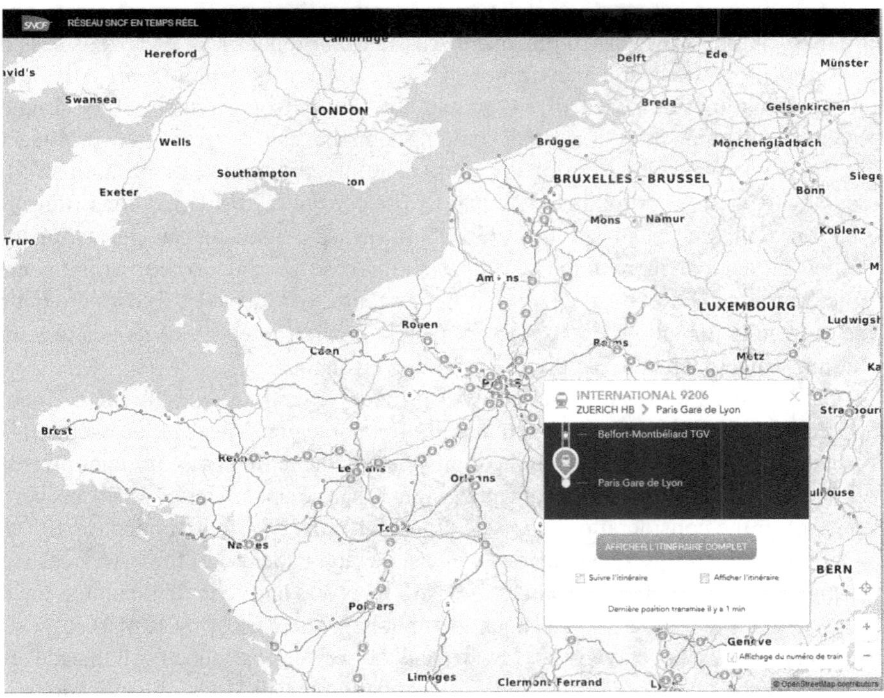

Figure 3.9 SNCF service for train tracking

[7]www.sncf.com

trains in the same time on a same track section" is performed by a decomposition of the track into fixed "blocks." Each train entering a block receives an authorization and a speed limit function of the track topology and the traffic on the line.

The detection of the train is a way of localizing the train and is typically performed, thanks to a track circuit. The same kind of solution is used for tramways. The track circuit detects the presence of trams and it operates on tramway signaling. Indeed, the train axles provoke a shunting in the electric circuit formed by the generator and receiver linked to the section of the rails. This system localizes the train punctually. The next positions of the train are computed based on this absolute point and an odometer (wheel sensor) until the next balise. However, such sensors suffer from the well-known bias of proprioceptive sensors, in particular caused by rail/road contacts and skating/slipper. This same equipment is also used to give speed limits onboard of the train. When the train is stopped between two balises there is no way to update its position.

3.3.2 The basics of Global Navigation Satellite Systems (GNSS)

Today, everybody can access to position services just by using a GNSS receiver. Satellite-based positioning systems are based on the propagation time measurement of signals emitted by satellites and a triangulation principle. The most famous of the GNSS is the American GPS, but a Russian constellation already exists, and the European Galileo and the Chinese Beidou are under deployment. All these constellations are composed of 24–30 satellites placed on circular orbits around the earth that transmit continuously ranging signals. The coverage is worldwide. The (basic) services are free for the users. The user computes its position itself. Accuracy offered is approximately equal to 7 m for the GPS in good conditions of reception. The European service EGNOS allows the user to use geostationary satellites, as a supplement to the GPS ones, to increase accuracy by correcting some of the common mode errors to reach 2 m of accuracy. Moreover, EGNOS signals send indicators that allow the user to compute a level of integrity of its positioning solution, which characterizes the risk of being erroneous.

Galileo, the European initiative, will provide several services, some of them still being in a definition stage, but that will offer some guarantees to the user that is of particular interest for safety applications. One is the authentication function that will avoid the use of external corrupted signal. These specifications will be very important in the considerations to use Galileo for railway safety applications.

For a transport user (passenger as well as operator), the use of these services are free, and only requires the installation of GNSS receivers on each mobile unit to track (bus, locomotive, wagon, or passenger smartphone). The cost of mass market receiver is very low, the coverage is worldwide, and the service continuous. All this makes this technology a very promising tool for localization for all kinds of applications!

3.3.3 Public transport experiences

Services mentioned in the introduction do not require very stringent performances in terms of accuracy or availability but new services are also tested in applied research projects that did not yet penetrate the market. These are mainly projects related to safety that requires very stringent performances and sometimes certification before being deployed on operational lines. This paragraph presents some examples, in particular in guided systems.

Like other transport modes, Bus Rapid Transit (BRT) systems take advantage of new intelligent systems to enhance their service offers. TEOR, the East-West BRT system serving Greater Rouen, in France, is an intermediate system between a classical bus line and a BRT. The role of this line is to bring into service a public transport system whose performance levels are equivalent to those of a tram line, at a lower cost than a tramway, and which meets the local population's expectations [44]. One of the specificities of the TEOR buses is that they are equipped with guidance system that provides vehicle accessibility for passengers and especially for the Person with Reduce Mobility (PRM) by reducing the horizontal gap between the vehicle and the platform.

In the project ANGO (2007–2010) that aimed at improving the performance of the TEOR bus, a complementary localization system has been added in order to provide an advanced driver assistance service (ADAS) for bus driver. Indeed a reduction of fuel consumption of 20% was expected by changing the driver way of driving, in particular by reducing the number and intensity of brakes and accelerations. Some companies propose training for eco-driving but Human Machine Interfaces can also help in real time by indicating to the driver the average speed to respect, depending on the bus position. Furthermore, smoothing the driving increases passengers' comfort and thus aims to increase public transport attractiveness. The solution was then to propose to the driver a driving profile. Such an innovation is directly in line with the objectives of reduced carbon emission and greener public transport.

For tramways, the need expressed by operators for the deployment of new information or signaling systems is to reduce the impact of the infrastructure to reduce maintenance costs, as expressed also for railway systems. An accurate positioning system is considered for tramway localization or freight locomotive tracking and developed in the CAPLOC project [45]. This project is a research project. It aims at showing the feasibility of an innovative positioning system that relies on the combination of image processing techniques and propagation knowledge. In this approach, image processing is used for obstacles detection that helps to analyze radio signal propagation conditions. Some similar approaches are published based on embedded 3D models of the cities, mainly addressed for car applications but that could absolutely be derived for such a transport mode. Indeed, a guided vehicle requires a smaller database than a car because of the non-changing environment crossed.

In railways, history plays an important role in the system conception. Indeed, each country in Europe developed its own infrastructures and equipment mainly because the trains were not crossing frontiers. The consequences are a huge variety of standards that concern as well power that operational rules or positioning balises. To answer interoperability request of railway systems, Europe decided, in 1996, to develop the ERTMS (European Rail Traffic Management System). ERTMS is composed of three levels: the two firsts are adopted by conventional existing systems, working with fixed blocks and equipment placed along the track. The deployment of ERTMS Regional, a level-3 like system, which limits the number of equipment on track, is now seriously envisaged in several Europeans projects like SATLOC or 3inSat.

One of the ideas of level 3 is to transfer most of the intelligence of the system onboard the train. In such a context, the potential of GNSS, including the European Galileo satellite navigation system, offers significant potential advantages. GNSS shall allow cost reduction (installation and maintenance), is not country-dependent, and can authorize a management based on moving blocks, thus shall increase safety and efficiency. Benefits include improved safety, management information, and increased use of existing infrastructure.

If a GNSS receiver alone is not seriously expected, a combination of a satellite-based receiver hybridized with other localization sensors and the use of a wireless telecommunications offer the possibility to develop and operate railway control systems at lower costs without impacting their high level of safety and answering the requirements of the ERTMS/ETCS.

In the United States, the PTC (Positive Train Control) is supported by a federal law that stipulates the use of GPS as compulsory for all the passengers intercity lines as well as suburban lines. The Russia recognized the GNSS as a cutting technology in its strategic map for the revival of infrastructures. In China, some rail lines are already operating with a GPS localization system. In Europe, with the arrival of Galileo, the European Commission, the European Space Agency (ESA), and also lots of diverse national institutions fund research projects on the topic to promote GNSS use since the end of the 1990s. However, no operational installation is working actually.

The first projects GADEROS or LOCOPROL (5th FP) showed that GNSS systems could answer the safe railway requirements being also compatible with the development in progress in ERTMS. A multisensor positioning system was chosen by GADEROS [46]. LOCOPROL chose an original use of satellite pairs [47]. Projects selected by the GJU (Galileo Joint Undertaking) pushed the development of equipment such as GIRASOLE (that provided a multimodes receiver for Galileo safety of life service) or GRAIL (and GRAIL2) that uses GNSS for an odometry function [48]. The more recent projects, funded by the GSA (European GNSS Agency), focus on operational demonstrations of the concepts. They target the low density traffic lines such as SATLOC on a Romanian line and 3inSat [49] in Sardinia. In SATLOC, the localization solution is composed of a GPS/EGNOS receiver and an odometer communication is performed by Euroradio and data transmission by a VPN tunnel. In 3inSat, a multiconstellation GNSS receiver is hybrid to embedded sensors (odometer, inertial central) and a digital map. Communication is performed, thanks to existing networks: cellular and satellite, without the deployment of dedicated network.

In parallel to these different actions, in March 2012, "EU decision makers and rail stakeholders agree that cost effective synergies between ERTMS and satellite technologies, such as GALILEO, can help make rail transport more efficient and reliable." That decision certainly marks time in the European march for

GNSS-based railway applications. Regional lines, with low traffic and where the cost of infrastructures is especially strong regarding the benefits, are the first targeted lines.

3.3.4 Challenges

3.3.4.1 Local errors – multipath effects

Railway environment suffers, like many other transport environments, of masking effects created by the surroundings: cuttings, urban areas with constructions, or vegetation. These obstacles for the EM waves reduce the level of accuracy of the GNSS receiver. The new constellation will offer new opportunities, and in particular Galileo's new signals should reduce the multipath effects. Multiconstellation receiver and the integration of advanced signal processing techniques are some of the research axes that will be explored in order to enforce robustness and reduce failures of GNSS-based solutions.

3.3.4.2 Tolerable hazard rate

Today, mainly used for non-safety services, GNSS localization will be integrated to future systems if we can prove the integrity (linked to the safety notion) of the positioning solution offered. EGNOS brings a first level of information, by giving some values about common mode errors. These data allow the user to compute a protection level considering the error margins on the measurement. However, as it is, EGNOS cannot satisfy the railway safety rules that ask for a stringent Tolerable Hazard Rate (THR). To face the THR challenge, the 3inSat project adds a new layer called EGNOS-R, local or regional, that will be verified, validated, and certified for railways applications. Its aim is to provide GPS and Galileo signal in space augmentation data compliant with SIL 4 for the localization of the train [50].

Moreover, in constricted environment, local errors are numerous and strongly dependent of the surroundings structure that make them difficult to model and take into account.

3.3.4.3 RAMS parameters

The exploitation of a new satellite-based system in a railway control and command system requires the proof that its RAMS parameters (Reliability, Availability, Maintainability, Safety) are, on the one hand, managed following the railway standards (EN50126 00) and, on the other hand, reach the same values than the existing systems. The goal is to control the dependability of the new products, to know the risks of using them, and these risks are focused by a European regulatory framework. The last European project began to include such analysis. Let us mention GaLoROI (FP7) that conducted to the development of a certifiable localization unit combining GNSS and an Eddy Current sensor. A RAMS analysis has been proposed based on a hybrid failure tree [51] to contribute to the certification objective.

3.3.4.4 Unavailability of satellite signals

The last challenge mentioned here is the absence of satellite signals in very dense urban environments or in masking areas like tunnels. GNSS-alone solutions are not appropriate here. The most common solutions developed rely on multisensor systems. The sensors are typically odometers, accelerometers, gyros, which can be initialized or updated by GNSS data when available. But other solutions can be exploited: the use of fix balises is still possible (transponders, RFID, visual landmarks, etc.), pseudolites (transmitters of GNSS-like signals) can replace non-received signals, or other radio signals like UWB are some of the solutions that could be installed in a near future.

3.3.5 Conclusion

The position data is now a key player in our everyday transportation uses. This position is useful for each individual traveler but also deeply for public transport managers. Moreover, new services based on "context awareness" or "ubiquitous computing" will benefit from this location information.

If some technical challenges like masking effects of satellite signals or multipath still need some work to be completely robust, the main frontier between existing applications and future ones is today the guarantee of performances required for safety.

3.4 Specific context of EM and propagation environments for railways and public transports

In the previous sections, we have seen that the deployment of wireless systems for communications and localization applications should answer to specific constraints such as robustness, reliability, high data rate, spectral efficiency, optimization of radio resource usages, and low energy consumption. Thus, the performance evaluation of wireless communication and localization systems requires radio channel models for numerous propagation environments: rural areas, urban areas, fast fading, Doppler effects, tunnels, cuttings, viaducts. We can mention for example some recent publications in the high-speed rail context to model propagation in cuttings [52] and viaduct [53]. Others deal with the propagation and channel modeling in the specific tunnel environment [54,55]. These channel models should also include *ad hoc* EM interferences and noise distributions. In Reference 56, the authors have modeled the impulsive noise due to the bad contact between the catenary and pantograph in the GSM-R band. The SECRET project (FP7) (SECurity of Railways against Electromagnetic aTtacks) aims at identifying and detecting the most probable EM attack scenario and the solutions that will permit the resilience of the wireless communication systems in the railway context (http://www.secret-project.eu/).

Furthermore, minimizing the energy consumption of railway vehicles is nowadays a real economic and technological challenge. Weight reduction using composites for the vehicle structure is one of the solutions currently considered. Indeed, composite materials offer high mechanical characteristics for an overall lighter structure. These composites are, however, not designed for an optimal EM characteristics and their impact on the performance of onboard wireless communication systems requiring a high level of reliability is not negligible. Equipping railway vehicles with wireless communication system is thus getting more complicated. Another constraint is the increasing number of wireless systems onboard, and antenna integration on such platforms is becoming a real challenge.

Recent research results in the field of wave–matter interaction and antenna technology suggest the design of wireless systems antennas accounting for the EM characteristic of the surrounding environment by using metamaterials. They are artificial EM composites which can be tailored for required properties. They provide, for instance, the same reflective properties as metals but with a much lower density in line with the weight reduction constraint of railway vehicles. Research on metamaterials is a promising and active field of research and its application to railway and public transport systems is new [26]. We can mention also recent developments of dedicated antennas with metamaterial [57] and methodologies for antenna positioning in tunnels [58].

3.5 General conclusion

This chapter was devoted to information and communications technology (ICT) for Intelligent Public Transport Systems. The aim of this chapter was to present to the readers a well-referenced contribution to state of knowledge and future trends. The first section was devoted to wireless communications. We described the main deployed technologies: beacons, bifilar communications, radio systems, etc. The future trends and ongoing research have been presented. The second section of the chapter focused on the localization function for public transport and railway applications. A brief description of the principles and the advantages and drawbacks of the systems was given. We focused on GNSS applications but we treat also the case where satellites are not available, particularly in the case of tunnels. Future trend was also presented. Finally, in the last section, we emphasize the needs for radio propagation models in various specific environments such as viaducts, cuttings, and tunnels. We refer also at new trends regarding the use of metamaterial for antenna design and placement.

In order to promote mobility and sustainable urban and railway systems, it is essential to enhance the quality, reliability, safety, and security of public transport while increasing accessibility and productivity in general. To do this, it is necessary to drastically reduce costs and deliver more innovative services to attract more customers to public and railway transport systems. Such objectives will be achieved through the widespread trade growing number of information between vehicles and infrastructure, inside vehicles and between vehicles. The connected car is a reality in the world of public and railway transport systems since a long time but this trend has increased with the widespread of fully automated driverless metro. We are witnessing the massive deployment of wireless communication systems at different scales along the infrastructure, railway stations, bus stops, inside vehicles, etc. To date, no technology is universal enough to meet all needs and all cases are operated in complex, heterogeneous environments. Moreover, spectrum is a scarce resource. Its use needs to be optimized. In this chapter, we have proposed an inventory of wireless technologies in the field of public and railway transport systems.

Tomorrow, the wireless communication systems must be available, robust, reliable, scalable, reconfigurable, and interoperable and have large transmission capacities to meet the diverse needs expressed for public transport in the context of mobility and heterogeneous systems. These systems must also be able to save energy and spectrum resources. Many research and developments are performed on the connected vehicle topic in the public transport field. They are done in the framework of cognitive radio, which requires a comprehensive and collaborative "system" approach between several scientific communities: software, information processing, signal processing, hardware, and electronics.

References

[1] Berbineau M., Chennaoui M., Gransart Ch., Afifi H., Bonnin J.M., Sanz D., Coutureau R., Duchange D. 'High data rate transmission for high speed trains, dream or reality? Technical state of the art and user needs'. *Les collections de l'INRETS*, synthèse No. 51, ISBN 978-2-85782-643-5, 2006

[2] Berbineau M., Ambellouis S. 'Etude d'un système de vidéo et audio surveillance sans fil… ou comment amener la continuité de service d'aide à la surveillance entre une flotte de véhicules et l'infrastructure?' *Les collections de l'INRETS*, Rapport No. 275, ISBN 978-2-85782-657-6, 2007

[3] IEEE standard method for CBTC performance and functional requirements, IEEE Standard 1474, 2004

[4] ERTMS/ETCS, Subset-037 ERTMS/ETCS – class 1 Euroradio functional interface specifications, UNISIG, July 2005. Available from www.era.europa.eu

[5] Cellmer J. Le réseau GSM-R de RFF, *REE*, No. 3, pp. 45–52

[6] Fitzmaurice M. 'Use of 2.4 GHz frequency band for CBTC data communication systems, Parson Corp., JCR06'. *IEEE Veh. Tech. Mag.* 2006

[7] Berbineau M. 'Les systèmes de télécommunication existants ou émergeants et leur utilisation dans le domaine des transports guidés'. *Les collections de l'INRETS*, synthèse No. 40, ISBN 2-85782-562-5, November 2001, 129p

[8] Eurobalise, FFFIS for Eurobalise, Subset-036, issue 3, Unisig, 2012

[9] 'Track circuit', in Wikipedia, the free encyclopedia, https://en.wikipedia.org/wiki/Track_circuit, updated on January 7, 2015

[10] ETSI technical report, *Electromagnetic compatibility and radio spectrum matters (ERM); spectrum requirements for urban rail systems*, 2012

[11] Heddebaut M., Boukour F., Fall B., Rivenq A. Perspective d'utilisation de l'ULB en exploitation ferroviaire, *REE* No. 5, 2013, pp. 102–109

[12] UIC, International Union of Railways, 'GSM-R Specifications', http://www.uic.org/spip.php?rubrique874, Published on July 30, 2010. Updated on May 22, 2015

[13] Sanz D., Almeida J.L., Lenormand R., Villeforceix B., Weller E., Feltrin E. 'Testing a DVB-S2 RCS link on the 66REE 4/2014 TGV IRIS320 (Inspection Rapide des Installations de Sécurité à 320 km/h) experimental train', *Toulouse Space Show (Space Appli)*, Toulouse, France, May 2010

[14] Sanz D., Ghannoum H., Villeforceix B. 'Delivering broadband internet access for high speed trains passengers using the new WiFi standard 802.11n for train-ground communications'. *InUIC HighSpeed 2010 Proceedings*; Beijing, China, December 2010

[15] Sethom K. *Gestion de la mobilité dans les réseaux sans fil de nouvelle génération*. PhD thesis, University of Paris; 2006. 135p

[16] Kassab M. *Layer-2 handover optimization for intra-technology and inter-technology mobility*. PhD thesis, Telecom Bretagne; 2008. pp. 13–23

[17] Kassab M., Bonnin J.M., Belghith A. 'Technology-integration framework for fast and low cost handovers. Case study: WiFi-WiMAX network'. *Journal of Computer Systems, Networks, and Communications*. 2010; Article ID 205786:21

[18] Khayat A., Berbineau M., Kassab M., Amine Abid M., Belghith A. 'LTE based communication system for urban guided-transport: A QoS performance study'. *Communication Technologies for Vehicles, Lecture Notes in Computer Science*. 2013;**7865**:197–210

[19] Aguado M. *Broadband communication architecture for train to ground communication services*. PhD Thesis, University of Bilbao

[20] Palicot J. Under direction of J. Palicot, De la radio Logicielle à la radio intelligente, *Hermes Sciences, Lavoisier*, @Institut Telecom et Lavoisier, Paris 2010, ISBN 978-2-7462-2598-5, ISSN 2109-8204, October 2010

[21] *Intelligent integration of railway systems*. InteGRail (2005–2009). Available from http://www.integrail.info/

[22] Berbineau M., Masson E., Cocheril Y., Kalakech A., Ghys J.-P., Dayoub I., *et al.* 'Cognitive radio for high speed railway through dynamic and opportunistic spectrum reuse'. *Proceedings TRA 2014*, Paris, 10p

[23] IFSTTAR, 'SYSTUF project, Telecommunication System for Future Urban Transport', http://systuf.ifsttar.fr/index-en.php, © 2012

[24] Wahl M. Les réseaux de terrain embarqués dans les transports guidés, *Les collections de l'INRETS*, synthèse No. 45, ISBN 2-85782-591-9, 2004

[25] Wahl M. Survey of railway embedded network solutions. Towards the use of Industrial Ethernet technologies, *Les collections de l'INRETS*, Synthèse Ref. S61, ISBN 2-85782-682-8, 2010

[26] Soulas C., Wahl M. Innovation dans les transports guidés urbains, *Hermès-Lavoisier, Collection Traité IC2, Traité Systèmes Automatisés*, 2010, 317p

[27] Nouvel F., Tanguy P., Pillement S., Pham H.M. 'Experiments of in-vehicle power line communications' in Almeida M. (ed.). *Advances in Vehicular Networking Technologies*. InTech; ISBN: 978-953-307-241-8

[28] Lienard M., Carrion M.O., Degardin V., Degauque P. 'Modeling and analysis of in-vehicle power line communication channels'. *IEEE Trans. Vehicular Technology*. 2008;**57**(2):670–9, DOI: 10.1109/TVT.2007.906372

[29] Van Den Abeele D., Wahl M., Berbineau M. Procédé de transfert de données d'alerte entre un véhicule ferroviaire en panne et un centre de contrôle, dispositif associé, ALSTOM, No. de dépôt FR 09 52875, 30/04/09, PCT MAI, 2010

[30] Elhillali Y., Tatkeu C., Deloof P., Sakkila L., Rivenq A., Rouvaen J.M. 'Enhanced high data rate communication system using embedded cooperative radar for intelligent transports systems'. *Transport research Part C.* 2010;**18**:429–39
[31] INTEGRAIL report, IGR-D-WIF-007-06 – Ultra Wide Band Application report
[32] Economou C., 'Innovation Programme 2 – Advanced Traffic Management and Control Systems', in Shift2rail – The Rail Joint Undertaking: http://www.shift2rail.org/ip2/, Updated on May 22, 2015
[33] Di Benedetto M.G., Giancola G. *Understanding UWB radio fundamentals.* Prentice Hall; 2004
[34] INTEGRAIL report, IGR-D-INR-014-05 – Embedded Middleware
[35] Flissi A., Gransart C., Merle P. 'A component-based software infrastructure for contextual transportation applications'. *Proceedings of the 5th International Conference on ITS Telecommunications (ITST 2005)*, Brest; 2005
[36] Bocquet A. 'Infrastructure logicielle multi-modèles pour l'accès à des services en mobilité'. PhD thesis in Informatics, USTL, December 2008
[37] Billion J., Van Den Abeele D. 'ICOM: a communication framework for interoperable European railways'. *Proceedings ITST 07, 7th International Conference on ITS-Telecommunications*, 6p
[38] Schmidt D.C., Parsons J. Overview of the OMG Data Distribution Service. Available from http://www.cs.wustl.edu/~schmidt/DDS.ppt
[39] Allemand L. 'Dossier "les promesses du Big Data" '. *La recherche*. 2013; **482**:27–42
[40] Gatin O., Isasi A. 'Track geometry condition monitoring system for non-intrusive measurements on commercial trains based on wireless sensor networks'. *Proceeding WCRR 2013*, Sydney, Australia
[41] Gatin O., Malouch N., Barro A. 'Wireless mesh and sensor networking for onboard event detection and video alarm transmission for BOSS project'. *Proceedings IEEE ITST 2009 Conference*, Lille. pp. 492–7
[42] Sondi P., Gantsou D., Lecomte S. 'A multiple-metric QoS-aware implementation of the optimized link state routing protocol'. *Int. J. Communication Networks and Distributed Systems.* 2014;**12**(4):381–400
[43] CENELEC EN 50129. *Railway Applications – Safety-related electronic railway control and protection systems.* European Committee for Electrotechnical Standardization; 1999
[44] Marais J., Pouyet V. 'Lateral guidance and longitudinal control for bus rapid transit – Focus on the positioning system based on GPS and RFID technology'. *In ITS World Congress*, Beijing, October 2007
[45] Marais J., Meurie C., Attia D., Ruichek Y., Flancquart A. 'Toward accurate localization in guided transport: combining GNSS data and imaging information'. *Transportation Research Part C: Emerging Technologies*, 2013. pp. 188–97
[46] Garcia-Prieto M.J., Gonzales Romero O., Guttierrez Dimingues R., Urech A. 'GADEROS project field trials underway'. *International Railway Journal and Rapid Transit Review.* 2004;**44**(6):38–9

[47] Marais J., Berbineau M., Frimat O., Franckart J.P. 'A new satellite-based fail-safe train control and command for low density railway lines'. *Proceedings of the TILT Seminar (Technological Innovations for Land Transportation)*; Lille, France, 2003

[48] Marradi L., Galimberti A., Foglia L., Zin A., Pecchioni C., Doronzo M., *et al.* GNSS for Enhanced Odometry: The GRAIL-2 results. Satellite Navigation Technologies and European Workshop on GNSS Signals and Signal Processing, (NAVITEC), 2012 6th ESA Workshop on (pp. 1–7)

[49] Rispoli F., Filip A., Castorina M., Di Mambro G., Neri A., Senesi F. 'Recent progress in application of GNSS and advanced communications for railway signaling'. *Radioelektronika (RADIOELEKTRONIKA), 2013 23rd International Conference*; IEEE, 2013. pp. 13–22

[50] Filip A., Rispoli F. 'SIL 4 compliant train location determination system based on dual-constellation EGNOS-R interface for ERTMS/ETCS'. *CERGAL 2014*; Dresde, 2014. pp. 109–14

[51] Nguyen T.P.K., Beugin J., Marais J. 'RAMS analysis of GNSS based localization system for the train control application'. *ITST 2013*, 5–7 November, Tampere, Finland

[52] He R., Zhong Z., Ai B., Ding J. 'Propagation measurements and analysis for high-speed railway cutting scenario'. *Electronics Letters*, 2011;**47**(21): 1167–8

[53] He R., Zhong Z., Ai B., Ding J. 'An empirical path loss model and fading analysis for high-speed railway viaduct scenario'. *IEEE Antennas and Wireless Propagation Letters*. 2011;**10**:808–12

[54] Masson E., Cocheril Y., Combeau P., Aveneau L., Berbineau M., Vauzelle R., *et al.* 'Radio wave propagation in curved rectangular tunnels at 5.8 GHz for metro applications, simulations and measurements'. *EURASIP Journal on Wireless Communications and Networking, SpringerOpen*. 2011;**202**:2011–202

[55] Hairoud S., Cocheril Y., Combeau P., Pousset Y., Berbineau M. 'WINNER model for subway tunnel at 5.8 GHz'. *Proceedings IEEE ITST 2012 Conference*, Taipei, Taiwan, November 2012, 5p

[56] Hassan K., Gautier R., Dayoub I., Berbineau M., Radoi E. 'Multiple-antenna based blind spectrum sensing in the presence of impulsive noise'. *IEEE Trans. Vehicular Technology*. 2013;99:10

[57] Addaci R., Seetharamdoo D., Rabah M.H., Berbineau M. 'Multi-band multi-antenna system for diversity and/or MIMO applications'. *International Journal of Research in Wireless Systems*. 2013;**2**(1):7–13

[58] Avella Castiblanco J., Seetharamdoo D., Berbineau M., Ney M.M. 'Electromagnetic modelling and definition of antenna specifications and positions for radio system deployment in confined environments'. *Institute of Physics (IOP) – Conference series – Materials Science and Engineering*. 2013; **44**(1):012010

Chapter 4

ITS and freight transport: stakes and perspectives

Corinne Blanquart[1] and Thomas Zeroual[2]

4.1 Introduction

Freight transport faces a dual challenge. It must satisfy the demands of globalised trade on the one hand and meet environmental requirements on the other. Innovation is key to taking up this dual challenge, both for transport policies and firms.

Transport policies are supporting massive investment in innovation and almost a quarter of Europe's private sector's research is in this area. For Europe and its member states, transport is a priority field of research and innovation because it is a major sector of the economy. Transport is considered to be vital for jobs, the economy and European integration. The increasing scarcity of oil and the increase in global warming are strengthening the commitment of public policies to fostering innovations to cope with these challenges.

Firms are also investing in innovation. In Europe, transport firms account for 10 of the 20 companies which commit the largest research budgets to innovation. The transport sector's performance makes it one of the most innovative in Europe.

Information and communication technology (ICT) is expected to make the most important contribution to the effectiveness and environmental viability of freight transport (Commission Européenne, 2007).

The term 'intelligent transport' denotes applications of ICT to the sphere of transport. The aim behind the greater presence of ICT is to implement logistics and integrated transport systems throughout the entire transport chain. This should make it possible to combine the qualities of each mode to provide shippers with the best service in terms of transport efficiency, price and environmental impact. The 2006 revision of the 2001 white paper is consistent with this, recommending 'co-modality' which it defines as 'the efficient use of different modes on their own and in combination' (Commission Européenne, 2006).

However, we are presented with a double paradox. First, transport – unlike nanotechnology, microelectronics or biotechnology – does not have the image of an

[1]IFSTTAR, Marne-la-Vallée Cedex 2, France
[2]ESCE, Paris, France

innovative sector that fosters high levels of inventiveness and creativity. Second, apart from the sector's image problem, we are inadequately informed about the effects of transport innovation. As Jacques Theys has stated with regard to transport innovations: '*the information we have on their cost, impact, potential market and difficulty of implementation... is – almost always – very fragmentary*'.

This chapter sets out to provide examples of innovations that implement ICT in freight transport before examining their suitability for the needs of shippers and the ability of transport undertakings to implement them. Ultimately, our purpose is to show that these technological innovations are a necessary, but not sufficient, condition for achieving efficient logistics and transport chains.

4.2 The role of ICT in extending the supply of transport services

It is widely accepted that the development and deployment of innovations plays a crucial role in increasing production and productivity. In transport, there is nothing new about interest in innovation either. It has nevertheless increased as a result of the impact of the concept of the knowledge economy, and can currently be witnessed, for example in France, in the setting up of competitiveness clusters, some of which are dedicated to transport and/or logistics.

In December 2008, the European Commission gave impetus to the development of intelligent transport by publishing its action plan for the deployment of intelligent transport system (ITS) in Europe. This set out to speed up and coordinate the deployment of ICT in the field of road transport and its interfaces with other transport modes. The Directive of July 2010 on the deployment of ICT, published in the Official Journal of the European Union on 7 July 2010, sets out a framework for supporting the coordinated and coherent deployment and use of ICT in the Union, and lays down general conditions for reaching this goal. The framework applies to ICT services and applications in the field of road transport and their interfaces with other transport modes.

The deployment of ITS affects road transport, due to the implementation of services in the areas of traffic management, freight management, road traffic information and the introduction of electronic tolling systems. But it also affects the other modes of transport, for example the European Rail Traffic Management System (ERTMS) whose main function is to monitor the spacing between trains and ensure a minimum is maintained. The development of ICT could also benefit air traffic and the safety of maritime traffic.

Some of the applications in question are made necessary by national or supranational regulations, one example being the digital tachygraph, and others are intended to meet the demands of fleet managers or the requirements of shippers. Some are useful for implementing multimodal transport operations, others improve traffic and infrastructure management.

With regard to freight, these technologies have the following roles to play:

- optimising logistics and the *organisation* of transport;
- optimising transport *operations* and routes;

- optimising urban freight transport and city logistics;
- limiting energy consumption, pollution and adverse impacts.

Each of these aspects will be examined in more detail below.

4.2.1 Prior to transport: ICT, optimisation of logistics and optimisation of the transport organisation

A TMS (Transport Management System) is a software package that covers all transport management activities, from the management of master data, offer creation, the management and dispatching of transport orders through to the invoicing of clients and subcontractors.

The functions of a TMS include:

- offer management;
- input of transport orders;
- Dispatching, providing a quick overview of scheduled transport activities and available vehicles;
- invoicing freight transport and services;
- the ability to be implemented alongside accounting applications and document management systems (DMS);
- management of insurance and other transport- and vehicle-related documents;
- monitoring of the budget allocated to the vehicle fleet.

Extended TMSs exist that can be used over the Internet. The benefits of these are as follows:

- clients can key in and track their consignments directly. They can also print out delivery forms and labels;
- like the clients, consignees can access the list of their shipments and find out the scheduled delivery date;
- the subcontractors, whether transport undertakings or subsidiaries, are able to feed data into the TMS: delivery dates and times, specific problems relating to consignments, ... for all the transport operations assigned to them.

In 2012, Logistiques Magazine listed 23 major TMS software packages, which shows that they are both very widespread and much in demand.

The function of APS (Advanced Planning and Scheduling) software is to organise the logistics chain. In response to the demand stated in advance by the firm, APS software analyses the capacities of resources (machines, workforce, materials, storage areas) and constraints in order to propose a detailed adaptable timetable that ensures optimal production.

It is active at every level:

- demand, it determines how many products need to be manufactured;
- purchasing, it checks the availability of raw materials and components based on the product's bill of material;
- production, it analyses constraints and the capacity to develop an optimal production plan;

- storage, it forecasts the areas needed;
- transport and distribution, it optimises the costs and guarantees the quality of the services delivered to clients.

APS automatically performs trade-offs between the anticipated demands of clients and the available resources.

An APS is not primarily a transport software system like a TMS, but it must take account of distribution when planning, so it nevertheless takes transport into account.

In 2012, Logistiques Magazine listed 10 major APS software packages. There are fewer APS than TMS because, like route optimisation software, they are difficult to implement.

4.2.2 ITS and transport and route optimisation

ICT has had a decisive impact on route optimisation. It takes account of a large number of management parameters: vehicle-kilometres travelled, truck filling rates, empty return rates, transport mode utilisation, batch sizes and reserve stocks.

These tools can also help to pool logistical resources, optimise and rationalise transport flows, improve truck filling rates and promote the use of modal transfers. Societal and regulatory pressures are thus creating a market for a new generation of Enterprise Resource Planning (ERP) and Supply Chain Management (SCM) software in which environmental constraints will figure from the design stage.

4.2.2.1 Route optimisation

Route optimisation helps firms use drivers' working time efficiently and provides direct improvements in productivity. It also has environmental benefits, as it minimises the vehicle kilometres travelled for deliveries. The route optimisation planning software packages that are currently available are highly effective. Interactive constrained optimisation software packages take account of clients' working hours, the capacity of vehicles, driving and resting times, etc. Routes are computed applying parameters that are appropriate for the firm. Personalisation modules make it possible to prevent the use of certain roads (motorways that are deemed expensive, roads which are too small for the trucks, etc.).

4.2.2.2 Quality of service and traceability

With regard to traceability, techniques are being developed to provide full traceability of transported products on a continuous basis and in real time by associating on-board geopositioning and radio-frequency identification (RFID) technologies that make it possible to display the positioning and the movement of objects throughout the transport process.

4.2.2.3 Pooling of freight and freight exchanges

Platforms also exist whose function is to identify freight. These freight exchanges are open to carriers and enable them to exchange freight and transport capacity and avoid empty return runs.

While pooling of this type exists between road transport undertakings, its implementation among shippers still poses difficulties. In addition, in response to the policy goal of developing co-modality, in particular by promoting modal transfer from the road to modes that are more environmentally friendly, the solutions put in place by the market stakeholders are often designed to meet the needs of a single carrier and do not always include all the links in the logistics chain.

However, it is of central importance to projects for urban distribution centres to have an operator in charge of an information system that gathers orders together, receives and prepares customers' packages and organises redistribution to the final recipients.

4.2.2.4 Towards the logistics Internet?

The Physical Internet is based on the global sharing of resources and logistics systems which can be seen as a type of globalised pooling. The Physical Internet aims to interconnect the various logistics services in order to create an open network of networks which brings together all the resources in the logistics chain. The idea behind interconnecting networks is to improve the efficiency of logistics systems by eliminating the fragmentation which is apparent today. This change in the organisation of logistics will have an impact on transport needs, the diversity of transport modes, storage facilities and, more generally, the performance of supply chains. The move towards a system of this type will inevitably lead to a thorough reorganisation of logistical practices and resources.

Interconnecting logistics networks requires technological changes with regard to the standardisation of packaging and information exchange. But above all it requires changes in the interactions between the members of the logistics chain.

Similarly, there are many obstacles in the way of the pooling of resources for multimodal networks.

4.2.2.5 Optimisation of traffic at the nodes of multimodal chains

Logistics must take account of the large range of transport modes used in freight transport. Managing multimodality and intermodality is therefore of fundamental importance. A number of ITSs are available that assist in the management and monitoring of multimodal centres (ports, airports, ferry terminals or railway stations). These deal with parking, operation and the interfaces between the various transport modes. In this case the task involved is managing dynamic environments.

In this connection, the InTraDE project (Interreg IVB) aims to design a virtual simulator to supervise traffic inside container ports together with two prototype safe, smart, environment-friendly vehicles. The innovative aspect consists of the ability of the proposed transport system to be adapted to any port infrastructure.

One of the problems that faces the development of port terminals is the management of traffic inside the port and achieving optimisation in the confined spaces that are available.

4.2.3 ITS and the optimisation of city logistics

Today, many projects focus on improving urban transport, due to the increasing presence of congestion, noise and atmospheric pollution in cities. Furthermore, freight transport is more and more concerned with environmental issues. Urban freight transport has therefore become an important topic within the wider area of freight transport research. Innovation in the areas of city logistics and urban freight transport can take several forms.

4.2.3.1 Limiting vehicle stationary time and facilitating deliveries

Delivery bays are scarce in urban areas, and when they do exist they are often occupied by other vehicles. Double parking is a persistent problem. Systems for the advance reservation and dynamic management of parking bays are therefore being developed. The information management system uses a combination of information technology, mobile communication devices and sensors embedded in the carriageway.

4.2.3.2 Making drivers' work easier: 'hands-free' vehicle access and egress

During their delivery rounds, drivers need to climb in and out of their vehicle all the time in order to perform loading and unloading operations. Drivers need to lock their vehicle as soon as they leave it to prevent the theft of the vehicle and or the goods it contains.

Considerable potential time savings can be made just by eliminating the use of keys, as is the case in the car industry, by using an electronic badge which unlocks the doors as soon as the driver comes near the vehicle.

4.2.3.3 RFID recognition for improved parcel tracking

In most transport undertakings that provide their drivers with a straightforward mobile phone, the drivers have to call their office at regular intervals to report their position, completion of a delivery or the beginning of a new task. When this is not the case, the operator has to spend a great deal of time trying to contact drivers by telephone to find out what they are doing, how their work is going and their position.

RFID technology provides a way of identifying parcels, tracking their movements and finding out their characteristics from a distance. Systems that are integrated within the vehicle speed up loading and unloading operations: the vehicle informs the driver about upcoming deliveries and automatically counts and checks stocks. Drivers can simply get on with their delivery rounds without spending time on administrative tasks.

4.2.3.4 Towards autonomous systems?

The vehicle that has been developed in the framework of the European FIDEUS (Freight Innovative Delivery of Goods in European Urban Spaces) project has been fitted with a geopositioning and communication system integrated within a pocket computer and will ultimately detect urban zones on its own and automatically switch over to appropriate operating conditions, particularly in relation to noise requirements.

4.2.4 ITS and the reduction of energy consumption and emissions

On-board computing systems set out to help to reduce energy consumption either by analysing driving behaviour or by guiding the driver to roads that are more suitable or less congested (every time a truck that is stuck in a traffic jam starts up again, it consumes 1.3 litres of fuel, which is equivalent to a consumption rate of 60 litres per 100 km). The majority of today's GPS devices have the capacity to analyse current traffic conditions and propose an alternative route in order to avoid traffic jams. This also applies to GPS devices that are dedicated to truck drivers.

To give one example, the European HeavyRoute project (Intelligent Route Guidance for Heavy Vehicles) has as its goal the development of an advanced route guidance system for trucks. This project sets out to provide a Europe-wide system that channels Heavy Goods Vehicles (HGVs) towards the safest routes and those that are the most cost-effective with regard to energy consumption. Heavy Route (HR) will use data on the characteristics of the road network for route planning and driver assistance purposes. The characteristics of the infrastructure will be used to compute journey times, vehicle operating costs, damage to roads and bridges, environmental impacts and accident risks.

A considerable number of on-board computing systems also transmit real-time journey data to fleet managers in order to compute fuel consumption and on that basis implement plans for cutting fuel consumption that are tailor-made for each driver. Examples include FleetBoard from Mercedes-Benz, the Dynafleet system from Volvo Trucks, Optifuel Infomax from Renault Trucks, Box from Scania France and the Telematics systems offered by DAF and Man.

Although many technological solutions exist and many services are available, firms nevertheless find it difficult to make a final decision in this strategically important area which is both complex and continually changing and where implementation often poses major difficulties.

4.3 What conclusion can we reach regarding the impact of ICT on transport services?

The potential impact of ITS has been evaluated both with regard to the research that has been done and the first phases of deployment. A considerable body of work has attempted to analyse impacts on the characteristics of trips in terms of cost, duration and the spatial distribution of flows (Haynes *et al.*, 2000; Taniguchi *et al.*, 2001). By employing different configurations it has frequently been possible to reduce travel times by as much as 20 per cent and improve network capacity by between 5 and 10 per cent. Many estimates place the resulting safety improvements at between roughly 10 and 15 per cent.

However, there are technical barriers that stand in the way of their implementation within the sector as a whole. For example, even if a large amount of data are available about freight logistics, they are neither interoperable, nor usable nor extractible, which means they cannot be exchanged between the systems.

In addition, the potential results depend on the mobilisation of the services. For this reason it is necessary to take account of the conditions under which these services are 'produced'. These are always the outcome not only of the obvious interaction between a shipper and a service provider, but also between a specific logistics organisation which in its turn is the outcome of the links between the members of the productive chain (Meunier and Zeroual, 2006). We shall show that what shippers demand from systems of this type varies a great deal and that the structure of the transport sector, which consists of a large number of Small and Medium Enterprises (SMEs), hampers take-up.

Thus, while these technological innovations are important means of improving the efficiency of transport, in many cases they need to be accompanied by innovations of other types if they are to realise their full potential.

4.3.1 The diversity of shippers' logistical needs

One of the difficulties in the freight transport sector stems from the diversity of the types of clients, freight to be carried and destinations, etc. This means that having a single transport solution and a single system is not an option.

Shippers make complex decisions and will clearly take into consideration the available infrastructure and services.

But firms' logistical and transport requirements are also strategic decisions that attempt to reconcile a very wide range of factors, stakeholders and management levels, including: constraints relating to microeconomic optimisation in terms of costs and transport and logistical processing time for the firm. Table 4.1 shows the importance of the different criteria of cost and time for shippers when selecting a mode.

The three most important criteria when selecting a mode for shippers are therefore the cost of transport, the reliability of delivery times and the duration of transport. It is important to note that the last of these is less important than the transport service provider meeting its commitments about delivery times. With regard to the increasing interactions between firms in the same productive chain, this decision concerning time already has impacts that extend beyond the firm. The reliability of transport times is essential to ensure the productive chain functions correctly, thereby allowing the firm to meet its productive commitments with respect to its partners in a context where just-in-time practices are constantly gaining ground.

It is for this reason that it is claimed that firms' transport decisions depend both on the context and meso-economic constraints that are imposed by ties with other stakeholders in the productive system (particularly the clients).

Needs will therefore differ in different contexts, as will the services that are mobilised. Figures 4.1 and 4.2 illustrate the diverse nature of the services that are performed in warehouses and hubs, respectively.

Table 4.1 The ranking of modal selection criteria for shippers
[Source: Guilbault and Soppé (2009); analysis of data from the ECHO[1] survey (INRETS, 2004).]

Selection criteria in decreasing order of mention in corrected no. of consignments/tonnes	Selection criteria (all modes)	
	No. of consignments (%)	Tonnage (%)
Transport cost	60	52
Reliability of delivery times	47	33
Duration	35	28
Conditions of access to the network	34	32
Flexibility	21	25
Working hours	20	14
Available frequencies	18	9
Security of goods	14	10
Availability of end-to-end transport	13	14
Quality of commercial contacts	12	10
Security of transport	7	10
Provision of associated logistical services	4	3
Regulatory constraints	3	2
Equipment available in advance	3	6
Legibility of supply	3	2
Supply of specialized equipment	2	10
Environmental friendliness	1	2
Management of own fleet	6	10

Figure 4.1 shows that the main service performed in warehouses is storage, which is a physical operation. Services that deal with information (stock management, computerised tracking of products, etc.) are performed in many fewer establishments. The functions performed in hubs are, incidentally, similar, as can be seen in Figure 4.2.

In addition, demand can also vary for the same type of operation. Analysis of the Echo survey for the Nord-Pas de Calais region for a number of information-related operations provides the results set out in Figures 4.3 and 4.4. They show the percentage of consignments which receive the services that we have selected during their journey.

These two examples of information-related operations (computerised tracking of consignments and electronic proof of delivery) show that deployment varies greatly from one sector to another. These two operations involve the traceability of

[1]This survey was conducted in 2004 on a sample of 3,000 firms, representing, after weighting, slightly fewer than 70,000 establishments at the national level. The sample includes firms with more than 10 employees in the sectors of industry and wholesale trading, mail order and salvaging. The sectors of agriculture, mining, building and civil engineering, other services and retail were excluded from the survey.

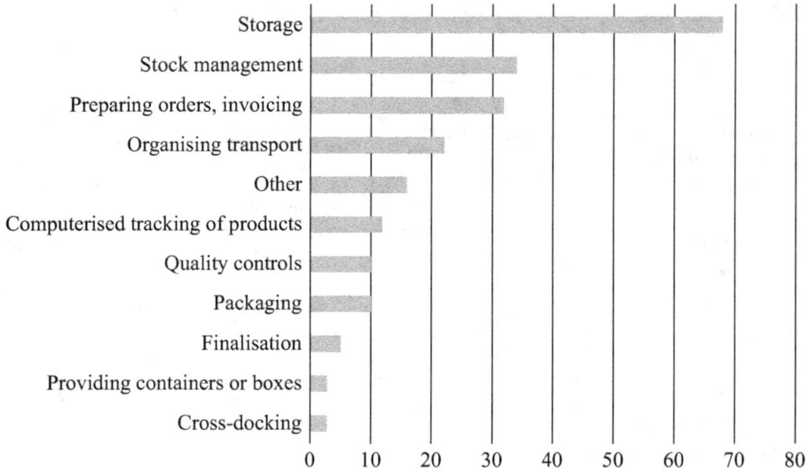

Figure 4.1 Services provided by warehouses (per cent of establishment)
[Source: ECHO (2004).]

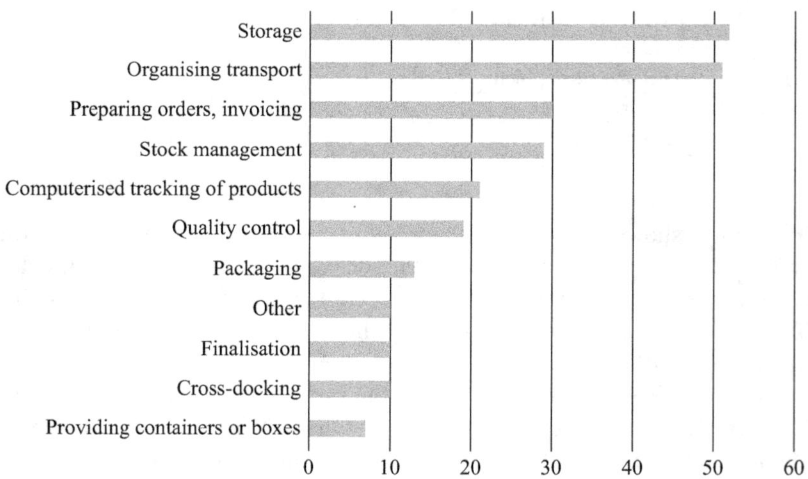

Figure 4.2 Services provided in hubs (per cent of establishments)
[Source: ECHO (2004).]

flows, which is defined as the ability to find out the history, the use, or the location of an entity from recorded data. This type of tracking is presented by firms as a sustainable logistics solution insofar as it improves the security of transported goods. Tracking operations of this type are essential for the smooth operation of the productive system as a whole and maintaining links between stakeholders. Logistics performance is thus linked to the reactivity and flexibility of the chain. It is important to be able to cope with unexpected events which might disrupt the

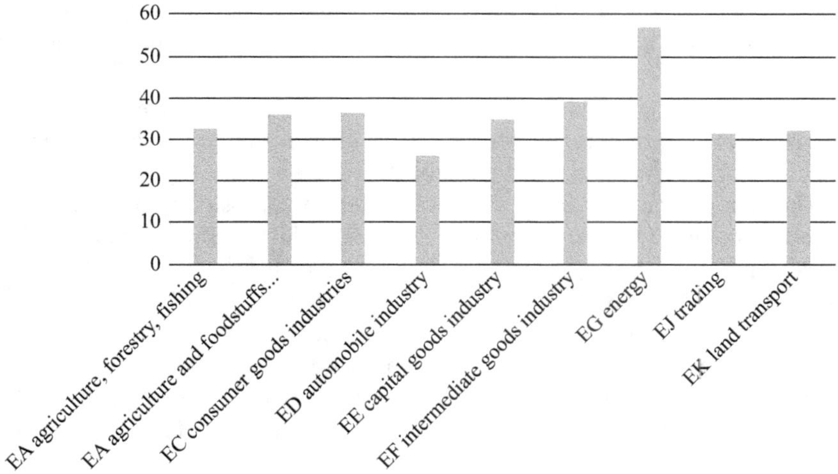

Figure 4.3 *Computerised tracking of consignments (per cent of consignments)*
[Source: ECHO (2004).]

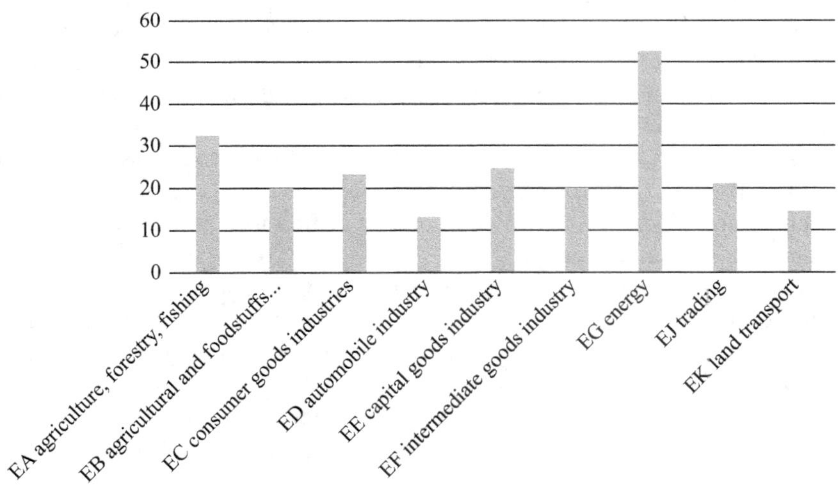

Figure 4.4 *Electronic proof of delivery (per cent of consignments)*
[Source: ECHO (2004).]

entire system. This is the area on which concerns about sustainability should focus. Thus, quality of service, in particular the reliability and forecastability of journey times, is both increasingly expected by shippers and an increasingly difficult issue in complex transport and logistics chains which contain a large number of links and stakeholders, and in which, in particular, some of the former are congested.

There is therefore not a single type but a range of logistics and transport strategies, as needs change depending on the context.

Consequently, the diverse nature of the strategic decisions made by firms with regard to transport and logistics opens up the possibility of diversity in the influence of ITS.

In addition, the needs of shippers will change in the future. The growing importance of e-commerce is changing transport demand and creating a need to provide suitable services that make use of ICT.

The courier firm Top Chrono offers its clients the possibility of appointments within a 30 minute or a 2-hour time slot. SLS-CGI (which specialises in the delivery of heavy parcels weighing over 30 kg) operates in the same way also making appointments, which it supplements with telephone calls. Growth in the number of pickup points and parcel delivery boxes also depends on the development of interfaces that make it possible to agree on an optimal delivery point for the goods.

In the case of parcel delivery boxes, the service is available 24/7 and real-time traceability is possible (using a barcode). This allows the level of stock in the box to be monitored in real time.

4.3.2 Innovations that must take account of the possibilities of adoption by the social system of transport

The implementation of innovations needs to consider them as part of a social system which, through a series of relationships between stakeholders, becomes stable in the form of a 'dependency path'. The deployment of radical innovation requires a change in the organisation of the system and the relationships between industry, state players, operators and users.

Thus, the main difficulty for the potential adopters is to assess the cost and benefits of these ICT's investments.

Small and Medium Enterprises in particular find it more difficult to valuate technical change than larger firms.

In view of the range of services and technologies that are available, whose differences and features are not always clear, and the proliferation of the solutions on offer, transport undertakings have real difficulties gaining a clear idea of whether or not a proposed solution meets their needs.

The situation is made even worse by the fact they have insufficient knowledge and understanding of the characteristics of the services and the related devices, and find it difficult to assess the real cost of purchase, use, maintenance and subscriptions, etc. These costs are often greatly overestimated. Furthermore, it is virtually impossible for these firms to make even an approximate estimate of the return on investment. Not only do they find the information available to them inadequate, but they also have doubts about the long-term utility and potential obsolescence of the proposed devices and services: is the technology mature?

4.3.3 Technological innovations which must be associated with innovations of other types

In the literature, the term innovation is often applied exclusively to technological innovation. The prime importance given to technological innovation by researchers may explain why it is currently the main focus for public policies. Unfortunately, this dominance can have negative impacts. For example, a report published by the French Economic Committee for Sustainable Development emphasises:

> the risk of focusing on technological innovations without distinguishing process, product and service innovations, as well as without considering their economic viability and the changes they imply. This logic leads to over-optimism in the emergence stage of new technologies and to severe corrections as soon as implementation difficulties appear. The internet bubble is an eloquent illustration (Crifo *et al.*, 2009).

Innovation is multifaceted. As McCormick (2002) has stated: 'the process of technological change is nothing less than a process of cultural transformation'. Non-technological innovations cover many areas and apply to organisation, processes and business models. The innovations that take place around ICT are often linked to innovative equipment. However, their implementation in the transport system requires additional organisational innovations which are frequently not recognised. For example, the FIDEUS project sets out to propose sustainable transport solutions for urban freight transport without sacrificing economic efficiency by:

- developing a complementary set of vehicles and equipment, specifically designed for urban delivery and collection;
- proposing a new approach to the organisation of urban logistics, involving co-ordinated use of vehicles of different types, and telematics systems to improve the management of delivery operations;
- providing tools and information that will be of practical assistance to local authorities in planning and implementing strategies for dealing with urban delivery traffic.

4.4 Conclusion: the implementation conditions for innovation

ICT combined with satellite positioning play a major role in all areas of land transport, whether improving the quality of safety, environmental aspects, etc. or providing new, more efficient, services.

The potential applications of ICT in the fields of freight transport and logistics appear to be numerous. Development of the corresponding services has been under way for many years, frequently as part of transport organisation services, and sometimes as free-standing services. The question is in what areas and for what type of services is the 'spontaneous' development of the services in question deemed to be inadequate necessitating some form of public intervention. In this case, what roles can the state and local authorities play to hasten the development of these services in order to bring about convergence between policy goals with regard to co-modality, the management of networks and the environment, and the needs of transport undertakings, logisticians and shippers.

Bibliography

CAS. *Le fret mondial et le changement climatique. Perspectives et marges de progrès, Rapport du groupe de travail présidé par Michel Savy*. La documentation Française; 2010

CEMT. *Le dumping social dans l'espace couvert par la CEMT. Application au cas du transport de marchandises*. UE, Bruxelles; 2002

CNT. *L'évolution sociale dans les transports terrestre, maritime et aérien en 2003–2004. Premier fascicule. L'évolution de la régulation sociale*. La Documentation française; 2005

Commissariat général au développement durable. *Cinq scénarios pour le fret et la logistique en 2040*. Rapport PREDIT; 2011

Commissariat général au développement durable. *Indice de production de service transport au troisième trimestre 2013*. 2014. Available from http://www.statistiques.developpement-durable.gouv.fr/fileadmin/documents/Produits_editoriaux/Publications/Chiffres_et_statistiques/2014/chiffres-stats485-ipst2013t3-janvier2014.pdf

Commission Européenne. *Livre Blanc: La politique européenne des transports à l'horizon 2010: L'heure des choix*. UE, Bruxelles; 2001

Commission Européenne. *Pour une Europe en mouvement – Mobilité durable pour notre continent. Examen à mi parcours du livre blanc sur les transports*. UE, Bruxelles; 2006

Commission Européenne. *Plan d'action pour la logistique du transport de marchandises*. UE, Bruxelles; 2007

Commission Européenne. *Un avenir durable pour les transports: vers un système intégré, convivial et fondé sur la technologie*. UE, Bruxelles; 2009

Commission Européenne. *EU energy and transport in figures*. Statistical Pocket-book. UE, Bruxelles; 2009

Commission Européenne. *Le réexamen du budget de l'UE*, Communication de la commission au parlement européen, au conseil, au comité économique et social européen, au comité des régions et aux parlements nationaux. UE, Bruxelles; 2010

Commission Européenne. *Document de travail des services de la commission accompagnant le rapport de la commission sur la politique de la concurrence de 2011*. UE, Bruxelles; 2012

Commission Européenne. *Proposition de règlement du Conseil portant création de l'entreprise commune Shift2Rail*. UE, Bruxelles; 2013

Crifo P., Debonneuil M., Grandjean A. *Croissance verte*; 2009. Available from http://www.developpement-durable.gouv.fr/IMG/pdf/03-10.pdf

Eurostat (ed.). *Measuring progress toward a more sustainable Europe*. Eurostat Statistical Book; 2007

Eurostat (ed.). *Panorama of transport*. Eurostat Statistical Books; 2009

Eurostat (ed.). *Energy, transport and environment indicators*. Eurostat Pocket Books; 2013

Guilbault M., Soppé M. Apport des enquêtes chargeurs: connaissance des chaînes de transport de marchandises et de leurs déterminants logistiques, actes INRETS n 121, les collections de l'INRETS; 2009

Haynes P., Shuckburgh E. Effective diffusivity as a diagnostic of atmospheric transport: 1, Journal of Geophysical Research: Atmospheres; 2000

March J. 'Exploration and exploitation in organizational learning'. *Organisation Science*. 1991;**10**(1):71–87

McCormick K. 'Veblen and the new growth theory: community as the source of capital's productivity.' *Review of Social Economy*. 2002;**60**(2):263–77

Meunier C., Zeroual T. *Transport durable et développement économique*. 2006. Available from http://%20developpementdurable.revues.org/3305

Mongo M. 'Les déterminants de l'innovation: une analyse comparative service/industrie à partir des formes d'innovation développées'. *Revue d'économie industrielle*. 2013;**143**(3):77–108

Morand P., Manceau D. 'Pour une nouvelle vision de l'innovation' in La Documentation Française (ed.). Paris; 2008

OCDE (ed.). *Les instruments économiques des politiques d'environnement en Chine et dans les pays de l'OCDE*. Paris; 1997

OCDE (ed.). *Perspectives des transports 2010. Le potentiel de l'innovation*. Paris; 2010

OCDE (ed.). *Politique énergétique et climatique: infléchir la trajectoire technologique*. Paris; 2013

Petersen M.S., Enei R., Hansen C.O., Larrea E., Obisco O., Sessa C. *et al. Report on transport scenarios with a 20 and 40 year horizon*. Final report, Copenhagen; 2009

Prades J., Mujica M., Revéret J.-P., Rubio-Ardanaz J.M. 'Vers une stratégiez de transport durable fondée sur le développement de l'innovation technologique'. *Esprit Critique*. 2005;**7**

SETRA 'Fret ferroviaire: quelles évolutions technologiques pour les wagons de marchandises'. *Revue Transports*. 2013;**25**:10–16

Savy M. 'Questions clefs pour le transport en Europe'. La Documentation Française. Paris; 2009

Steer Davis Gleave. *Impact assessment support study on the revision of the institutional framework of the EU railway system, with a special consideration to the role of the European Railway Agency*. Final Report. UE, Bruxelles; 2012

Salter A., Tether B. 'Innovation in Services: through the innovation glass of innovation studies'. Paper presented in Background Paper for Advanced Institute of Management (AIM) Research's Grand Challenge on Service Science; Tanaka Business School, London, 7 April 2006

Steinmueller W.E. 'Will new information and communication technologies improve the "codification" of knowledge?' *Industrial and Corporate Change*. 2000;**9**(2):361–76

Taniguchi E., Thompson R.G., Yamada T. *City Logistics. Network modelling and intelligent transport systems*. 2001.

Theys J. 'Quelles technologies futures pour les transports en Europe?' in CPVS (ed.) *Contribution au groupe – Technologies clés – de la Commission Européenne*; 2007

Tremblay D.-G. *Le rôle des institutions dans le processus d'innovation; l'apport de Thorstein Veblen*. Revue Interventions Economiques; 2007. Available from http://interventionseconomiques.revues.org/518

XERFI (ed.). *Le transport ferroviaire en France*. Xerfi **700**; 2013

Chapter 5

Energy-efficient and real-time databases management techniques for wireless sensor networks

Ousmane Diallo[1,2], Joel J.P.C. Rodrigues[1,3] and Mbaye Sene[4]

5.1 Introduction

Technological advances in recent years have allowed the maturity of wireless sensor networks (WSNs), which aim at performing environmental monitoring and data collection. This sort of network is composed of hundreds, thousands or probably even millions of tiny smart computers known as wireless sensor nodes, which are capable of sensing, processing and transmitting information about the environment in which they are deployed. Sensors are usually scattered around a given geographical environment for measuring physical data or performing event detection. These sensors organise themselves in *ad hoc* manner, without fixed infrastructure. Generally, the measured data is sent towards one or several sink nodes in order to be accessible by remote users through application-level gateway, for example, global sensor network [1–3]. To retrieve the data, users should use applications that provide supports of efficient query management techniques, which allow communication with the network [4–6]. An illustration of a WSN is given on Figure 5.1.

Sensor nodes are usually battery powered and are considered intelligent, equipped with sensors, a radio transceiver, a central processing unit (CPU) and some memory [7,8]. However, due to the small size and the requirements of low-cost nodes, the amount of available resources such as storage, computation power and particularly the energy into a node is generally very limited and the sensor network lifetime depends on the available energy in the nodes composing the network [9,10]. This available energy is consumed by sensing activities (data acquisitions from the environment), wireless communications (sending and receiving of packets), which are essential to form a WSN, and data processing,

[1]Instituto de Telecomunicações, University of Beira Interior, Covilhã, Portugal
[2]Department of Informatics, University of Assane Seck of Ziguinchor, Senegal
[3]King Saud University (KSU), Riyadh, Kingdom of Saudi Arabia (KSA)
[4]Department of Mathematics and Informatics, UCAD, Senegal

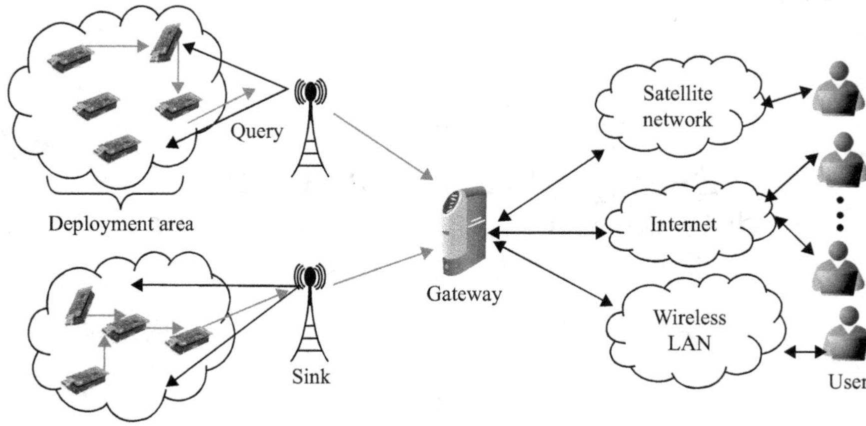

Figure 5.1 Illustration of a WSN architecture

which consists in some operations applied over data by smart sensors [11–13]. However, research into sensor hardware has shown that the energy depletion in the network is mainly due to the data communication tasks among the nodes [14,15]. Therefore, optimising sensor activities, mainly the wireless data communication activities, as a function of energy, is one of the main points of attention in many algorithms designed in database management for WSNs. One of the most used methods to save energy is to activate only necessary nodes and to put other nodes to sleep [16]. Some researchers have studied how a three-dimensional sensor field can be efficiently partitioned into cells in order to save energy [17].

Once sensors measure the physical data, the problem of data storing and querying arises. In fact, the sensors have small storage capacity and the ongoing interaction between the network devices and the environment results in huge amounts of data [18]. Therefore, in addition to data gathering [19] and data replication issues [20], in such applications, a database-oriented approach of WSNs has proven to be useful in order to manage the large amount of data generated by the sensors. This database-oriented approach has motivated the design of WSN data acquisition with two fundamental objectives [21]: like traditional database systems, a WSN database should provide SQL-like abstractions so that nodes can be easily programmed for simple data sensing and collection. Moreover, the data collection process should minimise the energy consumption in the network.

Sensor nodes can be placed anywhere to provide ubiquitous data access, resulting then on the increased use of sensor network systems in many areas of the knowledge, giving rise to several flavours of WSNs [22,23]. This maturity of WSNs involved the development of wide variety of applications, which may perform certain operations on collected sensor data. However, for certain applications, such as real-time applications, the sensor data must closely reflect the current state of the targeted environment [24]. Unfortunately, the environment changes constantly and the data is collected in discreet moments of time. As such, the collected data has a temporal validity, and as time advances, it becomes less accurate, until it does not reflect the state of the environment any longer. Thus, these applications must query and analyse the data in a bounded time in order to make decisions and to react efficiently. Among these applications, include intelligent transport system (ITS) applications [25–28], urban or environmental phenomena monitoring [29–31], tracking of actions on dynamic web pages [32,33], disaster management [34–36], military [37,38] and medical [39–42]. In this context, real-time database management techniques become tremendously important [43–45].

The aim of this chapter is to show how databases management techniques for WSNs are mature for managing the large amount of data generated by sensors in energy-efficient and real-time way by describing the requirements and characteristics for designing and implementing an energy-efficient and real-time database management technique for WSNs, as well as presenting and classifying some recent and relevant proposals in this area. Moreover, a discussion and open issues on energy-efficient and real-time database management techniques for WSNs are identified in order to facilitate further contributions.

The remainder of this chapter is organised as follows: Section 5.2 presents the main fields of application of WSNs, while the requirements and characteristics for designing energy-efficient and real-time databases techniques for WSNs, as well as recent and relevant protocols and techniques in this area are exposed in Section 5.3. Section 5.4 discusses essential techniques used on the studied approaches and proposes some open research issues. Finally, Section 5.5 concludes the chapter.

5.2 WSN-based applications

Sensor networks can be very useful in many applications for collecting and processing data from environments. So, with technological advances in sensor miniaturisation, the low cost, the large variety of available sensors, as well as the wireless communication support, wireless sensors increasingly develop in several fields of application. Among these fields of application, one can note the following areas.

5.2.1 Military applications

Like several information technologies, WSNs come mainly from military researches. Thus, autonomous WSNs are very essential components in the impetus of war systems centred on networks. They can be quickly deployed and used for the monitoring of battlefields in order to provide information such as the site, the number, the movement, and the identity of soldiers and vehicles, or even for the detection of nuclear, biological and chemical agents. Therefore, several projects were developed to either help military units in a battlefield or protect cities against attacks, such as terrorist attacks. The Distributed Sensor Network (DSN) project [38] financed by the Defense Advanced Research Projects Agency (DARPA) was one of the first projects having used the WSNs to collect distributed data. Another network named Wide Area Tracking System (WATS) [46] was developed by researchers of the national laboratory of Lawrence Livermore. The WATS network is composed of gamma ray detectors and neutrons to detect and track down nuclear devices and uses data aggregation techniques to send back the data towards a base station. Besides the WATS network, these researchers designed another network called Joint Biological Remote Early Warning System (JBREWS) [47] to prevent from biological attacks in a battlefield.

5.2.2 Monitoring applications

WSNs can be deployed in public structures for helping to give security of places and human beings and, in addition, reducing the devoted financial costs. Indeed, a WSN can be deployed with a purpose of supervising a small or wide geographical area to detect the appearance of phenomena or to measure a physical quantity (temperature, pressure, speed, etc.) [29,30]. For example, sensors can be included in dam-walls to calculate in real time the exerted pressure in order to control the water level if limits are reached. The same technique can also be used for other constructions such as bridges, railroad ways, mountain roads, buildings and other artworks to help detect cracks and deteriorations in structures due to an eventual seism or ageing of the structure. The deployment of a WSN of motion detection with a warning system can help to detect intrusions in targeted environment.

5.2.3 Environmental applications

Sensors can be placed anywhere there is data that should be collected, particularly in hostile environments. This leads to various types of applications, such as fire detection application with thermo-sensors [34,35]: sensor can be deployed in large forests in order to allow efficient intervention and to prevent from fire forests, volcano monitoring [36], other planets monitoring [30], in agriculture area [31,48]: sensors can be buried in the ground and the WSN can be questioned for knowing, for example, the state of the ground in order to sprinkle the driest sectors in priority.

5.2.4 Urban and home automation applications

With the pervasive computing [49] paradigm, sensors can be embedded into our daily living environment and provide sensor data for localisation and surveillance. Thus, in urban environments, sensors are already used in several monitoring and control applications, such as smart home [29,50,51] where sensors, embedded in devices, such as vacuum cleaners, microwave ovens, refrigerators, video tape recorders, can interact with other networks like Internet to allow locally or remotely control of domestic devices, intelligent buildings [52], industrial sites [53] to prevent from industrial risks such as the escape of toxic products (gases, chemicals, radioactive elements, oil, etc.), ITSs [25,26], etc. ITS applications can be divided into two groups, parking lots monitoring [27] and traffic monitoring and control [28,54] applications. And the successful development of these applications can efficiently improve the transportation systems with important economical and environmental impacts, such as reduction of traffic congestions, travel time, fuel consumption and air pollution [55].

5.2.5 Medical applications

With the growing needs in ubiquitous computing and recent progress in very-low-power wireless technologies, micro and nano technologies, mobile communications, human computer interface, there has been considerable interest in the development and application of wireless networks around humans. In the medicine area, it results on the development of new kind of network, known as either wireless body area network (WBAN), or Body Area Sensor Network (BASN) or in short Body Sensor Network (BSN) [56,57]. These sensor networks can be used to ensure a permanent monitoring of vital organs of human beings, thanks to micro-sensors installed either on clothes, on the body, or even under the human skin (glycaemia monitoring, cancer-detection, etc.) [58,59], and eventually using actuators that takes some specific actions according to the data they receive from the sensors or through interaction with users. For example, an actuator equipped with a built-in reservoir and pump administers the correct dose of insulin to give to diabetics based on the glucose level measurements [60]. They can also facilitate the diagnosis of certain diseases by taking physiological measurements such as heartbeat, blood pressure, body temperature, recording of prolonged electrocardiogram (ECG), thanks to sensors [40–42]. In addition, these networks can detect abnormal behaviours (falls from a bed, shock, cry, etc.) at the dependent people (handicapped or old person) [39,61].

5.2.6 Commercial applications

In commercial area, sensors can be integrated in the delivery and storage process. Then, the formed network could be used to know the position, the state and the direction of a product. This will allow, then, for a customer who awaits the reception of a product to have an opinion of delivery in real time, to know the current localisation of the product and to check if the production line is respected for edible products. For example, customer can check if the frozen chickens respected the required temperature during all transport [62].

Figure 5.2 tries to provide an illustration of different applications involved.

5.3 Energy-efficient and real-time databases techniques requirements

Like in traditional database management systems (DBMSs), sensor database managements try to create an abstraction between the end-users and the sensor nodes in order to allow the users to only concentrate on the needed data to be collected rather than bothering with the complexities of mechanisms deciding how to extract data from a network [63,64]. However, besides an abstraction-based design, an energy-efficient and real-time databases management techniques design is tremendously important due to the large amount of data generated by WSNs, the real-time requirements of data and tasks, and the resource limitations of WSNs. First, this section describes the techniques used to manage the large amount of generated data in an energy-efficient way, and second, presents the

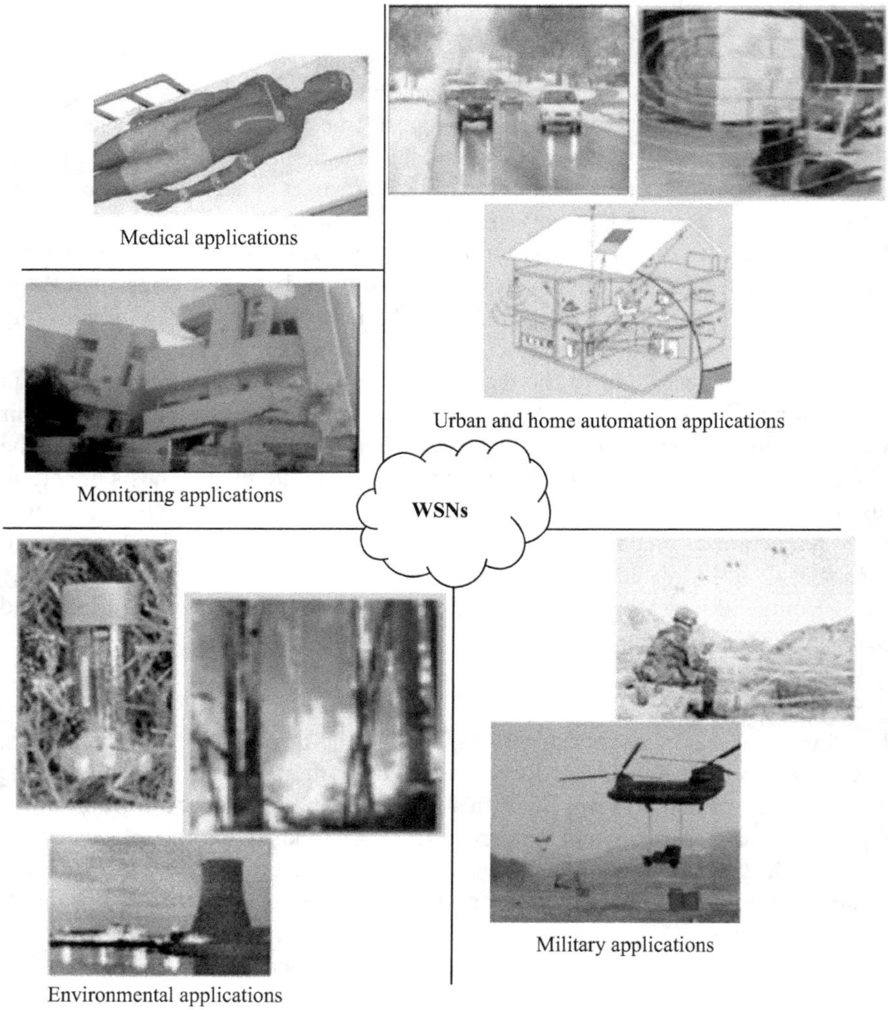

Figure 5.2 WSN-based applications

characteristics for designing a real-time database management technique. In addition, recent and relevant proposals are presented and classified into specific categories.

5.3.1 Energy-efficient databases management techniques

One important aspect that should be taken in account for the sensor data management is the energy management during treatments. In fact, the large amount of data generated by sensors greatly influences the sensor network lifetime, which particularly depends on the available energy in the nodes composing the network. The need to find better strategy has led to three types of approaches to data store and query in WSNs [65,66]: the *warehousing* approach, the *distributed* approach and the *hybrid* approach.

5.3.1.1 Databases management based on warehousing approach

The warehousing approach, again called external storage, proposes to centralise the data in a database server installed outside the sensor network. Then, the sensors act as only data collectors and periodically send the data gathered towards the central database where user queries are processed. As an example, the work in References 67 and 68, based on data from volcano monitoring, proposed a new indexing mechanism for centralised spatiotemporal databases with real-time constraints, named Po-Tree. Thus, fixed sensors scattered around the volcano periodically send the data towards a central database, later replicated to a data-warehouse, which is indexed by Po-Tree in order to retrieve most recent data. See Figure 5.3 for a visual illustration of the warehousing approach.

Other works, like Aurora [69] and RTSTREAM [70], proposed centralised data stream management systems (DSMSs) that provide real-time data stream processing capabilities. The large amount of sensor data stream comes continuously in a central repository managed by one of these real-time DSMSs that supports a real-time data stream query model, such as the Continuous Query Language (CQL) [71].

With this approach, the data access is more reliable; thus data from unreliable sensors will be always available in the centralised database server. However, it has some disadvantages: the transmission of the large amount of generated sensor data wastes resources and eventually can easily create a bottleneck on the central server.

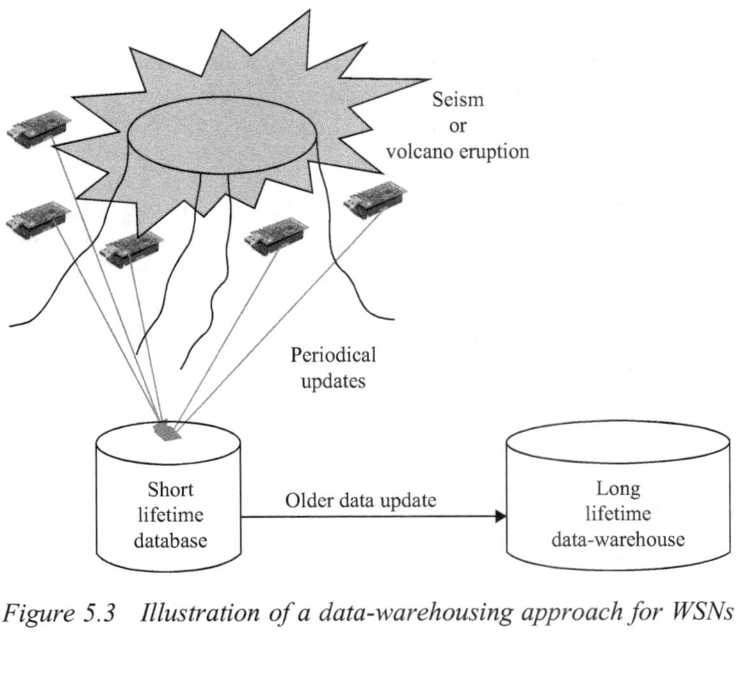

Figure 5.3 Illustration of a data-warehousing approach for WSNs

According to research into sensor hardware, data communication is one of the most energy-consuming operations, with each bit costing as much energy as about 1 000 instructions [14]. In addition, it is clear that this approach is unsuitable for real-time processing because it involves time delay for the results. The warehousing approach is the most-used approach in data storage and query processing for sensor network, and is greatly used in geographical information systems [72]. However, because of its drawbacks, researchers have proposed an alternative, which is the distributed approach.

5.3.1.2 Databases management based on distributed approach

The principal purpose of distributed database management techniques on WSNs is to support the management of the large amount of sensed data in an energy-efficient manner [65,73]. The distributed approach, again called local approach, proposes to locally keep the maximum of sensing data. Thus, unlike warehousing techniques that periodically transmit sensed data towards a centralised database, in distributed techniques the WSN forms a distributed database with sensed data, stored in sensor nodes, is in the form of rows with columns representing sensor attributes [13,74]. Some queries are disseminated, through a base station, into the network based on routing techniques as per References 75 and 76, and the sensors process them, thanks to their processing and storage capabilities. The sensors send their data to their parent nodes if they match the query requirements. The parent nodes combine this coming data with their own data and transmit to their parent nodes and so on until the data reaches the gateway. This approach that processes the data inside the sensor nodes themselves is commonly called in-network processing and it involves data reduction techniques [77,78], such as *data aggregation* [63,79], *packet merging, data compression techniques* [80,81], *data fusion* and *approximation-based techniques* [82,83]. As an illustration, see Figure 5.4.

There are four essential techniques to design a distributed data management system: *in-network processing, acquisitional query processing, cross-layer optimisation* and *data-centric data/query dissemination* [65,84].

In-network processing-based techniques

The in-network processing-based techniques [85–87] involve various types of operations that are traditionally performed on the database server, for instance, data aggregations and are used to process sensing values inside the network nodes in order to filter and reduce the large amount and needless data. The in-network processing technique can be divided into two sub-categories: *aggregation-based techniques* and *approximation-based techniques* [88]:

A. Aggregation-based techniques

The data aggregation techniques perform data aggregations (e.g. MAX, AVG) at intermediate nodes between the source nodes and the sink node. Since wireless communication involves more energy consumption than the sensing and processing operations, aggregating the data through intermediate nodes results in less data transmissions. As examples, COUGAR [64,74,89,90] and TinyDB [13,91] projects pioneer the use of sensor network as a distributed database. TinyDB is a distributed

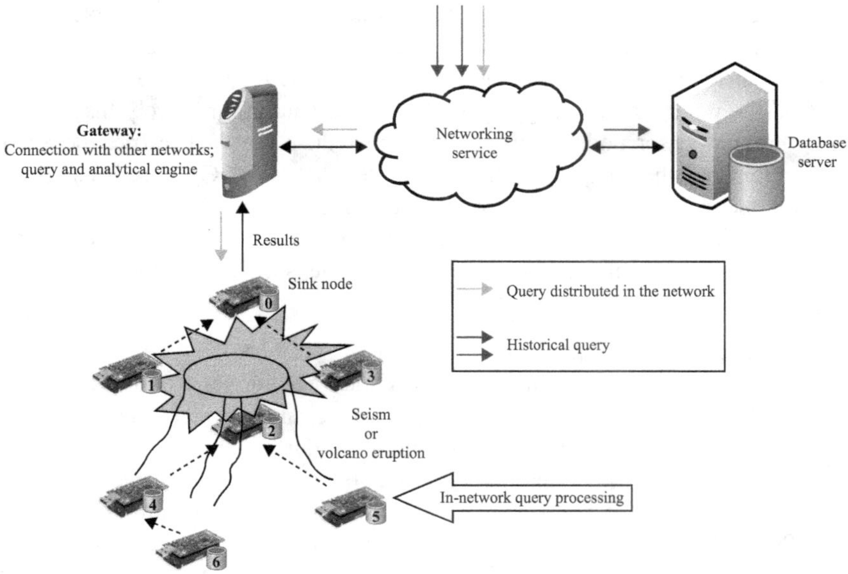

Figure 5.4 Illustration of a distributed database approach for WSNs

query processor for sensor networks that incorporate acquisitional techniques and was designed for networks based on the TinyOS operating system [92]. Through an interface, the user chooses what data he/she wishes to acquire, thanks to an SQL-like query. The query is parsed and optimised by a query processor at the base station and distributed into the network. The sensor nodes collect, filter and aggregate (e.g. MIN, MAX, SUM, COUNT, AVERAGE) the data and respond to the user query. Thus, when sensor readings flow up a communication tree called semantic routing tree (SRT), data is aggregated by intermediate nodes that contain relevant information for the query. This in-network aggregation reduces the large amount of data transmitted through the network, preventing, then, from bottleneck on the root node and energy wasting.

Cougar [90] is a platform for distributed query processing that uses a clustered technique to perform in-network data processing. A network is composed of several clusters, each of them managed by a cluster head. The child nodes that belong to clusters send periodically their readings to corresponding cluster heads, which then aggregate the received data and forward the computed result towards the Front End of the network. This Front End is a query optimiser, located at the gateway node, which generates optimised distributed query processing plans after receiving user queries. In addition, in this architecture, each node embeds a query layer [74,89] representing a query proxy between the network layer and the application layer, which processes queries. Additionally, Cougar performs packet merging by aggregating several packets into one. This reduces the power consumption, since sending multiple small packets is more expensive than sending one larger packet.

Among other examples, the works of References 93 and 94 can be quoted. These works propose an adaptive algorithm, called ADaptive AGgregation Algorithm for sensor networks (ADAGA), for performing in-network data aggregation. In ADAGA, the main idea is to aggregate the sensed data progressively that the data passes through sensor nodes. This will greatly reduce the data traffic, the energy consumption and the memory usage in the network. ADAGA also supports the packets replication in order to reduce the packets losses and allows the approximation of sensed values from collected data. Moreover, a data model for data streams and a declarative SQL-like query language named Sensor Network Query Language (SNQL) for WSNs are provided.

B. Approximation-based techniques

The approximation-based techniques use statistical techniques like approximation, linear regression, probabilities, etc. to approximate data or query answers. As an example, Barbie-Q (BBQ) [95] improves upon TinyDB and proposed a declarative query processing framework that uses a probabilistic model for answering queries about the actual state of the sensor network. Through an Structured Query Language (SQL)-like query that includes error tolerances and targeted confidence bounds, the user chooses what data he/she wishes to acquire. The query is parsed by the query processor and the model is used to estimate its answer on available sensor data with both spatial and temporal criteria. BBQ uses a time varying multivariate Gaussian model, and unlike TinyDB and COUGAR that interrogate all sensors every time a query is sent into the network, BBQ sensors are solicited to update the data for refining the model only if the model cannot satisfy the query with acceptable confidence.

Another example, among others, can be found in References 96 and 97. The research works in these last references, entitled distributed processing of probabilistic top-k queries in WSNs, proposed three suites of energy-efficient algorithms: sufficient set-based (SSB), necessary set-based (NSB) and boundary-based (BB). The main idea is, instead of transferring the large amount of sensor data from the network to the end-users, these algorithms return the subset of most relevant data answers efficiently with a constant round of data communications according to a probabilistic weight. These works allow to minimise the cost of retrieving all the data values and to only transfer relevant answers. In addition, for better minimising the communication and the energy consumption, this solution proposed also an adaptive algorithm that dynamically switches among the three algorithms based on their estimated costs.

Acquisitional query processing techniques

The acquisitional query processing techniques [13] attempt to reduce the number of sensor nodes participating in a given query processing in order to minimise the energy consumption in the network. This reduction is performed by expressing in the query when or what sensors should be in the sampling. As an example, TinyDB [13] is a distributed query processor for sensor networks that pioneer the use of sensor sampling, called as acquisitional query processing, in order to minimise the power consumption. TinyDB supports the management of metadata such as information about the costs of processing and delivering data, the necessary time and energy for that sampling. These metadata are periodically copied from the nodes to the root and used by the query optimiser for optimising the query processing. In addition, TinyDB extended the SQL language to formulate queries evaluated when the event specified in the request has occurred. The main purpose of that is to avoid the sending of measures that are not relevant of the query.

Among other examples that can be quoted, the research work in Reference 21 whose main purpose is, like most of research works on WSNs, the minimisation of the energy consumption. To reach this objective, it proposed an algorithm called pocket-driven trajectories (PDT) that bases on characteristics of monitoring queries for optimising the network lifetime. The main idea is the use of data acquisition techniques to optimise the communication overhead by sampling sensor nodes for processing a given query, after choosing efficient communication paths between the clustered sensor nodes and the base station. This eliminates the non-selected nodes in the data collection and, then, optimises the energy consumption. In addition, PDT is extended for context aware system and continuously adapts the data collection paths according to changes of node participation and environmental conditions.

Cross-layer optimisation techniques

In contrast to the classical computer networks, which allow independent design of different layers that results in strict boundaries between layers, the cross-layer optimisation [98,99] exploits interactions between different layers for improving various aspects of these networks, such as energy use efficiency and quality of service (QoS). For example, for improving the routing in WSNs, routing algorithms take care of various parameters, such as network connectivity, power availability on nodes and QoS parameters [100]. However, in classical computer networks, the routing is done by the network layer only considering the destination address of a packet. Among other examples, the ADAGA system [93,94], the work in Reference 101, and the research work in Reference 102 can be noted: ADAGA performs in-network data processing by adapting the data collecting and sending operations according to the remaining memory and energy of sensors, respectively. Thus, based on two parameters, Sense interval that specifies the interval between consecutive data collections and Send interval that specifies the interval between consecutive sending of packets, the main idea is to automatically adapt and update the values of these two parameters according to the available memory and power in the sensor nodes, respectively. This work allows the sensor nodes self-configurable, helps to reduce the power consumption and improves the memory availability.

In order to prevent from transmitting redundant data, the authors of Reference 101 proposed a distributed and self-organising scheduling algorithm (DOSA) that allows an in-network data aggregation, which is based on spatial and temporal correlations between sensor readings of neighbouring nodes. The DOSA algorithm automatically decides when a particular node should perform a correlating function. Thus, DOSA uses cross-layer information from the underlying MAC layer to detect eventual changes in the network topology and autonomously reassigns schedules of nodes.

The work in Reference 102 proposed a distributed query processing framework for WSNs composed of a compiler/optimiser, named Streaming NEtwork Engine (SNEE) [102–104], and a continuous declarative query language over sensed data streams, named SNEEql [105,106]. After receiving the SNEEql queries, SNEE optimises them by taking into account metadata, such as information about the network topology, the required energy and time, the cost of nodes sampling among other relevant parameters. After this step, SNEE creates, compiles and deploys an energy-efficient query evaluation plan, which will run into the participating nodes.

Data-centric data/query dissemination techniques

Generally, WSN applications are not interested in particular sensor nodes, but rather in data generated by sensors. Thus, the data-centric addressing scheme proposes to store collected data within the network based on attributes or types (e.g. geographic location, data type, timestamp). Queries for data according to particular attributes will be sent directly to the relevant nodes instead of performing flooding throughout the network; this improves routing protocols in terms of energy saving and latency [107–109]. Directed diffusion [107] is one of the first projects on data-centric data dissemination protocols designed specifically for WSNs. The sensor data is called events and the queries are called interests. Interests, events and responses of interests are represented by lists of '*attribute* = *value*'. The interests (e.g. interest for a specific data in the network, such as, 'detection of abnormal temperature in a given region') are injected into the network through one sink node arbitrarily chosen, which broadcasts the interest to its neighbours, which in turn broadcasts it to their neighbours, and so on. Every node receiving an interest compares it with its detected events, and if there is a correspondence, it sends back data towards the nodes from which it received the interest, until the data reaches the sink node.

Like previous works that investigate to improve the network lifetime, the dynamic data-centric routing and storage (DDCRS) [110] mechanism minimises the communication overhead by dynamically determining the locations of the data-centric nodes according to the location of multiple sink nodes and the data collecting rate. The DDCRS scheme consists of two phases: a static phase which consists of defining a data-centric node, called home data-centric responsible for storing sensed data and future replying to sink nodes, based on hash table. And the dynamic phase, which handles data storage and delivery when the locations of the data-centric nodes change according to the locations and the frequencies of reply of multiple sink nodes.

The distributed approach mainly based on in-network processing can provide several advantages, such as energy saving (processing the data inside the network greatly reduces the amount and size of communicated data), quasi-real-time query processing (query processing inside the devices themselves means that the most current data will be acquired and the data will certainly arrive to the end-users with their temporal validity), support of long-running queries and instant queries processing. However, with this approach the cost of data access is not negligible because it is necessary to distribute queries through the sensor network.

Moreover, although there is a significant advance in ubiquitous computing, the conditions for using sensor networks still restrict the use of sensor as data storage systems. Indeed, the improvement of sensors in terms of operating time, power calculation and memory capabilities does not cancel the risks of failure related to natural conditions. For more details about the distributed approach, readers can see Reference 65.

Regarding the advantages and inconvenience of these two approaches, certain applications try to profit from both distributed and warehousing approach for improving the data management in WSNs. This last approach called *hybrid approach* can be seen in the research works of Reference 66. These works combine statistical modelling techniques with the distributed approach to optimise the real-time query processing. Thus, sensor readings are stored in both the devices and a central repository located on the gateway and the query answers are estimated based on data in the central repository. If the data is not sufficiently fresh to answer one query, then the query is processed in the network and the data in the central repository is updated. This allows finding a compromise between the cost of query and transmission and help to real-time processing.

5.3.2 Real-time databases management techniques

Traditionally, the real time is considered as the respect of temporal constraints, which means that the data has temporal validity and the transactions must be run in well-defined time intervals [111,112]. Thus, a real-time database management system (RT-DBMS) must process transactions and ensure that the logical consistency of data is not violated, as well as the temporal validity of data and the temporal constraints or deadlines of transactions [43,44,113]. The temporal constraints help to maintain the temporal consistency between the current state of the targeted environment and the state reflected by the content of the database. The temporal consistency can be measured in two ways [114]:

- Absolute consistency, which deals with the need to maintain the view representing the state of the targeted environment consistent with the real state of the environment,
- Relative consistency, which concerns data derived from other ones.

The following sections give a brief overview of the characteristics of data and transactions in real-time databases management for WSNs and some proposals on this field. For more information, readers can see References 113 and 114.

5.3.2.1 Real-time data constraints

The data generated by sensors generally changes along the time in unexpected way, representing temperature values, moving objects, etc. and unfortunately the data is collected in discreet times, providing then to the collected data a temporal validity characteristic. In this context, the data should be controlled so that it reflects as closely as the current state of the targeted environment.

To include temporal constraints, a real-time data d can be modelled by $d = (d_{\text{value}}, d_{\text{timestamp}}, d_{\text{avi}}, d_{\text{DE}})$, where d_{value} represents the real data value of the data d, $d_{\text{timestamp}}$ indicates the instant when the data was captured from the environment, d_{avi} is the absolute validity interval (avi) that denotes the time interval following the timestamp during which the data is considered valid and d_{DE} is the imprecision or data error (DE), which refers to how the current state of the targeted environment may differ from the measured data [114].

According to Reference 114, the absolute and relative consistency for real-time data presented in the previous sub-section can be formally defined as follows: a set of data used to derive a new data item constitutes a relative consistency set. Each set R is associated with a relative validity interval denoted by R_{rvi}. Given a real-time data item $d = (d_{\text{value}}, d_{\text{timestamp}}, d_{\text{avi}}, d_{\text{DE}})$ as defined earlier in this section and $d \in R$, d has a correct state if:

1. d_{value} is logical consistency, that is, satisfies all integrity constraints.
2. d is temporally consistent:

   - Absolute consistency: $\left(\text{current\_time} - d_{\text{timestamp}}\right) \leq d_{\text{avi}}$ (5.1)
   - Relative consistency: $\forall d' \in R, \left|d_{\text{timestamp}} - d'_{\text{timestamp}}\right| \leq R_{\text{rvi}}$ (5.2)

For example, let us consider a system with two data: temperature and pressure with $\text{Temperature}_{\text{avi}} = 5$, $\text{Pressure}_{\text{avi}} = 10$, $R = \{\text{temperature, pressure}\}$, the relative validity interval $R_{\text{rvi}} = 2$. If the current_time is equal to 100, then one has with respect to the two conditions in the previous sub-paragraphs:

- The temperature = (347, 96, 5) and pressure = (50, 97, 10) have temporal consistency.
- However, the temperature = (347, 96, 5) and pressure = (50, 92, 10) are not temporally consistent (the relative consistency is not met).

The data error on d can be calculated by:

$$d_{\text{DE}} = 100 * \left|\frac{d_{\text{CurrentValue}} - d_{\text{UpdateValue}}}{d_{\text{CurrentValue}}}\right| \% \tag{5.3}$$

5.3.2.2 Real-time transactions constraints

The methods used in traditional DBMS are generally not suitable for RT-DBMS. However, they can be used and adapted to take into account the constraints of RT-DBMS, which involve that the transactions should be executed before their deadlines and that these transactions should have access to valid data within limited validity periods. This has implications for how to design these systems. In particular, the atomicity, consistency, isolation and durability (ACID) properties of transactions for traditional databases were reviewed and adapted to the real-time databases context. Thus, the atomicity may be relaxed. It is only applied to the sub-transaction that wants to deal with completely data consistency. Since timeliness is more important than correctness, in many situations, correctness can be traded for timeliness. The isolation allows transactions to communicate with others to better perform control functions. In real-time databases, not all data must be permanent and some of them are temporal [113–115].

Transactions in an RT-DBMS can be classified either according to their real-time constraints or their type of data access. According to their real-time constraints (deadlines missing), transactions in an RT-DBMS are classified into three categories: hard deadline transactions (strict and critical), firm deadline transactions (strict and non-critical) and soft deadline transactions (non-strict). And according to the type of data access, two types of transactions were proposed in the literature [43,44]: *update transactions* and *user transactions*. Update transactions, also called sensor transactions, are write transactions that obtain the current state of the targeted environment and periodically update the real-time sensor data in the database. This kind of transactions can be also executed sporadically to update the derived data in order to reflect the state of the real world. Derived data are the data computed using sensor data. For example, the lane of an aircraft is derived from the location and the altitude data items. User transactions, representing user queries, arrive aperiodically. They may read or write non-real-time data, but only read real-time data.

In order to comply with temporal constraints, real-time transactions must have also these following attributes: (i) liberation time that represents the moment on which all the resources for the transaction processing is available, (ii) computing time that indicates the execution time needed for the transaction, (iii) maximum time, which indicates the maximum time limit for the transaction execution, and (iv) the periodicity that refers to the frequency with which the transaction happens [115].

For example, the research work in Reference 115 proposed real-time database management techniques in WSNs that use distributed approach and solve the conflict problems that arise between read and write transactions processing in real time. The main idea is: each collected data X has these properties: a value (X_{value}), a timestamp ($X_{timestamp}$), an avi (X_{avi}), the imprecision on the data (X_{imp}) and the limit of imprecision of the data (X_{lim_imp}). Considering a read transaction on a data X as $R(X)$ and a write transaction on a data Y as $W(Y)$, if transactions wish to run on the same data object, a concurrency control technique is performed in order to allow conflicting transactions to execute concurrently in a way that their scheduling does not result in an imprecision that is higher than the one associated to the data. First, the algorithm verifies the temporal validity of the data based on Equation (5.1). If the data is still temporally valid, the concurrency control protocol uses negotiation functions (NFs) that base on these parameters defined earlier in this paragraph on data and transactions to provide compromise between logical and temporal consistency. Thus, when a read transaction is running and a write transaction is invoked, the NF is:

$$\mathrm{NF}(R(X), W(Y)) = |X_{\text{value}} - Y_{\text{value}}| \leq X_{\text{lim\_imp}} - (X_{\text{imp}} + Y_{\text{imp}}) \tag{5.4}$$

If the NF returns true, the imprecision on X is updated by the value defined as:

$$X_{\text{imp}} = X_{\text{imp}} + Y_{\text{imp}} + |X_{\text{value}} - Y_{\text{value}}| \tag{5.5}$$

When a write transaction is running and a read transaction is invoked, the NF is:

$$\mathrm{NF}(W(Y), R(X)) = |Y_{\text{value}} - X_{\text{value}}| \le Y_{\text{lim\_imp}} - X_{\text{imp}} \tag{5.6}$$

If the NF returns true, the imprecision on X is updated by the value defined as:

$$X_{\text{imp}} = X_{\text{imp}} + |X_{\text{value}} - Y_{\text{value}}| \tag{5.7}$$

When a write transaction is running and another write transaction is invoked, the NF is:

$$\mathrm{NF}(W(X), W(Y)) = |X_{\text{value}} - Y_{\text{value}}| \le X_{\text{lim\_imp}} - X_{\text{imp}} \tag{5.8}$$

If the NF returns true, the imprecision on X is updated by the value defined as:

$$X_{\text{imp}} = X_{\text{imp}} + |X_{\text{value}} - Y_{\text{value}}| \tag{5.9}$$

5.4 Discussion and open issues

The integration of real-time applications and WSNs raised challenges to handle real-time data storage and query in energy-efficient way. In fact, in real-time applications data and tasks have temporal constraints. Moreover, since in WSNs energy is one of the most crucial resources and the ongoing interaction between the network devices and the environment results in large amounts of data causing great waste of energy if not well handled, it is challenging to design energy-efficient and real-time databases management techniques for WSNs. Although the researchers have proposed several good techniques, many of them are limited because of the underlying architecture used, either warehousing or distributed, or complex algorithms used that can have a negative impact in either the temporal constraints or the energy constraints.

Building techniques suitable for this kind of applications is very difficult and the data management techniques used in traditional databases are not generally suitable for sensor networks because of their specificities. Most of the proposals are based on the warehousing approach, which is a centralised approach that requires to periodically updating the central database. This can cause, among other drawbacks, delay on the response time and great waste of energy. On the other hand, other proposals focus on the distributed approach that, mainly based on in-network processing, provides help on energy saving and the temporal constraints. However, according to the unpredictable and generally harsh environment, there may be sudden failures of sensor devices. This can lead to information loss that greatly influences the result analysis or even the system blocking. In addition, in this approach the time delay is distance-sensitive and depends on the depth of the nodes containing the required information and the cost of data access is not negligible because it is necessary to distribute queries through the sensor network. Some other studies used complex algorithms mainly to satisfy temporal constraints with sometimes the lack of mechanism aware of the energy wasting. Moreover, in some of these algorithms the waiting or aborting of the transactions is determined by the temporal validity of the data to be accessed, yet when one uses temporal constraints directly, one can make the mistake of performing wait or abort the transactions too many times without having variations that justify these operations. This can lead to performance slowdown when one knows that, based on this study, in real-time systems, for some applications, the accuracy of results may be sacrificed under some limit to reduce the response time.

Regarding advantages and disadvantages of the proposed techniques one can opt for a hybrid approach which will allow one or the other solution depending on the requirements of the application that is the source of queries. The proposed framework could also profit from data reduction techniques such as data aggregation, packet merging, data compression techniques, data fusion and approximation-based techniques. Indeed, the approximation-based techniques are useful in many applications which do not require the exact readings, resulting in the form of energy and latency minimisation. However, in order to get more good results in terms of energy savings, both aggregation and approximation techniques could be used in combination where possible. The WSNs are application specific; so one technique could not be always efficient for all types of applications. Therefore, it is desirable to have a generalised DBMS so that all various applications can customise and tailor it according to their needs.

5.5 Conclusion

This chapter has described the characteristics and the different steps to design energy-efficient and real-time databases management techniques for WSNs, as well as presented and classified several recent and relevant proposals in this area. In addition, several essential techniques are discussed in order to provide efficient databases management techniques suitable for real-time applications based on WSNs.

Acknowledgements

This work has been partially supported by *Instituto de Telecomunicações*, Next Generation Networks and Applications Group (NetGNA), Covilhã Delegation, Portugal, by the Visiting Professor Program at King Saud University, and by National Funding from the FCT – *Fundação para a Ciência e a Tecnologia* through the UID/EEA/50008/2013 Project.

References

[1] Shu L., Hauswirth M., Cheng L., Ma J., Reynolds V., and Zhang L. 'Sharing worldwide sensor network'. *International Symposium on Applications and the Internet, IEEE*, 189–192, 2008, DOI: 10.1109/SAINT.2008.22

[2] Aberer K., Hauswirth M., and Salehi A. 'Global sensor networks'. *Technical report LSIR-REPORT-2006-001*, http://lsir.epfl.ch/research/current/gsn/, 2006

[3] Aberer K., Hauswirth M., and Salehi A. 'Infrastructure for data processing in large-scale interconnected sensor networks'. *In Proceedings of the 8th International Conference on Mobile Data Management*, Mannheim, Germany, May 7–11, 2007

[4] Callaway E.H. 'Wireless sensor networks: architectures and protocols'. CRC Press, 2004

[5] Baronti P., Pillai P., Chook C.V., Chessa S., Gotta A., and Hu F.Y. 'Wireless sensor networks: A survey on the state of the art and the 802.15.4 and ZigBee standards'. *Journal of Computer Communications*, 30(7), 1655–1695, 2007

[6] Oliveira L.M.L. and Rodrigues J.J.P.C. 'Wireless sensor networks: a survey on environmental monitoring'. *Journal of Communications (JCM)*, 6(2), 143–151, 2011, DOI: 10.4304/jcm.6.2.143-151

[7] Akyildiz I.F., Su W., Sankarasubramaniam Y., and Cayirci E. 'Wireless sensor networks: a survey'. *Computer Networks*, 38, 393–422, 2002

[8] Nayak A. and Stojmenovic I. 'Wireless sensor and actuator networks: algorithms and protocols for scalable coordination and data communication'. Wiley-Interscience, 2010

[9] Lin K., Chen M., Zeadally S., and Rodrigues J.J.P.C. 'Balancing energy consumption with mobile agents in wireless sensor networks'. *Future Generation Computer Systems*, 28(2), 446–456, 2012, DOI: 10.1016/j.future.2011.03.001

[10] Lin K., Rodrigues J.J.P.C., Ge H., Xiong N., and Liang X. 'Energy efficiency QoS assurance routing in wireless multimedia sensor networks'. *IEEE Systems Journal*, 5(4), 495–505, 2011, DOI: 10.1109/JSYST.2011.2165599

[11] Akyildiz F.I., Melodia T., and Chowdhury R.K. 'A survey on wireless multimedia sensor networks'. *Journal of Computer Networks*, 51(4), 921–960, 2007

[12] Cho S., Shih E., Ickes N., and Min R. 'Physical layer driven protocol and algorithm design for energy-efficient wireless sensor networks'. *Proceedings of ACM MOBICOM'01*, Rome, Italy, pp. 272–287, July 2001, DOI: 10.1145/381677.381703

[13] Madden S.R., Franklin M.J., Hellerstein J.M., and Hong W. 'TinyDB: an acquisitional query processing system for sensor networks'. *ACM Transactions on Database Systems (TODS)*, 30(1), 122–173, 2005

[14] Culler D., Estrin D., and Srivastava M. 'Overview of sensor networks'. *IEEE Computer, Special Issue in Sensor Networks*, August 2004

[15] Pottie J.G. and Kaiser J.W. 'Wireless integrated network sensors'. *Communication of the ACM*, 43(5), 51–58, 2000

[16] Sendra S., Lloret J., Garcia M., and Toledo J.F. 'Power saving and energy optimization techniques for wireless sensor networks'. *Journal of Communications*, 6(6), 439–459, 2011, DOI: 10.4304/jcm.6.6.439-459

[17] Matsumae S. 'Energy-efficient cell partition of 3d space for sensor networks with location information'. *Network Protocols and Algorithms*, 1(2), 85–98, 2009

[18] Silva F.A., Braga T.R.M., Ruiz L.B., and Nogueira J.M.S. 'Tecnologia de Nós Sensores Sem Fio'. *Contrôle & Instrumentação*, 92, 76–87, 2004

[19] Shu L., Zhang Y., Min G., Wang Y., and Hauswirth M. 'Cross-layer optimization on data gathering in wireless multimedia sensor networks within expected network lifetime'. *Accepted in Springer Journal of Universal Computer Science (JUCS)*, 2009

[20] Zhu C., Shu L., Hara T., Wang L., Nishio S., and Yang L.T. 'A survey on communication and data management issues in mobile sensor networks'. *Wireless Communications and Mobile Computing*, November 10, 2011, DOI: 10.1002/wcm.1219

[21] Umer M., Kulik L., and Tanin E. 'Optimizing query processing using selectivity-awareness in wireless sensor networks'. *Journal of Computers, Environment and Urban Systems*, 33(2), 79–89, 2009

[22] Neves P.A., Rodrigues J.J.P.C., Chen M., and Vasilakos A.V. 'A multichannel architecture for IPv6-enabled wireless sensor and actuator networks featuring PnP support'. *Journal of Network and Computer Applications*, 2011, DOI: 10.1016/j.jnca.2011.03.033

[23] Oliveira L.M.L., Rodrigues J.J.P.C., de Sousa A.F., and Lloret J. 'Network access control framework for 6LoWPAN networks'. *Sensors*, 13(1), 1210–1230, 2013, DOI: 10.3390/s130101210

[24] Diallo O., Rodrigues J.J., and Sene M. 'Real-time data management on wireless sensor networks: A survey'. *Journal of Network and Computer Applications*, 35(3), 1013–1021, 2012

[25] Dias J.A., Rodrigues J.J.P.C., Isento J.N., Pereira P.R.B.A., and Lloret J. 'Performance assessment of fragmentation mechanisms for vehicular delay-tolerant networks'. *EURASIP Journal on Wireless Communications and Networking*, 195, 2011

[26] Tubaishat M., Zhuang P., Qi Q., and Shang Y. 'Wireless sensor networks in intelligent transportation systems'. *Wireless Communications and Mobile Computing, Wiley-InterScience*, 9, 287–302, 2009, DOI: 10.1002/wcm.616

[27] Wang Y., Zhou G., and Li T. 'Design of a wireless sensor network for detecting occupancy of vehicle berth in car park'. *In PDCAT'06: Proceedings of the Seventh International Conference on Parallel and Distributed Computing, Applications and Technologies, IEEE Computer Society*, Washington, DC, USA, 2006. pp. 115–118

[28] Chowdhury M., Wang K.-C., Fries R., Ma Y., and Bagaria D. 'A distributed wireless sensor network system for transportation safety and security'. *In Sensors, and Command, Control, Communications, and Intelligence (C3I) Technologies for Homeland Security and Homeland Defense VI*, Vol. 6538, Carapezza E.M. (ed.). SPIE: Orlando, FL, USA, 2007

[29] Mainwaring A., Culler D., Polastre J., Szewczyk R., and Anderson J. 'Wireless sensor networks for habitat monitoring'. *Proceedings of the 1st ACM International Workshop on Wireless Sensor Networks and Applications*. New York: ACM Press; 2002. pp. 88–97

[30] Ulmer C., Yalamanchili S., and Alkalai L. 'Wireless distributed sensor networks for in-situ exploration of Mars'. *Work in Progress for NASA, Technical Report (Citeseer)*, 2003. Verifions l'ordre des auteurs
[31] Carneiro J.P. 'Environmental monitoring in wind farms based on WSANs'. *Instituto Superior Tecnico*, 2009
[32] Yunyue Z. and Dennis S. 'Statstream: Statistical monitoring of thousands of data streams in real time'. *Proceedings of the 28th International Conference on Very Large Data Bases. VLDB Endowment*, 2002. pp. 358–369
[33] Chen J., DeWitt D.J., Tian F., and Wang Y. 'NiagaraCQ: A scalable continuous query system for internet databases'. *Proceedings of the 2000 ACM SIGMOD International Conference on Management of Data.* New York: ACM Press; 2000. pp. 379–390
[34] Kechar B., Houache N., and Sekhri L. 'Using wireless sensor networks for reliable forest fires detection'. *The 4th International Conference on Ambient Systems, Networks and Technologies (ANT 2013), the 3rd International Conference on Sustainable Energy Information Technology (SEIT-2013).*
[35] Liyang Y., Wang N., and Meng X. 'Real-time forest fire detection with wireless sensor networks'. *Proceedings of International Conference on Wireless Communications, Networking and Mobile Computing*, 2, 10.1109/WCNM.2005.1544272, 1214–1217, September 2005
[36] Lorincz G. and Welsh M. 'Deploying a wireless sensor network on an active volcano'. *IEEE Computer Society*, 2006
[37] Akyildiz I., Su W., Sankarasubramaniam Y., and Cayirci E. 'A survey on sensor networks'. *In IEEE Communication Magazine*, 40(8), 102–114, 2002
[38] Chee-Yee Chong and S.P. Kumar, 'Sensor networks: evolution, opportunities, and challenges,' *Proceedings of the IEEE*, 91(8), 1247–1256, 2003
[39] Lorinez K., Malan D.J., Fulford-Jones T.R.F., Nawoj A., Clavel A., Shnayder V., *et al.* 'Sensor networks for emergency response: Challenges and opportunities'. *IEEE Pervasive Computing*, 3(4), 16–23, 2004
[40] Caldeira J.M.P.L., Rodrigues J.J.P.C., Garcia J.F.R., and de la Torre I. 'A new wireless biosensor for intra-vaginal temperature monitoring'. *In Sensors Journal, MDPI*, Basel, Switzerland, ISSN 1424-8220, 10(11), 10314–10327, November 2010, DOI: 10.3390/s101110314
[41] Rodrigues J.J.P.C., Caldeira J., and Vaidya B. 'A novel intra-body sensor for vaginal temperature monitoring'. *In Sensors Journal, MDPI*, Basel, Switzerland, ISSN 1424-8220, 9(4), 2797–2808, 2009
[42] Diallo O., Rodrigues J.J.P.C., Sene M., and Jianwei N. 'Real-time query processing optimization for cloud-based WBAN'. *Inform. Sci., Special Issue on Cloud-assisted Wireless Body Area Networks*, 284, 84–94, 2014, http://dx.doi.org/10.1016/j.ins.2014.03.081
[43] Idoudi N., Duvallet C., Sadeg B., Bouaziz R., and Gargouri F. 'How to model a real-time database?' *Proceedings of 12th IEEE International Symposium on Object-oriented Real-time distributed Computing (IEEE ISORC'2009)*. Tokyo, Japan: IEEE, 2009. pp. 321–325

[44] Idoudi N., Louati N., Duvallet C., Bouaziz R., Sadeg B., and Gargouri F. 'Framework to model real-time databases'. *International Journal of Computing & Information Sciences*, 7, 1–11, 2009

[45] Diallo O., Rodrigues J.J.P.C., Sene M., and Lloret J. 'Simulation framework for real-time database on WSNs'. *Journal of Network and Computer Applications*, 39, 191–201, 2014, http://dx.doi.org/10.1016/j.jnca.2013.07.001

[46] Gosnell T.B., Hall J.M., Ham C.L., Knapp D.A., Koenig Z.M., Luke S.J., *et al.* 'Gamma-ray identification of nuclear weapon materials'. *Lawrence Livermore National Lab. Livermore CA, USA: s.n.* Technical report DE97053424, February 1997

[47] Brown M.J. 'Users guide developed for the JBREWS project'. *Los Alamos National Laboratory of California University*. Technical report LA-UR-99-4676, 1999

[48] 'PODS: a remote ecological micro-sensor network project'. Available from http://www.pods.hawaii.edu/

[49] Estrin D., Culler D., Pister K., and Sukhatme G. 'Connecting the physical world with pervasive networks'. *IEEE Pervasive Computing*, 1(1), 5969, 2002

[50] http://www.intel.com/pressroom/archive/releases/20040316corp.htm

[51] Petriu E.M., Georganas N.D., Petriu D.C., Makrakis D., and Groza V.Z. 'Sensor-based information appliances'. *IEEE Instrumentation Measurement Magazine*, 3(4), 31–35, 2000

[52] Lynch J.P. and Loh K. 'A summary review of wireless sensors and sensor networks for structural health monitoring'. *Shock and Vibration Digest*, Sage Publications, 38(2), 91–128, 2005

[53] Bonivento A., Carloni L.P., and Sangiovanni-Vincentelli A. 'Platform-based design of wireless sensor networks for industrial applications'. *In DATE '06: Proceedings of the Conference on Design, Automation and Test in Europe*, 3001 Leuven, Belgium, 2006, pp. 1103–1107. European Design and Automation Association

[54] Coleri S.C.S. and Varaiya P. 'Sensor networks for monitoring traffic'. *In 42nd Allerton Conference on Communication, Control and Computing*, 2004

[55] Guitton A., Skordylis A., and Trigoni N. 'Correlation-based data dissemination in traffic monitoring sensor networks'. *In IEEE Wireless Communications and Networking Conference (WCNC 2007)*, Hong Kong, March 11–15, 2007

[56] Poon C.C.Y., Zhang Y.T., and Bao S.D. 'A novel biometrics method to secure wireless body area sensor networks for telemedicine and mhealth'. *IEEE Communications Magazine*, 44(4), 73–81, 2006

[57] 'IEEE 802.15 WPAN Task Group 6 Body Area Networks' [online]. Available from http://www.ieee802.org/15/pub/SGmban.html

[58] Rodrigues J.J.P.C., Reis N., Moutinho J.A.F., and de la Torre I. 'Breast alert: an on-line tool for predicting the lifetime risk of women breast cancer'. *J. Med. Syst.*, 36, 1417–1424, 2012, DOI: 10.1007/s10916-010-9603-z

[59] Rodrigues J.J.P.C., Pedro L.M.C.C., Vardasca T., de la Torre-Díez I., and Martins H.M.G. 'Mobile health platform for pressure ulcer monitoring with electronic health record integration'. *Health Informatics Journal* 19(4), 300–311, 2013, DOI: 10.1177/1460458212474909 jhi.sagepub.com

[60] Latre B., Braem B., Moerman I., Blondia C., and Demeester P. 'A survey on wireless body area networks'. *Journal of Wireless Networks*, 17, 1–18, 2011

[61] Lopes I.C., Vaidya B., and Rodrigues J.J.P.C. 'Towards an autonomous fall detection and alerting system on a mobile and pervasive environment'. *Telecommun. Syst.*, 2011, DOI: 10.1007/s11235-011-9534-0

[62] Beydoun K. 'Conception d'un protocole de routage hiérarchique pour les réseaux de capteurs'. Ph.D. Thesis, 2009. pp. 32–33

[63] Madden S., Franklin M., Hellerstein J., and Hong W. 'TAG: a tiny aggregation service for ad-hoc sensor networks'. *ACM Symposium on Operating System Design and Implementation, OSDI 2002*, Boston, MA, USA, 2002. pp. 131–146

[64] Bonnet P., Gehrke J., and Seshadri P. 'Towards sensor database systems'. *2nd International Conference on Mobile Data Management*, Hong Kong, China, January 8–10, 2001

[65] Diallo O., Rodrigues J.J.P.C., Sene M., and Lloret J. 'Distributed database management techniques for wireless sensor networks'. *IEEE Transactions on Parallel and Distributed Systems*, August 19, 2013, http://doi.ieeecomputersociety.org/10.1109/TPDS.2013.207

[66] Diallo O., Rodrigues J.J.P., and Sene M., 'Real-time query processing optimization for wireless sensor networks'. *International Journal of Sensor Networks (IJSNet), InderScience Publishers, ISSN (Online): 1748-1287 – ISSN (Print):* 1748-1279 (in press)

[67] Noël G., Servigne S., and Laurini R. 'Real-time spatiotemporal data indexing structure'. *Proceeding of 2004 AGILE 7th Conference on Geographic Information Science*. Heraklion: Citeseer, 2004. pp. 261–268

[68] Noël G., Servigne S., and Laurini R. 'The Po-tree: a soft real-time spatio-temporal data indexing structure'. *Developments in Spatial Data Handling. SDH 04, 11th International Symposium on Spatial Data Handling.* Springer Verlag, 2004. pp. 259–270

[69] Abadi D.J., Carney D., Çetintemel U., Cherniack M., Convey C., Lee S., *et al*. 'Aurora: a new model and architecture for data stream management'. *The VLDB Journal (Springer)*, 12(2), 120–139, 2003

[70] Wei Y., Son S.H., and Stankovic J.A. 'RTStream: Real-time query processing for data streams'. *IEEE Symposium on Object-Oriented Real-Time Distributed Computing (ISORC'06)*. Gyeongju, Korea: IEEE Computer Society, 2006

[71] Arasu A., Babu S., and Widom J. 'The cql continuous query language: Semantic foundations and query execution'. *The International Journal on*

Very Large Data Bases (VLDB), Springer-Verlag, New York, Inc., 15(12), 121–142, 2006

[72] Catalina C. and Rodriguez G. 'Qualité des données capteurs pour les systèmes de surveillance de phénomènes environnementaux'. *Thesis Report, National Institute of Applied Sciences of Lyon*, June 2010

[73] Cheng L., Chen Y., Chen C., Ma J., Shu L., Vasilakos A.V., and Xiong N. 'Efficient query-based data collection for mobile wireless monitoring applications'. *The Computer Journal*, 53(10), 1643–1657, 2010

[74] Yao Y. and Gehrke J. 'Query processing in sensor networks'. *In First Biennial Conference on Innovative Data Systems Research (CIDR 2003).* Asilomar, CA, January 5–8, 2003

[75] Li C., Zhang H., Hao B., and Li J. 'A survey on routing protocols for large-scale wireless sensor networks'. *Sensors*, 11(4), 3498–3526, 2011

[76] Al-karaki J.N. and Kamal A.E. 'Routing techniques in wireless sensor networks: a survey'. *IEEE Wireless Communications*, 11(6), 6–28, 2004

[77] Kacimi R. 'Techniques de conservation d'énergie pour les réseaux de capteurs sans fil'. Ph.D. dissertation, Dept. Math., Info. and Telecom., INPT Univ., Toulouse, France, September 2009

[78] Dini C. 'Les Réseaux Capteurs Sans Fil Avec Access Sporadique Au Noeud-puits'. Ph.D. dissertation, Info. Eng., Haute Alsace Univ., France, December 2010

[79] Fasolo E., Rossi M., Widmer J., and Zorzi M. 'In-network aggregation techniques for wireless sensor networks: a survey'. *IEEE Wireless Communications*, 14(2), 70–87, 2007

[80] Tang C. and Raghavendra C.S. 'Compression techniques for wireless sensor networks'. Chapter 10 in Wireless Sensor Networks (C. Raghavendra, K. Sivalingam, T. Znati, Eds.), Kluwer Academic Publishers, Boston, USA, pp. 207–231, 2004

[81] Dini C. and Lorenz P. 'Primitive operations for prioritized data reduction in wireless sensor network nodes'. *4th International Conference on Systems and Networks Communications, ICSNC 2009*, Porto, Portugal, 2009. pp. 274–280

[82] Xu J., Tang X., and Lee W.C. 'A new storage scheme for approximate location queries in object tracking sensor networks'. *IEEE Transactions on Parallel and Distributed Systems (TPDS)*, 19(2), 262–275, 2008

[83] Debono C.J. and Borg N.P. 'The implementation of an adaptive data reduction technique for wireless sensor networks'. *IEEE International Symposium on Signal Processing and Information Technology (ISSPIT)*, Sarajevo, December 16–19, 2008

[84] Chatterjea S. and Havinga P. 'A taxonomy of distributed query management techniques for wireless sensor networks'. *International Journal of Communication Systems*, 20(7), 889–908, 2007

[85] Demirbas M., Lu X., and Singla P. 'An in-network querying framework for wireless sensor networks'. *IEEE Trans. on Parallel and Distributed Systems (TPDS)*, 20(8), 1202–1215, 2009
[86] Girighar A. and Kumar P.R. 'Toward a theory of in-network computation in wireless sensor networks'. *IEEE Communications Magazine*, 44(4), 98–107, 2006
[87] Andreou P., Zeinalipour-Yazti D., Chrysanthis P.K., and Samaras G. 'In-network data acquisition and replication in mobile sensor networks'. *Distributed and Parallel Databases*, 29, 87–112, 2011
[88] Ahmad M.B., Islam M.H., Asif M., and Aziz S. 'A short survey on distributed in-network query processing in wireless sensor networks'. *Proceedings of first International Conference on Networked Digital Technologies (NDT)*, IEEE, Ostrava, pp. 541–543, 2009, DOI: 10.1109/NDT.2009.5272100
[89] Yao Y. and Gehrke J. 'The Cougar approach to in-network query processing in sensor network'. *ACM SIGMOD Record*, 31(3), 9–18, 2002
[90] Bonnet P., Gehrke J., and Seshadri P. 'Cougar: The sensor network is the database'. Available from www.cs.cornelle.edu/database/cougar/, 2001 [Accessed 25 January 2013]
[91] Madden S., Hong W., Hellerstein J., and Stanek K. 'TinyDB: a declarative database for sensor networks'. Available from http://telegraph.cs.berkeley.edu/tinydb/overview.html [Accessed 25 January 2013]
[92] Berkeley U. 'TinyOS'. Available from http://www.tinyos.net/, 1999
[93] Brayner A., Lopes A., Meira D., Vasconcelos R., and Menezes R. 'An adaptive in-network aggregation operator for query processing in wireless sensor networks'. *The Journal of Systems and Software*, 81(3), 328–342, 2008
[94] Brayner A., Lopes A., Meira D., Vasconcelos R., and Menezes R. 'Toward adaptive query processing in wireless sensor networks'. *Signal Processing*, 87(12), 2911–2933, 2007
[95] Deshpande A., Guestrin C., Madden S., Hellerstein J.M., and Hong W. 'Model-driven acquisition in sensor networks'. *Proceedings of VLDB*, Toronto, Canada, 2004
[96] Ye M., Liu X., Lee W., and Lee D.L. 'Probabilistic top-k query processing in distributed sensor networks'. *IEEE 26th International Conference of Data Engineering (ICDE)*, March 1–6, 2010, Long Beach, California, USA. pp. 585–588
[97] Ye M., Lee W., Lee D.L., and Liu X. 'Distributed processing of probabilistic top-k queries in wireless sensor networks'. *IEEE Transactions on Knowledge and Data Engineering*, 25(1), 76–91, 2013, DOI: 10.1109/TKDE.2011.145
[98] Radunovic B. 'A cross-layer design of wireless ad-hoc networks'. Thesis, 2005
[99] Mendes L.D. and Rodrigues J.J. 'A survey on cross-layer solutions for wireless sensor networks'. *Journal of Network and Computer Applications*, 34(2), 523–534, 2011

[100] Oliveira L.M.L., de Sousa A.F., and Rodrigues J.J.P.C. 'Routing and mobility approaches in IPv6 over LoWPAN Mesh Networks'. *International Journal of Communication Systems*, 24(11), 1445–1466, 2011, DOI: 10.1002/dac.1228

[101] Chatterjea S., Nieberg T., Meratnia M., and Havinga P. 'A distributed and self-organizing scheduling algorithm for energy-efficient data aggregation in wireless sensor networks'. *ACM Transactions on Sensor Networks (TOSN)*, 4(4), 1–42, 2008

[102] Galpin I., Brenninkmeijer C.Y., Gray A.J., Jabeen F., Fernandes A.A., and Paton N.W. 'SNEE: a query processor for wireless sensor networks'. *Distributed and Parallel Databases*, 29, 31–85, 2011

[103] Galpin I., Brenninkmeijer C.Y., Gray A.J., Jabeen F., Fernandes A.A., and Paton N.W. 'An architecture for query optimization in sensor networks'. *IEEE 24th International Conference on Data Engineering (ICDE 2008)*, Cancun, Mexico, April 7–12, 1439–1441, 2008, DOI: 10.1109/ICDE.2008.4497582

[104] Galpin I., Brenninkmeijer C.Y., Jabeen F., Fernandes A.A., and Paton N.W. 'Comprehensive optimization of declarative sensor network queries'. *21st International Conference on Scientific and Statistical Database Management (SSDBM 2009)*, Lecture Notes in Computer Science, 5566, 339–360, 2009

[105] Brenninkmeijer C.Y., Galpin I., Fernandes A.A., and Paton N.W. 'A semantics for a query language over sensors, streams and relations'. *25th British National Conference on Databases, BNCOD 25*, Cardiff, UK, July 7–10, 2008. Lecture Notes in Computer Science, 5071, 87–99

[106] Brenninkmeijer C.Y., Galpin I., Fernandes A.A., and Paton N.W. 'Validated cost models for sensor network queries'. *Sixth International Workshop on Data Management for Sensor Networks (DMSN '09)*, Lyon, France, August 24, 2009

[107] Intanagonwiwat C., Govindan R., Estrin D., Heidemann J., and Silva F. 'Directed diffusion for wireless sensor networking'. *IEEE/ACM Transactions Networking*, 11(1), 2–16, 2003

[108] Shenker S., Ratnasamy S., Karp B., Govindan R., and Estrin D. 'Data-centric storage in sensor nets'. *ACM SIGCOMM Computer Communication Review,* 33(1), 137–142, 2003

[109] Diao Y., Ganesan D., Mathur G., and Shenoy P. 'Rethinking data management for storage-centric sensor networks'. *Third Biennial Conference on Innovative Data Systems Research (CIDR'07)*, Asilomar, CA, USA, January 2007. pp. 22–31

[110] Chang C., Sheu J., Chang S., and Chen Y. 'A frequency-aware data-centric mechanism for wireless sensor networks'. *Wireless Communications and Mobile Computing 2010*, 10(8), 1078–1101, 2010

[111] Lam Y.K. and Kuo W.T. 'Real-time database systems: architecture and techniques'. Springer: the Netherlands, 2001

[112] Christophe B. 'Petite introduction théorique au temps réel'. *GNU Linux Magazine France*, 82, 76–79, 2006

[113] DiPippo L.C. and Wolfe V.F. 'Real-time databases'. *Journal of Database Systems Hand-book*, Multiscience Press (Citeseer), 1997

[114] Ramamritham K. 'Real-time databases'. *Journal of Distributed and Parallel Data-bases (Springer)*, 1(2), 199–226, 1993

[115] Chagas L.D., Lima E.P., and Neto P.F.R. 'Real-time databases techniques in wireless sensor networks'. *Proceedings of the Sixth International Conference on Networking and Services*, IEEE, Cancun, pp. 182–187, 2010, DOI: 10.1109/ICNS.2010.32

Chapter 6

Proactive safety – cooperative collision warning for vehicles

Chung-Ming Huang[1], *Shih-Yang Lin*[2], *Alexey Vinel*[2] *and Magnus Jonsson*[2]

6.1 General description

Telematics is an interdisciplinary technology that combines telecommunications, vehicular technologies, road transportation, road safety, electrical engineering, and computer science to provide applications and services for the purpose of comfort and safety enhancement. From the timing point of view, the driving safety can be classified into two domains: (1) active safety and (2) passive safety. Passive safety systems are used to reduce damage and protect passengers and drivers when an accident occurs. Common passive safety systems include airbags, seatbelts, whiplash injury lessening systems, and energy absorbing steering column. Active safety systems are used to prevent accidents before they occur. An example of active safety system is the collision warning/avoidance system. It basically collects/detects neighboring vehicles' motion states to compute potential collision between vehicles.

Based on future technology, cooperative active safety systems emerge. Vehicles can exchange their information between each other through wireless communication [1], for example, over a vehicular *ad hoc* network (VANET), for cooperative purposes such as collision warning/avoidance. In a project named smart intersection, a collision avoidance system based on the concept of active safety was developed by Ford and the US government [2]. The system collects a vehicle's information like Global Positioning System (GPS) coordinates, velocity, and heading and delivers it through wireless communication to other vehicles in order to prevent accidents and congestion before vehicles arrive to an intersection.

To understand the details of cooperative collision warning (CCW), this chapter exposes main factors that affect the accuracy of CCW, challenges of CCW, communication techniques for cooperative safety, and collision prediction techniques. CCW systems are also introduced in detail. Moreover, we present

[1]Department of Computer Science and Information Engineering, National Cheng Kung University, Taiwan, Republic of China
[2]School of Information Science, Computer and Electrical Engineering, Halmstad University, Sweden

some existing safety-related techniques and systems that are developed by automobile manufacturers.

6.2 Main factors of CCW

6.2.1 Positioning

Some collision warning systems use infrared or radar to measure neighboring vehicles' position and motion for collision prediction. In the CCW system, positions are exchanged between vehicles through wireless communication; hence, positioning accuracy is an important factor affecting the collision prediction. With larger position errors, the probability of collision prediction error will increase too. Currently, GPS is widely used for various purposes, for example, mountain climbing, navigation, timing synchronization, distance measurement, and collision prediction. Although a GPS receiver is cheap and convenient to position a user's location by receiving at least four GPS satellite signals, the positioning error is about 15 m.

In order to enhance the positioning accuracy, differential GPS (DGPS) is developed, which provides positioning error within 10 cm. The principle of DGPS is to use the reference location of a fixed base station to correct the positioning error of the user. Since the precise location of the base station is known, we can use the difference between the base station and user to create a correction vector for positioning error correction.

6.2.2 Vehicular communication

A few years ago, the WLAN standard IEEE 802.11 was extended with the IEEE 802.11p amendment to support vehicular communication on the PHY and MAC levels using the 5.9 GHz frequency band, that is, Dedicated Short-Range Communications (DSRC) [3]. The standard is complemented separately for the US (by IEEE) and Europe (by ETSI) covering, for example, message types and send rates. Some related standards are made, such as the wireless access in the vehicular environment (WAVE) standards, which include the IEEE 1609.X protocols [4–8]. In both variants, there is multichannel support, where safety messages shall be transmitted over a special control channel. The communication can involve road-side units (RSUs), but only a subset of all road segments can be covered by RSUs in practice. Even in the case of RSU coverage, the communication is assumed to be more or less in *ad hoc* mode, that is, forming a VANET, although some coordination, like service announcements, can exist.

6.2.3 Collision prediction

Vehicle collision means two or more vehicles colliding with each other in terms of the vehicles' trajectories in both space and time domains. Here we classify collision prediction as the following types.

6.2.3.1 Trajectory-based prediction

Trajectory prediction utilizes the vehicles' position and motion state to predict an intersection point in the space domain. There will be an intersection point between moving vehicles if their relative course is bigger than 0 and smaller than 180°. Given two points and vectors for two moving vehicles, that is, points representing the vehicles' coordinates and vectors representing the vehicles' heading and velocity, we can easily calculate the intersection point based on points and vectors as vehicle collision prediction. However, it is difficult to predict the intersection point if the vehicles' moving states are changed. Furthermore, intersection point prediction is not enough to judge whether the vehicles will collide or not because the vehicles may arrive to the intersection at different time instances. Hence, a CCW system needs more information to make the prediction.

6.2.3.2 Timer-based prediction

Timer-based prediction basically utilizes the Estimated Time of Arrival (ETA) for collision prediction. Given a point being an intersection point and another point with a vector facing the intersection point, we can calculate an estimated time for the vehicle to arrive at the intersection. If two vehicles are predicted to arrive at the intersection at the same time, that is, the same ETA for two vehicles, it means that the two vehicles could be collided.

6.2.3.3 Probability-based prediction

Considering both space and time domains, trajectory- and timer-based predictions can be judged whether collision will be happening or not. To realize a CCW system, more considerations are needed. The alternative way to realize a CCW system is probability-based collision prediction. A probability prediction that is proposed in Reference 9 considers the driver reaction time, headway distance, relative velocity, and acceleration as an index for collision prediction. Based on the probability prediction, unnecessary warning messages can be filtered.

6.3 CCW challenges

CCW is one of the proactive techniques that advances driving safety. Drivers can get collision warning messages before crash to have more reaction time for avoidance of an oncoming crash if the CCW system operates well. In order to provide higher accuracy of the collision warning messages, the CCW algorithms/systems face some challenges, which are addressed as follows.

6.3.1 Unnecessary calculation

In a cooperative scenario, vehicles periodically broadcast their beacons with information such as coordinates, coursing, and speed. Since a CCW system predicts potential crashes depending on received messages coming from other vehicles, a vehicle can only predict collisions with vehicles it has received information from recently. A vehicle will receive many other vehicles' related information, especially at an intersection. It is unnecessary to calculate predictions based on all received messages because some of them are, for example, not from vehicles within range of potential collision. Based on limited computing resource, the collision prediction may be out of date if all received messages are processed.

6.3.2 Vehicle's motion state changing

Basically, a CCW system predicts a potential collision based on some calculation results, for example, time and/or distance to arrival at an intersection, or trajectory prediction. Then, the collision calculation algorithm utilizes the produced information as input data to calculate the collision prediction. Normally, vehicles do not keep the same velocity and coursing unless they are driving on a straight highway. The challenge is to get correct collision prediction when vehicles change their velocity and/or coursing. In an intersection scenario, vehicles may keep the same coursing when approaching the intersection, but their velocity may change. They may speed up, slow down, or keep the same velocity toward the intersection. Moreover, vehicles may change their coursing at an intersection or curvy road. Collision prediction may have errors if the vehicles' motion states are not considered. Hence, collision warning messages may fail in warning drivers if the collision prediction errors are not considered. In order to enhance the accuracy, vehicles' motion state changing should be considered in a CCW algorithm/system even if these considerations will increase the CCW's complexity.

6.3.3 Positioning error

Positioning error is a distance error caused by the GPS sampling rate, transmission delay between vehicles, and computing delay. It is a distance difference between a positioned point and the actual point. For example, assuming a vehicle gets its GPS position P1 at time T1. After a while, its GPS position should be P2 at time T2, but according to its GPS sampling rate, the GPS sampling point will be P3 at time T3. In this scenario, the positioning error occurs at time T2. Since two continuous GPS positions were sampled at time T1 and T3 for positions P1 and P3, respectively, the vehicle will use position P1 as its position point at time T2. Hence, the positioning error is equal to P2 subtracted by P1.

The transmission delay between two vehicles and the computing delay will also generate positioning error. Assume a vehicle A gets its GPS position P4 and broadcasts it immediately at time T4 without any delay. Because of transmission and computing delays, vehicle B uses position P4 as vehicle A's position at time T5. However, the position of vehicle A at time T5 should be P5; hence the positioning error of vehicle A is equal to P5 subtracted by P4.

Packet loss is another issue resulting in having old GPS position data. According to our field trials [10], the probability of packet loss is non-negligible in many environments. Even though there are methods to cope with packet loss in vehicular communication (see, e.g., [11]), such methods are hard to implement in a non-centralized way.

Even if the positioning error may be small often, it is an important factor affecting the accuracy of the collision prediction. Considering these conditions, a CCW system should consider positioning errors to enhance the accuracy of collision prediction and warning.

6.3.4 Different types of vehicles

CCW algorithms usually utilize GPS coordinates to represent the vehicles' locations. Assume the GPS receiver is placed in the middle of a vehicle. If we have an omnidirectional collision prediction algorithm, the length and width of the vehicles need to be considered. Since vehicles are rectangular in shape, the distance from the center of a vehicle to the bumper and the distance from the center of a vehicle to the side of the vehicle differ. Furthermore, two approaching vehicles may be of different sizes, for example, a sedan and a truck. Their lengths and widths will affect the accuracy of collision prediction and need to be considered in a CCW algorithm.

6.4 Communication techniques for cooperative safety

An important factor for proactive safety systems and applications is vehicular communication capability. Cooperative manners are normally performed relying on information exchanged between vehicles. In Reference 12, the authors indicate that inter-vehicle real-time information communication is the key point to solve or reduce accidents in a CCW system. One of the pressing challenges in vehicular communication for safety applications is to maintain acceptable communications. The question is how often and how far the information should be broadcasted to its neighbors. To tackle the question, a transmission control protocol that adapts the communication rate and power control is proposed in Reference 13.

The purposes of rate and power control are to decide how frequently a vehicle should broadcast its own state information and how far the state information should be broadcasted, respectively. Basically, the transmission rate control mechanism calculates a probability based on the distance of neighboring vehicles toward its own position in a Euclidean sense. If the distance is smaller than the threshold, transmission from the vehicle is not allowed; otherwise, the probability is computed. Besides, the transmission power control mechanism is adjusted according to the observed channel status. If the channel occupancy is higher than a maximum threshold, minimum transmission range is used; if the channel occupancy is lower than a minimum threshold, maximum transmission range is used; otherwise, the transmission range is computed. Actually, it is more dangerous if a vehicle does not obtain the state of its nearby neighbors, and it could therefore temporarily give up faraway vehicles. The power control mechanism achieves the goal to reduce interference of faraway vehicles and increase successful reception probability of nearby vehicles.

6.5 CCW techniques

6.5.1 Unsignalized intersection

Unsignalized intersection represents an intersection without traffic signal lights, which is a place where two or more roads join and cross each other. The driver can choose to drive on the same road or join the other when arriving at the intersection. Since the intersection does not have traffic signal lights, vehicles may fall into danger when they are crossing the intersection simultaneously. Some collision warning/avoidance approaches are designed to solve the problem [14].

As reported in Reference 14, the authors have designed a risk assessment architecture based on the trajectory prediction manner, in which the architecture supports standalone and cooperative approaches. In the architecture, a data fusion model is used to integrate data that comes from proprioceptive and exteroceptive sensors, and communication. In order to consider errors and imprecisions on the positions, environment state models are designed to process the fused data. After that, a collision detection procedure is performed to detect potential dangerous situations based on trajectory prediction. The trajectory prediction is performed using a linear Kalman filter, which can filter the noise of GPS position, estimate the position of the vehicle, estimate the motion model of the vehicle, and estimate the positioning errors. Finally, a risk management entity displays predicted results based on estimation of dangerous situations.

Considering DGPS, the authors of Reference 15 have explored the engineering feasibility of a CCW system based on future-trajectory prediction. In Reference 16, four functional blocks constituting the CCW system are used: (1) positioning, which means a positioning system should provide position accuracy with an error within 1 m, timing and update rate of the system should be fast enough, and its reliability should be able to operate under all normal operational conditions; (2) communication, which maintains reliability, quality, and bandwidth of vehicular communication between vehicles; (3) situation awareness, which integrates relevant information that may come from its devices (e.g., positioning system, Controller Area Network (CAN)) or communication system for potential collision determination; (4) collision determination, which makes decisions by receiving relevant information. In the collision determination, the system selects target vehicles that may collide in trajectory conflicts and computes their future trajectories at first. After that, it identifies the corresponding time, location, and vehicle ID for vehicles where the distance between target vehicles and itself is smaller than a threshold of 2–3 s into the future. Finally, it computes the associated probability and persistency of the collision to minimize false alarms and to tolerate errors.

A DSRC-based cooperative collision prediction and warning mechanism was proposed in References 16 and 17. It utilizes related information such as location and velocity calculation for the Safety Headway to assist drivers to have more reaction time to avoid collision. Four warning levels were defined, and they were determined by the distance between a host and its neighboring vehicles, and driving patterns.

Another CCW system based on the distance calculation between approaching vehicles for omnidirectional protection was proposed in Reference 18. A similar collision warning algorithm named intersection collision warning system (ICWS) was proposed in Reference 19. The basic idea is to calculate the time for which vehicles reach the intersection. Two vehicles will be judged to collide if the time difference is equal to or smaller than a threshold. In Reference 20, the authors utilize future-trajectory prediction to construct a robustness and reliability CCW system. These kinds of cooperative mechanisms were operating relying on vehicles' position and information exchanging between vehicles. The authors of Reference 21 analyze factors of communication and position for a CCW system, which indicates that the accuracy of position and the wireless communication are two important factors for the success of a CCW system.

In our previous research on cooperative systems, the early collision warning algorithm (ECWA) [22] and the collision pre-warning algorithm (CPWA) [23] were proposed considering more factors in order to enhance the accuracy of the collision warning prediction. These two algorithms provide 360° collision warning prediction, and they can be applied in various conditions, for example, multi-intersections and intersection with different intersection angles. The basic idea of them is that each vehicle periodically broadcasts its related information such as position information (i.e., GPS coordinates), coursing, and speed through DSRC. Each vehicle itself calculates whether a collision can happen or not when neighboring vehicle's related information is received. If two approaching vehicles are judged to collide, some collision warning messages will be shown to the driver several seconds before, where the timing can be determined by the drivers themselves.

6.5.2 Acceleration and turning

Vehicle motion contains some states such as to go forward and backward, turn left and right, and keep the same state. It is more difficult to predict the vehicle's trajectory for collision warning when its state is changing.

Our previous study proposed a vector-based cooperative collision warning (VCCW) [24] algorithm based on ECWA and CPWA to correct errors in the collision prediction. In VCCW, acceleration is used to correct calculation errors caused by velocity and/or direction changing. For a detailed description of the VCCW algorithm, please refer to the next section.

6.6 Our VCCW algorithm

Here we introduce our VCCW algorithm. The VCCW system architecture is introduced first. Next, some techniques used in the VCCW, such as information exchanging, collision calculation, and judgment rule and alerting, are respectively explained in detail.

6.6.1 Architecture

VCCW contains three parts: information exchanging, collision calculation, and judgment rule and alerting. Let each vehicle be equipped with a positioning device (e.g., GPS or DGPS), an accelerometer, and a vehicular communication module (e.g., WAVE), which provide the GPS position, acceleration, and capability of inter-vehicle communication, respectively. Information exchange collects the host vehicles' foregoing related information and exchanges it with neighboring vehicles, periodically. The collision prediction starts calculating potential collisions between the host vehicle and the target vehicle when the related information is received. According to the prediction results, judgment rule and alerting will show warning messages if it is judged that two approaching vehicles will collide.

6.6.2 Information exchanging

At first, the host vehicle collects its own related information (called MyRI) which contains position information (e.g., GPS coordinates) and acceleration, periodically. MyRI is broadcasted in a default interval (e.g., one time per second) via inter-vehicle communication. Although we can frequently broadcast MyRI to reduce the positioning error, it is not a good idea because frequent broadcasting will result in higher probability of packet collision. If packets collide, vehicles cannot receive neighboring vehicles' related information to predict potential collision between the vehicles. All vehicles perform the same procedure, so that vehicles will receive neighboring vehicles' related information (called TargetRI).

When TargetRI is received, some conditions can be filtered to save computing resources. For example, it is unnecessary to perform the collision calculation if the distance between the target vehicle and the host vehicle is large, the velocity of the target vehicle is low, or the target vehicle is leaving from the host vehicle. Therefore, we design a filter to save computing resources. Before introducing the filter, we first illustrate the relative bearing and relative course. Relative bearing is a direction that points out a specific object, and relative course is a coursing that points out the target vehicle's moving direction. Referring to Figure 6.1a, from the view point of host vehicle H, the relative bearing of target vehicle T is 240°, and the relative course of target vehicle T is 330°. On the other hand, referring to Figure 6.1b, from the view point of host vehicle T, the relative bearing of host vehicle H is 90°, and the relative course of host vehicle H is 30°.

Based on the relationship of relative bearing and relative course, a filter is designed. If the relationship between the host vehicle and the target vehicle satisfies the following four conditions, it means that the target vehicle will probably collide with the host vehicle. In such a case, the collision calculation procedure will be triggered. Otherwise, the message from the target vehicle will be filtered out.

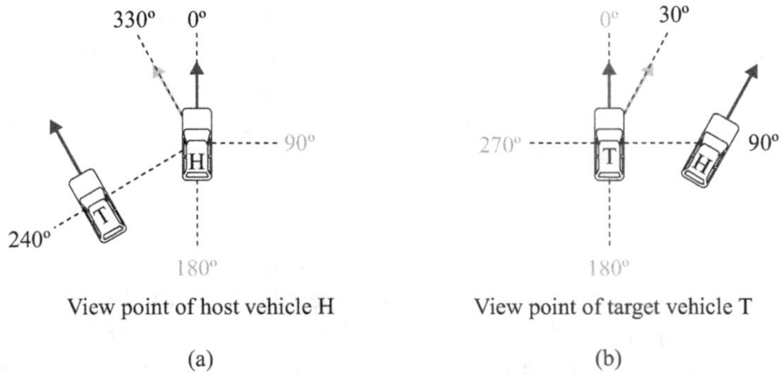

Figure 6.1 Illustration of relative bearing and relative course

1. The relative bearing of vehicle T is 0°–90° and the relative course of vehicle T is 180°–360°.
2. The relative bearing of vehicle T is 270°–360° and the relative course of vehicle T is 0°–180°.
3. The relative bearing of vehicle T is 180°–270° and the relative course of vehicle T is 0°–90°.
4. The relative bearing of vehicle T is 90°–180° and the relative course of vehicle T is 270°–360°.

After some unnecessary messages are filtered out, two RI messages, including MyRI and TargetRI, are used to perform the collision calculation procedure.

6.6.3 The collision calculation

6.6.3.1 Basic technique of collision calculation

The procedure of the collision calculation computes whether the collision between host and target vehicles will occur or not based on the concept of Closest Point of Approach (CPA) [25], and send warning messages to drivers depending on the judgment rule. The judgment rule compares some parameters, such as Time to Closest Point of Approach (TCPA) and Distance to Closest Point of Approach (DCPA). Note that DCPA is a distance that denotes the shortest distance between two approaching vehicles after TCPA time units, and TCPA is a time unit that denotes how many time units does two approaching vehicles need to reach the shortest distance (DCPA). If collision is judged not to occur, the procedure returns to the waiting state for the next oncoming RI messages; otherwise, the warning message is shown to the driver depending on the predefined warning levels, for example, 3, 6, or 9 s before the oncoming collision occurs.

Assume host vehicle H and target vehicle T are approaching the same intersection. Let $D_{\mathrm{H,T}}$ denote the distance between vehicle H and T, and V_i denote the vector of vehicle i, that is, V_{H} and V_{T} represent the velocity and moving direction of host vehicle H and target vehicle T, respectively. Symbols $\mathrm{TCPA_{H,T}}$ and $\mathrm{DCPA_{H,T}}$ denote time and distance to the CPA between vehicle H and T, respectively. When the collision calculation is started, reversing vector V_{H} and then adding it to vector V_{T} can get a new point called NP. After that, draw the line $L_{\mathrm{T,NP}}$ from vehicle T, pass the point NP until the line is over vehicle A. Then, draw another line which passes vehicle H and is perpendicular to the line $L_{\mathrm{T,NP}}$. Finally, we can obtain a point called CPA on the two lines intersected. The distance between CPA and host vehicle H is called $\mathrm{DCPA_{H,T}}$. After point CPA is obtained, $\mathrm{TCPA_{H,T}}$ can be obtained by using Equation (6.1).

$$\mathrm{TCPA_{H,T}} = \mathrm{Distance_{CPA,T}}/\mathrm{Distance_{NP,T}}$$

$$\text{where} \begin{cases} \mathrm{TCPA_{H,T}} = \text{Time to CPA of vehicles H and T} \\ \mathrm{Distance_{CPA,T}} = \text{Distance between point CPA and vehicle T} \\ \mathrm{Distance_{NP,T}} = \text{Distance between point NP and vehicle T} \end{cases} \tag{6.1}$$

After calculation, we can say that vehicle H and vehicle T will collide after $\mathrm{TCPA_{H,T}}$ time units if $\mathrm{DCPA_{H,T}}$ is equal to zero. If $\mathrm{DCPA_{H,T}}$ is not equal to zero, we can say that the minimum distance between vehicle H and vehicle T will be $\mathrm{DCPA_{H,T}}$ after TCPA time units.

6.6.3.2 Velocity changing

The aforementioned technique of collision calculation cannot get accurate collision prediction when the vehicles' motion states are changed, for example, velocity changing. Here, acceleration is used to tackle the problem.

Let Acc_i denote the acceleration of vehicle i, that is, $\mathrm{Acc_H}$ denotes vehicle H's acceleration. The acceleration can have a positive or negative value. It has a positive value when the vehicle is accelerating, while it is negative if the vehicle is decelerating. If the vehicle keeps the same velocity, the acceleration will be 0. When the collision calculation is started, in each round, let V_i be equal to V_i plus Acc_i. After that, the reversed vector V_{H} is added to vector V_{T} to find the first temporary point called TP_i, where i means ith round. If the distance between vehicle H and vehicle T ($D_{\mathrm{H,T}}$) is larger than the distance between point TP_i and vehicle T, the algorithm will be repeated. The repetition of n rounds will be stopped if the total distance of the round is larger than $D_{\mathrm{H,T}}$. Finally, a similar way is used to find the CPA point in the last round.

After that we can find the CPA point, which is an intersection between a line passing vehicle H and being perpendicular to a line between the last two temporary points, for example, TP_{n-1} and TP_n (assuming there is a total of n rounds). Once CPA is found, $TCPA_{H,T}$ can be obtained using Equation (6.2).

$$TCPA_{H,T} = (Distance_{CPA,TP_{n-1}}/Distance_{TP_n,TP_{n-1}}) + n - 1$$

$$\text{where} \begin{cases} n = \text{Times to repeat} \\ Distance_{CPA,TP_{n-1}} = \text{Distance between point CPA and the temporary point } (TP_{n-1}) \\ Distance_{TP_n,TP_{n-1}} = \text{Distance between temporary point } TP_n \text{ and temporary point } TP_{n-1} \\ TCPA_{H,T} = \text{Time to CPA of vehicles H and T} \end{cases} \quad (6.2)$$

6.6.3.3 Curve environment

Curve condition is another problem affecting the accuracy of collision calculation. Referring to Figure 6.2, let the steering angle of vehicle T be obtained from the accelerometer and be denoted as α_T, while the heading of vehicle T is denoted as C_{Tj} in the jth round. The value of α_T is positive when the steering angle is clockwise; otherwise, α_T is negative. Next, the following three steps are performed: (1) finding the vector point (denoted as VP_{Tj}) of the target vehicle T in the jth round, (2) finding the temporary point (denoted as TP_j) in the jth round, and (3) repeating n rounds.

Details of these three steps are as follows:

- Step 1: Finding the vector point of target vehicle T.

$$\left.\begin{array}{l}
\text{If } (0 <= C_{Tj} < 90)\ \{ \quad /* \text{ First Quadrant } */ \\
\quad VP_{Tj}(x) = T_x + \sin(C_{Tj}) \bullet \text{velocity} \\
\quad VP_{Tj}(y) = T_y + \cos(C_{Tj}) \bullet \text{velocity}\} \\
\text{If } (90 <= C_{Tj} < 180)\ \{ \quad /* \text{ Fourth Quadrant } */ \\
\quad VP_{Tj}(x) = T_x + \sin(C_{Tj}) \bullet \text{velocity} \\
\quad VP_{Tj}(y) = T_y - \cos(C_{Tj}) \bullet \text{velocity}\} \\
\text{If } (180 <= C_{Tj} < 270)\ \{ \quad /* \text{ Third Quadrant } */ \\
\quad VP_{Tj}(x) = T_x - \sin(C_{Tj}) \bullet \text{velocity} \\
\quad VP_{Tj}(y) = T_y - \cos(C_{Tj}) \bullet \text{velocity}\} \\
\text{If } (270 <= C_{Tj} < 360)\ \{ \quad /* \text{ Second Quadrant } */ \\
\quad VP_{Tj}(x) = T_x - \sin(C_{Tj}) \bullet \text{velocity} \\
\quad VP_{Tj}(y) = T_y + \cos(C_{Tj}) \bullet \text{velocity}\}
\end{array}\right\} \quad (6.3)$$

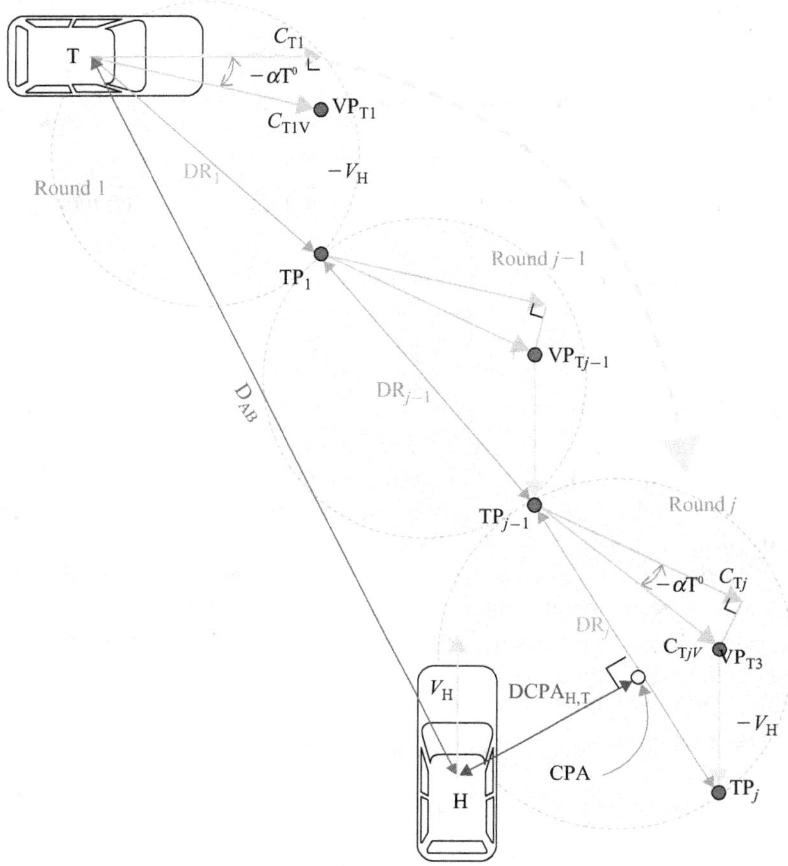

Figure 6.2 Illustration of the CPA calculation

Referring to Equation (6.3), let T_x and T_y be the coordinates of vehicle T. The coordinates of vector point VP_{Tj} in the jth round can be obtained using Equation (6.3), which is dependent on vehicle T's heading C_{Tj} in the jth round. For example, assuming the heading of vehicle T is 60° ($C_{Tj} = 60$) and the velocity of vehicle T is 15 m/s, the coordinates of the X axis for vector point $VP_{Tj}(x)$ is equal to $T_x + \sin(60)*15$. Using Equation (6.3), the vector point VP_{Tj} can be obtained, and the heading to the vector point VP_{Tj} is denoted as C_{TjV}.

- Step 2: Finding the temporary point in each round.

  Next, reverse vector V_H and then add it to VP_{Tj} to get a temporary point called TP_j, where j means the jth round.
- Step 3: Repeating n rounds.

  Let DR_j be the distance between TP_j and TP_{j-1}, where j means the jth round. Repeat the same procedure depicted in the first and second steps until the total distance, that is, $DR_1 + DR_2 + \cdots DR_j$ is larger than the distance between host vehicle H and target vehicle T. Finally, a similar way is used to find the CPA point in the last round. After point CPA is found, $DCPA_{H,T}$ can be obtained. Moreover, $TCPA_{H,T}$ can be obtained using Equation (6.2).

6.6.4 Judgment rule and alerting

Based on the result of the collision calculation, we can know (i) whether collision between two vehicles may occur or not and (ii) when it will happen. There are two conditions needed to be considered.

The first condition is about timing. It means, how long before the collision should a warning message be sent. For example, the warning messages should not be sent to drivers 30 s before the collision. Hence, we design three warning levels, which can be set by the user himself/herself. Three classified levels are high, middle, and low warning levels. The low warning level, which sets the action time to be less than 3 s, could minimize the number of warning messages alerting to the driver. It also means that the warning message should be sent to drivers 3 s before the collision. The middle warning level sets the action time to be less than 6 s, while the high warning level sets it to 9 s.

The second condition is warning distance. Warning distance determines a circular coverage around a car, which is used for detecting whether some collision will happen or not. Assuming a GPS receiver is located in the center of the vehicle, the warning distance is expressed in Equation (6.4).

$$\text{Warning distance} = (\text{Length of the car}/2) \tag{6.4}$$

Now, we can finalize the judgment rule. According to the judgment rule, a warning message is sent to the driver if $DCPA_{H,T}$ is smaller than the warning distance and $TCPA_{H,T}$ is smaller than the warning level, which is expressed in Equation (6.5).

$$\begin{aligned} &\text{If } ((DCPA_{H,T} \leq \text{Warning Distance}) \ \& \ (TCPA_{H,T} \leq \text{Action Time}))\{ \\ &\quad \text{Send(Warning Message)}\} \end{aligned} \tag{6.5}$$

6.7 CCW systems

6.7.1 Vector-based cooperative collision warning

6.7.1.1 Device description

The VCCW system was implemented on a notebook platform, which contains four components. They are (1) the GPS receiver: providing the GPS position coordinates; (2) G-sensor (accelerometer): providing the acceleration information; (3) the WAVE communication module: providing V2V communication based on the IEEE 802.11p protocol and the WSM protocol and (4) 3.5G wireless device: providing connection to Internet such that the E-MAP information can be shown on the VCCW user interface.

6.7.1.2 User interface

Figure 6.3a illustrates the user interface of the VCCW system. In Figure 6.3a, the host vehicle is always kept in the center, which is marked with the blue arrow. All neighboring vehicles are marked with red arrows. The license plate should be input in the MyID box; hence it can be encapsulated in the MyRI packet. The ComPorts box provides the COM ports selection, which depends on the connection with GPS receiver. The user can disable the E-MAP in the MAP option box. If the option of the MAP is disabled, the relative positions of neighboring vehicles are shown on the user interface, such as that depicted in Figure 6.3b. The Coordinate option provides two coordinate systems for the GPS receiver, which are WGS84 and EPSG:900913. What kind of coordinate system should be used is determined by the GPS receiver. The PICorrection option provides the function to correct the problem of position interruption, which can be enabled or disabled. A driver or user can select different warning levels in the WarningLevel option. Additionally, the CarInformation box should be chosen and inputted before the system is started. The VCCW system connects to the WAVE box through the IP address that was entered in the DSRC IP box. Finally, users can select a suitable distance value in the WarningDistance box. If the VCCW system detects an oncoming collision, warning messages with voice warning will be shown to drivers.

6.8 State of the art

6.8.1 Automobile manufacturers – collision avoidance system

Collision avoidance systems are designed to reduce or avoid an accident. Some sensing techniques are used to be aware of the states of neighbors, such as radar, laser, or camera sensors for collision detection. Once the collision is oncoming, the systems may provide warnings to the driver. Here we introduce some active systems that are developed by automobile manufacturers.

The full directions of the active system Pre Sense Plus is developed by Audi. It contains five systems. (1) Side Assist, which provides warnings if vehicles behind are approaching beside the vehicle. (2) Lane Assist, which provides warnings if the vehicles depart the lane. (3) Pre Sense Front, which provides warnings if the system detects the distance between itself and vehicle in front is too small. In addition, the brake assist system will activate to reduce the vehicle speed if a collision will happen. (4) Pre Sense Rear, which performs in a similar way as the Pre Sense Front. (5) Night Vision Assistant, which provides the capability of night vision for objects in front of the vehicle up to 300 m away. It will show a pedestrian on the monitor if the pedestrian is in front of the vehicle.

Bavarian Motor Works (BMW) provided the ConnectedDrive system in 2013, which integrates information, amusement, and driving assist system. For the safety-related capability, it contains (1) Active Cruise Control system, which integrates

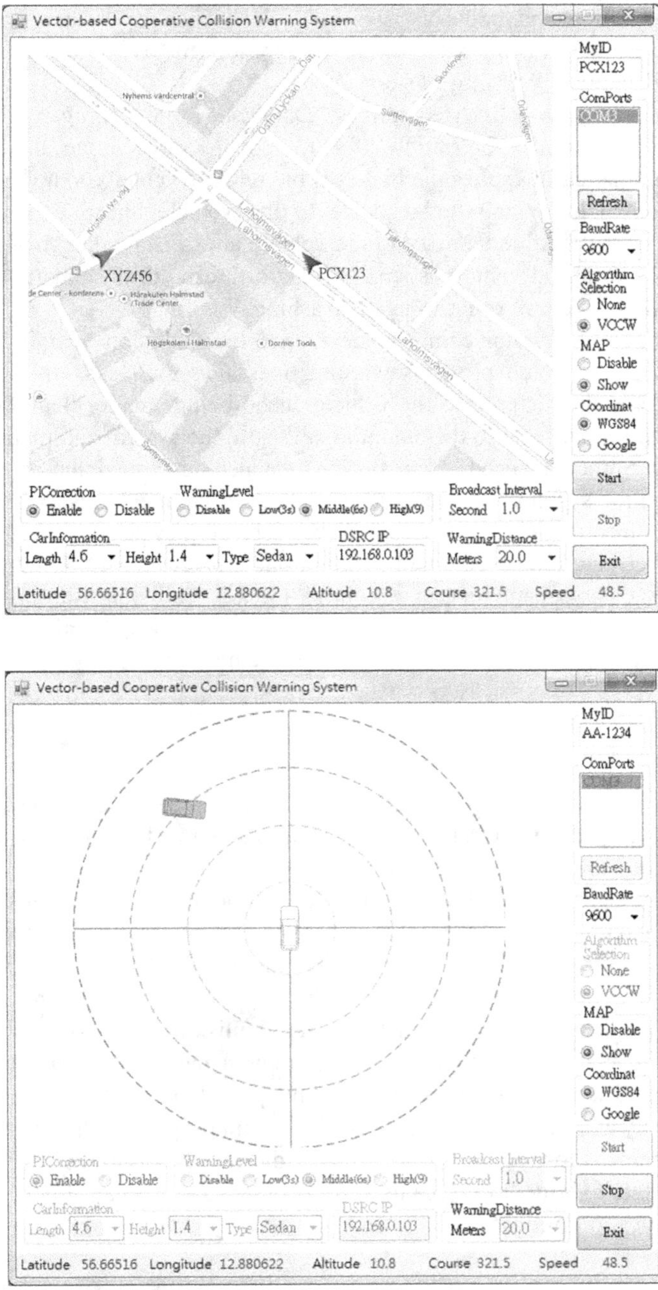

(a)

(b)

Figure 6.3 The user interface of the VCCW system

collision warning and brake assist system, lane change warning, and start-stop system; (2) Night Vision with person recognition and Dynamic Light Spot; and (3) Lane Departure Warning System.

Considering proactive techniques, Daihatsu Motor company developed a Smart Assist system in the middle of 2013. For the safety issue, the Smart Assist system automatically applies the brakes if the relative velocity to a vehicle ahead is higher than 30 km/h and the distance to the vehicle ahead is smaller than a threshold; if the relative velocity to the vehicle ahead is smaller than 30 km/h and the distance to ahead vehicle is smaller than a threshold, the Smart Assist system will stop the vehicle in order to avoid crashing.

In 2009, Ford Motor company developed the Collision Warning with Brake Support system, which provides warning messages on Head Up Display if the vehicle senses the distance to the vehicle ahead being smaller than a threshold. If the driver does not react to the potential collision, the system automatically applies the brakes. Some advanced safety systems are also developed accordingly, such as Curve Control system, Lane Keeping system, Adaptive Cruise Control system, Blind Spot Information system, and Active City Stop system.

Another automotive company, Volvo, has also developed some active safety systems, such as adaptive cruise control (ACC), collision warning with brake support (CWBS), emergency break light (EBL), blind spot information system (BLIS), and intelligent driver information system (IDIS).

6.9 Conclusion

In this chapter, we have presented some proactive safety techniques. A specific proactive safety technique called CCW has been addressed. A CCW system detects a potential collision between two vehicles in a cooperative manner, and is periodically exchanging some relevant information with neighboring vehicles for collision prediction. Vehicles can change their motion state, for example, speed up, slow down, or change its direction, to avoid oncoming danger. We have discussed challenges that may affect the accuracy of the collision prediction. We overcome the three identified challenges to have an active safety collision prediction algorithm called VCCW. The VCCW system has also been implemented and tested.

We have also discussed some existing proactive safety systems that were developed by automobile manufacturers. The main purpose of the various proactive safety systems is to try to avoid or mitigate dangerous situations. The cooperation between vehicles cannot only be used for driving safety enhancement, but also for traffic congestion mitigation.

References

[1] Dar, K., Bakhouya, M., Gaber, J., Wack, M. 'Wireless communication technologies for ITS applications'. *IEEE Communications Magazine*, 2010; **48**(5):156–162

[2] Ford Motor Company. *Smart Intersection* [online]. 2008. Available from http://ophelia.sdsu.edu:8080/ford/10-21-2008/about-ford/news-announcements/press-releases/press-releases-detail/pr-ford26rsquos-new-smart-28611.html

[3] IEEE Std. 802.11p-2010, Part 11: *Wireless LAN Medium Access Control (MAC) and Physical Layer (PHY) specifications: Amendment 7: Wireless Access in Vehicular Environment*. July 2010

[4] IEEE Std. 1609.0-2013: *IEEE Guide for Wireless Access in Vehicular Environments (WAVE) Architecture*. December 2013

[5] IEEE Std. 1609.1-2006: *IEEE Trial-Use Standard for Wireless Access in Vehicular Environments (WAVE) – Resource Manager*. October 2006

[6] IEEE Std. 1609.2-2013: *IEEE Standard for Wireless Access in Vehicular Environments – Security Services for Applications and Management Messages*. April 2013

[7] IEEE Std. 1609.3-2010: *IEEE Standard for Wireless Access in Vehicular Environments (WAVE) – Networking Services*. December 2010

[8] IEEE Std. 1609.4-2010: *IEEE Standard for Wireless Access in Vehicular Environments (WAVE) – Multi-channel Operation*. February 2011

[9] Yag, G., Yang, W., Weigle, M.C., Olariu, S., Rawat, D. 'Cooperative collision warning through mobility and probability prediction'. *Proceedings of the IEEE Intelligent Vehicles Symposium*; San Diego, CA, June 2010. pp. 1172–1177

[10] Böhm, A., Lidström, K., Jonsson, M., Larsson, T. 'Evaluating CALM M5-based vehicle-to-vehicle communication in various road settings through field trials'. *Proceedings of the 4th IEEE Workshop on User MObility and VEhicular Networks*; Denver, USA, October 2010. pp. 613–620

[11] Böhm, A., Jonsson, M., Kunert, K., Vinel, A. 'Context-aware retransmission scheme for increased reliability in platooning applications'. *Proceedings of the 6th International Workshop on Communication Technologies for Vehicles*; Offenburg, Germany, 2014. pp. 30–42

[12] Xing, J.G., Liu, S.F. 'A robust cooperative collision warning algorithm for unreliable wireless communication channel'. *Proceedings of the International Conference on Optoelectronics and Image Processing*; Haiko, China, November 2010. pp. 112–115

[13] Huang, C.L., Fallah, Y.P., Sengupta, R. 'Adaptive intervehicle communication control for cooperative safety systems'. *IEEE Network*. 2010;**24**(1): 6–13

[14] Ammoun, S., Nashashibi, F. 'Real time trajectory prediction for collision risk estimation between vehicles'. *Proceedings of the IEEE Intelligent Computer Communication and Processing*; Cluj-Napoca, Romania, August 2009. pp. 417–422

[15] Tan, H.S., Huang, J. 'DGPS-based vehicle-to-vehicle cooperative collision warning: engineering feasibility viewpoints'. *IEEE Transactions on Intelligent Transportation Systems*. 2006;**7**(4):415–428

[16] Young, C.P., Chang, B.R., Chen, S.Y., Wang, L.C. 'A highway traffic simulator with dedicated short range communications based cooperative collision

prediction and warning mechanism'. *Proceedings of the IEEE Intelligent Vehicles Symposium*; Eindhoven, the Netherlands, June 2008. pp. 114–119

[17] Young, C.P., Chang, B.R., Lin, J.J., Fang, R.Y. 'Cooperative collision warning based highway vehicle accident reconstruction'. *Proceedings of the 8th International Conference on Intelligent Systems Design and Applications*; Kaohsiung, Taiwan, November 2008. pp. 561–565

[18] Sebastian, A., Tang, M., Feng, Y., Looi, M. 'Multi-vehicles interaction graph model for cooperative collision warning system'. *Proceeding of the IEEE Intelligent Vehicles Symposium*; Xi'an, China, June 2009. pp. 929–934

[19] Miller, R., Huang, Q. 'An adaptive peer-to-peer collision warning system'. *Proceeding of the IEEE 55th Vehicular Technology Conference*; Birmingham, USA, May 2002. pp. 317–321

[20] Huang, J., Tan, H.S. 'Error analysis and performance evaluation of a future-trajectory-based cooperative collision warning system'. *IEEE Transactions on Intelligent Transportation Systems*. 2008;**10**(1):175–180

[21] Huang, J., Tan, H.S. 'Impact of communication reliability on a cooperative collision warning system'. *Proceedings of the IEEE Intelligent Transportation Systems Conference*; Bellevue, USA, October 2007. pp. 355–360

[22] Huang, C.M., Lin, S.Y. 'An early collision warning algorithm for vehicles based on V2V communication'. *International Journal of Communication Systems*. 2012;**26**(6):779–795

[23] Huang, C.M., Lin, S.H., Yang, C.C., Chou, C.H. 'A collision pre-warning algorithm based on V2V communication'. *Proceedings of the 4th International Conference on Ubiquitous Information Technologies & Applications*; Fukuoka, Japan, December 2009. pp. 1–6

[24] Huang, C.M., Lin, S.H. 'Cooperative vehicle collision warning system using the vector-based approach with dedicated short range communication data transmission'. *IET Intelligent Transport Systems*. 2014;**8**(2):124–134

[25] Subramanian, A., Christopher, J. 'Closest-point-of-approach join for moving object histories'. *Proceedings of the 22nd International Conference on Data Engineering*; Atlanta, GA, April 2006. pp. 86–95

Chapter 7
Electronic toll collection systems in Europe

*Mike Hayward**

7.1 Introduction

7.1.1 Why collect road tolls?

Tolls have been collected on roads in England since at least the twelfth century. For many centuries, tolls were the principal means of financing road improvements and maintenance (Plate 7.1).

However, the railway age during the nineteenth century in England resulted in turnpike trusts being abandoned and roads falling under the control of local authorities.

In the twentieth century, continental Europe began to adopt tolls again as a method of financing the construction of sections of motorways. The first examples of this were in Italy and Greece in the 1920s, followed by France, Spain and Portugal from the 1950s onwards.

One of the main advantages of road tolls are that they can be used as a revenue stream to secure private sources of finance. The burden of road finance on taxation can be reduced, and the cost of road maintenance and operation can be directly linked to road use.

In this way road construction has been progressed relatively quickly by countries without adding massive amounts to the public sector debt.

Nowadays, many European countries have examples of private concession contracts funded by road tolls in order to finance major new road infrastructure. Many of these concessions have installed electronic toll collection (ETC) systems to improve the efficiency of their operations.

Additionally, during the early part of the twenty-first century, a new trend has emerged in Europe – the introduction of nationwide systems for the collection of road use charges for heavy vehicles. These systems were introduced by European countries concerned by the cost of maintaining roads being used by increasing volumes of heavy vehicles travelling across the European Union. These new systems adopted 'free-flow' technology for ETC which did not require vehicles to stop at toll plazas (Plate 7.2).

The approach of taxing heavy vehicles electronically was first adopted by Switzerland, with Germany, Austria and Czech Republic introducing free-flow electronic charging systems shortly afterwards.

*Transport Logic Ltd, UK

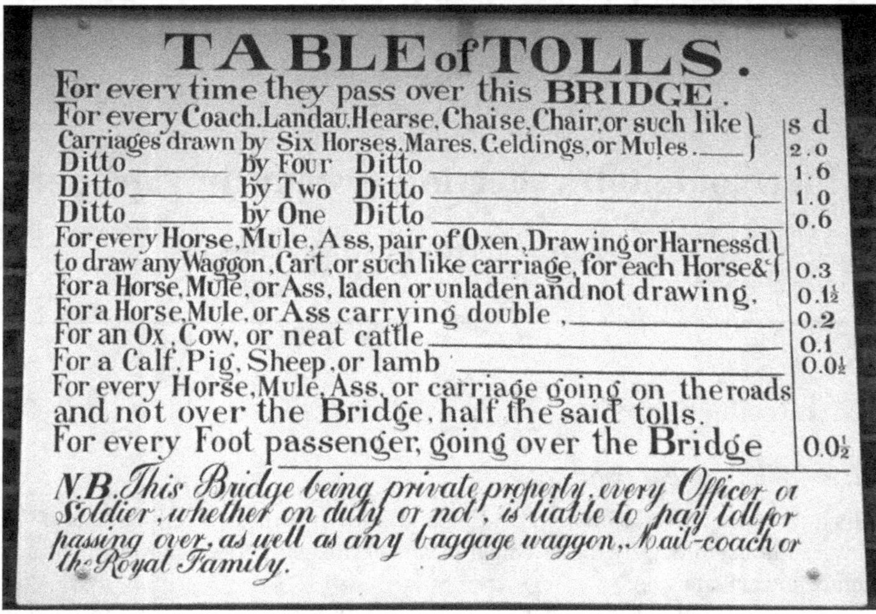

Plate 7.1 Table of Tolls, Ironbridge, England
[Source: Photo provided by Ironbridge Gorge Museum Trust Library and Archives]

Plate 7.2 GoBox on-board unit for heavy vehicle free-flow electronic tolling in Austria
[Source: Photo provided by ASFiNAG]

Plate 7.3 A typical mixed manual and electronic toll plaza at Dartford River Crossing near London, England
[Source: Photo provided by SANEF]

The European Directive on the charging of heavy goods vehicles[1] required Member States to introduce fairer charging systems for the use of road infrastructure. Fairer in this context meaning that the charging system should follow certain principles, namely 'user pays' and 'polluter pays'. The objectives were to achieve a reduction in the negative impacts of road use whilst avoiding double taxation and without imposing additional burdens on vehicle operators.

The Directive does not require the introduction of electronic methods of road charging; nevertheless, several Member States have decided that the best way to achieve the requirements of the Directive is through the use of an ETC system (Plate 7.3).

So in summary, ETC has been introduced in Europe to meet two broad needs. The first, to improve the efficiency of toll plazas on privately operated toll roads, and the second, a public policy objective to link road taxes to road use and in particular distance travelled and vehicle emissions.

7.2 Motorways, tunnels and bridge tolls

The first operational ETC system in Europe was introduced in Norway in 1986. The system was introduced alongside traditional manual toll collection booths.

Drivers were for the first time able to pay without the need for a manual or cash transaction using a dedicated electronic tag fixed to the inside of the vehicle's windscreen. The ETC tag uses dedicated short-range communications (DSRCs) to verify the vehicle and open the toll barrier. The toll is then deducted from the user's account automatically.

The dominant technical solution for ETC within the European Union is the use of microwave tags. As a first step towards interoperability of services in Europe, the 5.8 GHz frequency band has been adopted as the standard for electronic tolling systems.

ETC systems are popular with motorway operators because they have the following advantages compared to manually operated toll collection:

- lower dwell time at toll booths,
- less delay for motorists approaching toll plazas,
- lower costs per transaction,
- less revenue leakage/more secure.

7.2.1 France

Tolls have been in operation in France since 1955. At the time of writing, France has 14 separate motorway concession companies plus a further 4 operating companies, with over 9 000 km of motorway in operation in total.

France's motorway concession companies and operators have introduced two ETC services which allow a user to travel using a single DSRC on-board unit throughout France. Interoperability of the on-board unit is guaranteed across all motorways and users receive only one invoice. The TIS[a]-Liber-t service for light vehicles has been in operation since 2000 and the TIS-PL service for heavy vehicles since 2005.

The Liber-t service is offered by 11 different service providers, has over 5 million subscribers and accounts for 42% of the total light vehicle transactions at French toll plazas.[2]

France has over 4 500 toll lanes equipped with ETC equipment, and each Liber-t on-board unit registers on average 213 transactions per year. Of the ETC equipped lanes, 450 are 'non-stop' lanes where drivers do not have to stop to allow the barrier to open.

The TIS-PL service is offered by five service providers, has over 640,000 subscribers and accounts for 85% of the total heavy vehicle toll transactions at French toll plazas. TIS-PL is available at 2 364 lanes at toll plazas.[3] The system is fully compliant with the requirements of European Directive 2004/52/EC on interoperability of ETC systems.

The French ministry has developed a nationwide free-flow electronic charging system known as 'Ecotaxe' applying to heavy vehicles when driving on selected national class roads. The system is separate and in addition to, but also interoperable with, the TIS-PL ETC for heavy vehicles, which is installed at French motorway plazas. This system was due to be introduced in 2014 but at the time of writing, its introduction has been postponed by the French government and is not in operation. The system which was fully developed and implemented, used an on-board unit type combining the DSRC charging system used in TIS-PL with a satellite Global Navigation Satellite System (GNSS) receiver-based system for use

[a]TIS: Télépéage Intersociétés.

on roads within the TIS-PL system. The system required on-board units to report position data to a back office system where charge data was calculated. The system was to be offered by service providers known as 'Société Habilité Télépéage (SHT), which would also act as service providers for and providing interoperability with the existing TIS-PL service on the concession-operated motorway network.

7.2.2 *Italy*

Italy can claim to have introduced the first section of modern tolled motorway in the world in 1925 (Milan – Varese).

Another first was the introduction of the first ETC system on a motorway in 1990. This system, known as Telepass, now provides an interoperable electronic payment system across all tolled motorways in Italy. Italy has 25 concession motorway operators, and a single interoperability agreement exists between them ensuring Telepass is accepted across the whole national motorway network of approximately 6 000 km. Telepass has over 9 million subscribers and accounts for some 60% of total toll payments.[4]

Telepass uses a slightly different standard for its DSRC on-board unit to roadside infrastructure communications, using an European Telecommunications Standards Institute (ETSI) standard instead of the more recent CEN TC 278 derived standard. On-board units used in the Telepass system, therefore have to comply with ETSI ES 200674-1 and its related Technical Reports. The European Decision 2009/750/EC on the definition of the European electronic toll service and its technical elements requires that on-board units for a European-wide service must support both the ETSI standard and the CEN TC 278 standard. When the decision was enacted, Italy was allowed to keep its roadside infrastructure in line with ES 200674-1 due to the investment already made by its toll chargers.[b]

7.2.3 *Spain*

Spain also has an interoperable ETC system (via-T) which is accepted by 30 concession companies across a network of over 3 000 km of motorways.[5]

The system has been in operation since March 2003 and uses 5.8 GHz DSRC according to CEN standards and the 'PISTA' transaction model. It can be used by all vehicle classes at either dedicated ETC lanes or mixed ETC/manual lanes.

A unique feature of the via-T system is that there are over 150 'service providers' from which users can choose to obtain an on-board unit and tolling account. This is because the system uses the retail banking network as service providers who provide users with accounts and also issue approved on-board units directly from the manufacturers. This arrangement also enables the existing bank clearing system to be utilised for clearing transactions between the acquiring bank and the service providers. Banks use their existing customer relationship channels for transaction statements, invoices and payments. Banks define their own commercial conditions for users such as the price for acquiring an on-board unit, annual fees, etc. The same on-board unit can be used in different vehicles.

[b]See European Commission Decision of 6 October 2009 on the definition of the European electronic toll service and its technical elements (2009/750/EC).

In addition, transport companies are able to choose from five non-financial issuers which are already active in the commercial haulier fleet market. These organisations have direct relationships with each concession company and invoice users directly.

Since launch, over 1 million via-T on-board units have been distributed. There are nearly 1 800 equipped lanes, 15.6 million via-T transactions occur each month accounting for 34% of the total toll transactions.

There are several examples of interoperability between the via-T system in Spain and with other countries:

- between the Spanish operator Bidegi of the A8 motorway and ASF the operator of the A63 and A64 in France,
- a two-contract on-board unit accepted in France's TIS-PL service issued by three Spanish service providers,
- a pilot service offering the user full interoperability between the French TIS-PL service and motorways operated by Abertis,
- with the ViaVerde service in Portugal.

7.3 Free-flow charging and city congestion charges

A further innovation was the introduction of free-flow systems which could be used in locations where it was not practical or economical to install toll plazas. These free-flow systems are able to read both DSRC on-board units and vehicle number plates. The larger the proportion of vehicles fitted with on-board units, the smaller the number of vehicles for which vehicle licence plates have to be read and correlated to a toll payer's details. In some systems, such as those used for city congestion charges in London and Stockholm, vehicle licence plates are the only method of vehicle identification. These systems rely on pre-registration of a large proportion of users to enable a white list of vehicle licence numbers to be generated, plus a close link with an accurate vehicle registration database in order to identify non-registered users after the toll payment period has expired. Again, Norway was the pioneer, introducing (single lane) electronic free-flow toll collection in Trondheim and Oslo in 1989.[6] Multilane free-flow (MLFF) systems were developed in the mid-1990s which could be used without the need for toll plazas with 'channelised' traffic.

7.3.1 Norway

In 2002, Norway created a national system of free-flow ETC called Autopass. The system combined ETC with video tolling. Users without electronic on-board units are charged by means of pre-registering their vehicle licence plate numbers with the system.

Norway's Autopass ETC system is now used to pay tolls on the roads of 53 tolled road and tunnel operating companies in Norway. The Autopass system is free-flow and all vehicles equipped with an Autopass on-board unit pass without stopping through toll plazas and free-flow gantries.

7.3.2 Ireland

Ireland's motorways have been constructed through public-private partnerships, where the government utilises private sources of funding to reduce public sector borrowing requirements. Sections of motorway are operated under concession by private contractors, which use the toll revenue used to secure financing for construction and maintenance of the road. A contractual obligation was also placed on the road operators to introduce ETC, to issue tags and to accept tags of other issuers. To facilitate interoperability, the National Road Administration established an Information Exchange Agent with the role of clearing all transaction information between operators. This resulted in a nationally interoperable system of electronic toll payment by 2007.

In 2007, the National Road Administration removed a toll plaza on the M50 Dublin Ring in order to alleviate severe queues and replaced it with a free-flow system of ETC at this location (Plates 7.4, 7.5 and 7.6).

Since the introduction of the M50 free-flow system in 2008, the number of drivers using tag accounts has increased dramatically. As part of establishing the M50 free-flow system, independent service providers were introduced in the market. The majority of tags in Ireland are now issued by three service providers, and the number of tags issued by individual road operators is now quite low.

The M50 free-flow system uses a combination of DSRC tags and number plate recognition to charge the toll. Users are able to choose either a tag account from a service provider, or registration for a video charging account. Users who do not

Plate 7.4 The M50 free-flow electronic toll charging station
[Source: Photo provided by SANEF]

Plate 7.5 The M50 free-flow electronic toll charging station
[Source: Photo provided by SANEF]

Plate 7.6 DSRC tag readers installed in M50 free-flow system
[Source: Photo provided by SANEF]

register for either a tag account or a video account are allowed to settle the toll the following day. An account is automatically opened for users not registering in order to reduce the number of unregistered users in the system.

7.3.3 Portugal

At the time of writing there are 21 separate tolled motorway and bridge concessions in Portugal. A single interoperable ETC service provider – ViaVerde – exists and is operated on behalf of all of the concession companies. Portugal was the first country in the world to introduce ETC throughout its motorway network. The Portuguese ViaVerde service has one of the highest rates of ETC usage with over 3.5 million subscribers. The ViaVerde service is now accepted at over 300 toll plazas throughout Portugal, as well as 108 petrol stations, 113 car parks and 11 MacDonald's McDrive outlets (see ViaVerde figures at www.viaverde.pt). Portuguese motorway operators have recently pioneered the widespread introduction of free-flow ETC for light vehicles on motorways.

7.3.4 England

In November 2014, electronic free-flow tolling was introduced at the Dartford River Crossing near London, England. The Dartford River Crossing comprises two two-lane tunnels and the four-lane QE2 bridge. It links the M25 motorway north and south of the River Thames and is one of the busiest sections of the UK's road network handling over 135,000 vehicles per day (Plates 7.7 and 7.8).

7.4 National schemes for electronic tolls for heavy vehicles

Once the operation of ETC was successfully demonstrated, it was not long before other applications for the technology were developed. In particular, in Europe, the potential was spotted for ETC systems to be introduced in locations where there were previously no toll booths or toll plazas. In this way, toll collection would be made possible for reasons other than financing motorways, tunnels or bridges. In particular, as a method of linking road charges more widely to vehicle use and environmental performance, particularly for heavy vehicles which impose far greater maintenance costs than light vehicles.

Switzerland was the first country to introduce a nationwide electronic charging system applied to heavy goods vehicles.

Six EU Member States have since introduced free-flow ETC systems for heavy vehicles:

- Austria
- Germany
- Czech Republic
- Slovakia
- Poland
- Hungary

Plates 7.7 and 7.8 Free-flow electronic toll gantries at Dartford River Crossing near London, England
[Source: Photos provided by SANEF]

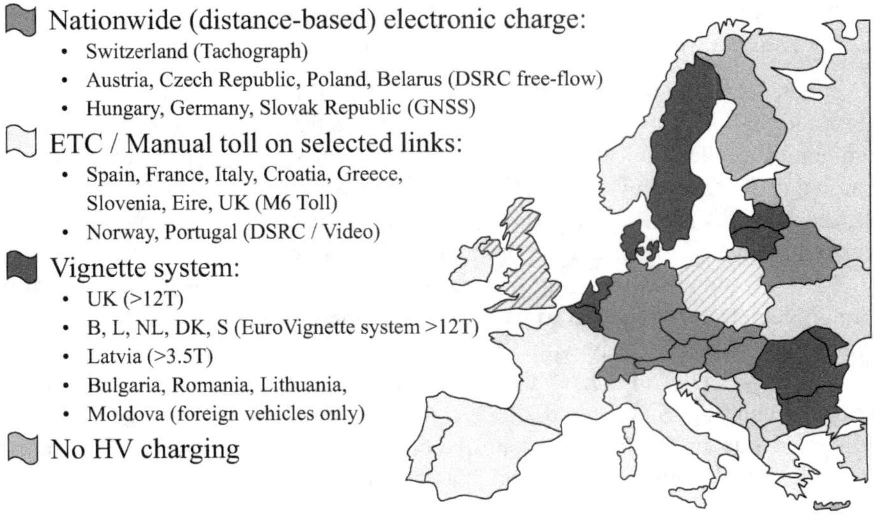

Figure 7.1 How heavy vehicles are charged in Europe

In addition, Portugal and Norway now require payment of tolls for heavy vehicles electronically, and France, Spain, Italy and Ireland also allow (but do not require) motorway tolls to be paid electronically (Figure 7.1).

7.4.1 Switzerland

Switzerland operates three types of road charge:

(i) a vignette for vehicles up to 3.5 tonnes applicable on most national highways,
(ii) a 'performance-related' charge (LSVA) for Heavy Goods Vehicles (HGVs) over 3.5 tonnes used to transport goods on the Swiss road network,
(iii) a lump-sum heavy vehicle charge (PSVA) for buses, coaches and certain other classes of heavy vehicle.

The vignette for light vehicles has been in operation since 1985 and is applicable to 1 801 km of roads. Revenue from the vignette is reserved for expenditure on national roads.

The distance-related charge for heavy vehicles was approved by referendum and replaced the previous flat rate fee in 2001. The charge operates on all roads in an effort to encourage goods transport by rail.

The charge is calculated based on the distance travelled, the emission rating (Euro class) and the weight of the vehicle. The weight of the vehicle corresponds to the lower of:

- the maximum permissible laden weight of the towing vehicle plus maximum permissible laden weight of the trailer or, for an articulated vehicle, the unladen weight of the towing vehicle plus maximum permissible laden weight of the semi-trailer; or

- the maximum permissible laden weight of the vehicle train; or
- the national weight limit in Switzerland (40 tonnes).[7]

All goods vehicles above 3.5 tonnes registered in Switzerland are fitted with an electronic on-board unit. The on-board unit records the distance travelled which is communicated electronically via DSRC to a central system which calculates the charge due. Distance travelled is measured from an electronic odometer sensor and verified by GPS data. Detachment of a trailer is recorded by a sensor and also declared manually. On entry, the on-board unit will declare the vehicle's mileage by DSRC communication. For non-equipped vehicles, this is done manually. Foreign vehicles are required to pay on exiting the country, and Swiss registered vehicles are required to pay within 60 days.

Foreign vehicles also have the option to pay using a manual system using a card. On entering Switzerland, the vehicle will declare its mileage then pay the charge due on leaving the country. Each set of journeys between entering Switzerland and leaving Switzerland is assessed separately.

Enforcement of the charge is achieved by a combination of automatic checks and manual checks at the roadside. The on-board unit also contains a number of features to protect against fraud. For example, opening or disconnection of the on-board unit is logged, an internal power supply allows operation whilst disconnected and the status of the on-board unit is visible outside the vehicle.

Odometer and GPS data are compared and vehicle licence plates are recorded at free-flow gantries.

7.4.2 Austria

In Austria, a government-owned group of companies ASFiNAG is responsible for the construction, maintenance and operation of all motorways and express roads including tunnels. ASFiNAG is fully financed by road tolls. In Austria, light vehicles are obliged to pay for a 'Toll Sticker', and pay special tolls for tunnels. Heavy vehicles are subject to a national distance-based charge 'Go-Maut' which is collected by means of a mandatory electronic on-board unit, the 'Go-Box' (Plate 7.9).

Introduced in 2004, the Go-Maut system uses DSRC communications between the on-board unit and 5.8 GHz microwave beacons mounted on free-flow toll gantries. The toll is collected automatically from the users account when a transaction is registered at the roadside.

The tariff varies according to the number of axles of the vehicle, with different charges being levied for vehicles with either two, three or four or more axles. Since 2010, tariffs also have been differentiated according to vehicle emission class with different tariffs for:

- Classes III or below,
- Classes IV and V,
- Class VI and electric vehicles.

The on-board unit does not require a connection to the vehicle's power supply or GPS antenna, and is mounted using a simple self-adhesive mounting on the inside of the vehicle's windscreen. Drivers are warned of a malfunction of the on-board

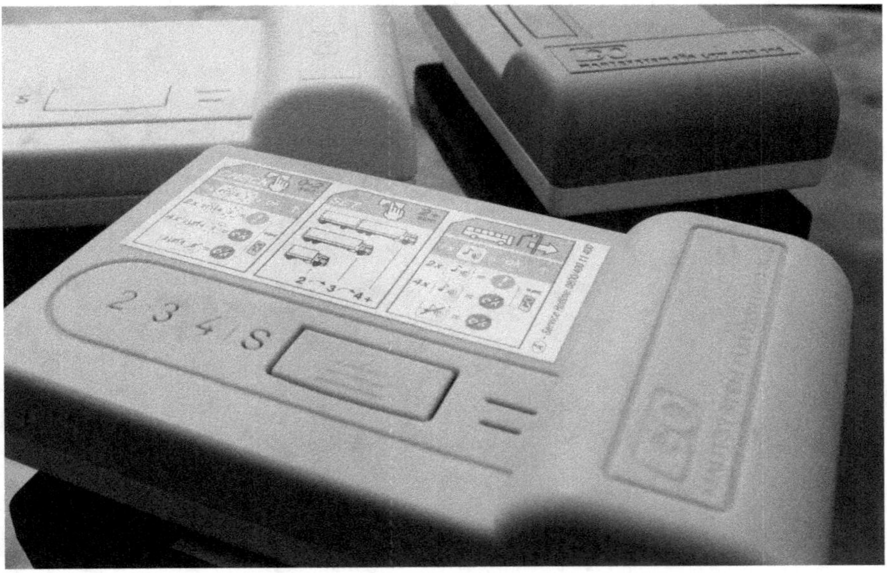

Plate 7.9 Austrian Go-BOX
[Source: Photo provided by ASFiNAG]

unit by a means of a simple, non-distracting, human–machine interface (HMI) which issues four beeps when a malfunction occurs.

Enforcement is by means of mobile patrols with specially equipped vehicles, and stationary enforcement equipment – number plate recognition cameras and laser scanner vehicle classifiers – mounted at some free-flow gantry sites. Vehicles which are found not to have a valid on-board unit correctly fitted and operating are checked by enforcement staff and penalty notices issued (Plate 7.10).

ASFiNAG has led in the area of creating interoperability between systems of neighbouring countries. It has established the Toll2Go service with Germany which allows on-board units from the German LKW-Maut system to be used in Austria. It has also recently established an interoperable service together with the EasyGo group of toll chargers in Scandinavia, allowing users to pay tolls in Denmark, Sweden, Norway and Austria using one contract and one on-board unit.

7.4.3 Germany

In 2005, a new form of free-flow ETC was introduced by the German government for the first time in the world.

The system used what was then an innovative combination of satellite positioning and GSM data communications.

The system, known as the LKW-Maut System, applies to all vehicles over 12 tonnes and levies a charge based on vehicle emission class, number of axles and distance travelled. The GPS capability of the on-board unit allows the road sections used by the vehicle to be determined through map-matching techniques. Once it has

Plate 7.10 Permanent enforcement station
[Source: Photo courtesy ASFiNAG]

determined the road section, the on-board unit then generates the charge for entering the section.

Enforcement is achieved by a combination of mobile patrols and fixed gantries. Data from on-board units can be read by roadside equipment via a special DSRC link known as a CCC link (Compliance Checking Communications).

Unlike other countries, heavy vehicle on-board units are not mandatory in Germany. Drivers have the option of paying the toll using a manual system which requires them to register their journey prior to travelling on the road network. To do this, they go to a payment terminal and obtain a ticket covering the sections of road network on which they intend to travel. If their vehicle is detected on section of road for they have not registered payment, an enforcement penalty will be issued.

The introduction of the LKW-Maut system has delivered measurable positive environmental impacts. For example, the fact that the toll rate is 45% more expensive for the highest polluting vehicles has resulted in a substantial rise in the adoption of less-polluting vehicles. The proportion of trucks running empty has also reduced by 10% since introduction, and the number of containers carried by train has increased by 7%.[8]

7.4.4 Czech Republic

Since January 2007, the Czech Republic has operated a compulsory free-flow ETC system for heavy vehicles.

The ETC system applies to vehicles above 3.5 tonnes, and charges are calculated based on distance travelled. The system uses a MLFF DSRC technology similar to the Austrian Go-Maut system already described above.

7.4.5 Slovakia

Slovakia introduced an electronic charging system for heavy vehicles in 2010 applying to all vehicles over 3.5 tonnes.

The system initially applied to a road network of 600 km of motorways and 1 800 km of first-class roads, but has since been expanded to include all classified roads in Slovakia adding approximately another 15 000 km. Vehicles above 12 tonnes in weight are not permitted to use class 2 or class 3 roads.

The system uses a mandatory GNSS-based on-board unit although different in specification to the German unit. Charges are levied when vehicles pass 'virtual gantry' locations on the road network (Plate 7.11).

The on-board unit required in Slovakia does not require a permanent connection to the vehicle's electrical system and is installed on the vehicle's windscreen in a similar way to a DSRC on-board unit. The on-board unit recognises tolled sections of road and beeps to confirm to the user when a 'virtual' gantry has been passed.

Plate 7.11 GNSS Onboard unit used in Slovakia
[Source: Photo provided by Siemens AG]

The unit communicates with the toll charger's back office via an on-board GSM modem.

It also includes a microwave DSRC link for enforcement purposes, and this link can be used at charging gantries in other countries (subject to approval by the relevant toll chargers).

7.4.6 Poland

In July 2011, Poland joined the group of countries equipped with a free-flow electronic toll system (Plate 7.12).

The system uses the same DSRC technology as is used in Austria and Czech Republic. However, in Poland, tolls on motorways are collected at toll plazas in addition to the free-flow ETC mode used when travelling on express roads and national roads (in cases where charges are levied on these roads). In the Polish system, heavy vehicles use the DSRC on-board unit to raise the toll barrier at motorway plazas on motorways operated by the national highway authority GDDKiA. This means that from the users' point of view, the Polish ETC system effectively appears as an integrated version of the Austrian and French ETC systems.

The heavy vehicle charge in Poland is being applied to wider part of the road network as roads are upgraded to motorway or expressway standard. The charge

Plate 7.12 A Polish free-flow DSRC charging gantry
[Source: Photo provided by Kapsch Telematic Services, Poland]

also applies on some other roads considered to be potential diversion routes. The roads to which the charge applies at the time of writing are listed on the following webpage: www.viatoll.pl.

On-board units are mandatory for all vehicles with a total permissible gross vehicle weight of over 3.5 tonnes including trailer when towing. This means that vehicles below 3.5 tonnes may be charged when towing. Tariff classes vary according to vehicle emission class and weight (Plates 7.13 and 7.14).

The on-board unit operates in a similar way to the Austrian and Czech systems, although there are small differences in the use of the HMI due to the differences in vehicle classes between the countries. At the current time, work is ongoing to develop a method of encoding the vehicle class into the on-board unit so that it is correctly recognised in all countries. This method will support a regional European ETC service known as REETS. It is intended that REETS will be offered by service provider companies which are independent from the national toll chargers. This will allow users to open a 'roaming' account and receive one toll charge invoice for travelling throughout a defined region including several different countries. Ultimately, the European Commission expects such a service to be available throughout the European Union so that users do not have to set up individual accounts with every toll charger when travelling from country to country. This service is known as EETS.

Plate 7.13 A Polish viaBOX on-board unit
[Source: Photo provided by Kapsh Telematic Services, Poland]

Plate 7.14 A permanent enforcement gantry in Poland
[Source: Photo provided by Kapsch Telematics Services, Poland]

The introduction of the viaTOLL system has produced many benefits for Poland. An older paper-based vignette system has been replaced by a fairer distance-based electronic charge for road use. The initial cost of the system was recovered in the first 15 months of operation.

A wider benefit has been through the acceleration of vehicle fleets upgrading to newer less-polluting vehicles. In Poland, the toll rate for the most polluting classes of vehicle is double that of cleaner vehicles. Truck manufacturers have used this information within their own marketing campaigns.

An ETC service known as viaAUTO is also provided by GDDKiA. The viaAUTO on-board unit is currently accepted at motorway toll plazas on state-operated motorways but not on concession-operated motorways.

7.4.7 Hungary

A distance-based ETC system was introduced in Hungary from July 2013. It covers 6 501 km of roads of different classes and applies to approximately 20% of the overall length of the road network in Hungary.

The charge applies to vehicles with a gross weight above 3.5 tonnes. The tariff structure varies according to the number of axles in a similar way to Austria. The tariff is also differentiated according to Euro emission class of the vehicle with different tariffs applying to:

- Euro I class or below,
- Euro II class,
- Euro III class or higher.

The system in Hungary is based on a system of toll declaration. The toll declarations are produced either by route tickets which are purchased at terminals (in a similar way to the German LKW-Maut occasional user journey ticket system) or by on-board units.

An innovation introduced by Hungary is the use of existing GPS data sources from vehicle fleet management systems. This means that instead of fitting an additional GPS on-board unit, users can opt to allow their vehicle fleet management service provider (if approved) to provide journey data for toll declaration purposes.

Enforcement in Hungary is achieved by a combination of fixed gantries and mobile patrols equipped with licence plate recognition systems, vehicle classification scanner and DSRC beacons.

7.5 Interoperability

Many countries in Europe have successfully established ETC systems which provide an efficient means of collecting both motorway tolls and distance-based charges for heavy vehicles. However, the emergence of interoperable services across Europe has been slower than was anticipated when the interoperability Directive was introduced in 2004.

Efforts have been made over a long period of time to develop and agree necessary technical standards and contractual conditions which would allow commercial European-wide services to emerge in the market. The concept of the European Directive on interoperability is that it should be possible for a road user to travel on any tolled road in Europe using a single on-board unit and pay through a single invoice from a single service provider.

Individual countries like France, Italy, Spain, Portugal, Ireland and Norway have all achieved interoperability within their own borders. As described earlier, there are also examples of limited interoperability where on-board units are accepted by motorway concessions in neighbouring countries.

Currently, there is no single service provider offering a pan-European service. However, there are several examples of technical interoperability between countries. Technical interoperability means that it is possible to use one on-board unit in two or more countries but the user still has to enter into multiple toll payment contracts. In technical interoperability, the service provider does not guarantee to pay the toll on behalf of the user. An example of technical interoperability is the Toll2Go service in which a subscriber in the German LKW-Maut system is able to elect to use his on-board unit in Austria. However, the user will receive a separate invoice from ASFiNAG.

A step beyond technical interoperability has been achieved in Scandinavia through the EasyGo cooperation of toll chargers. In EasyGo, toll chargers in Norway, Denmark and Sweden have agreed that on-board units issued by any toll charger or by an approved service provider will be accepted by all other toll chargers in the region. Toll transactions occurring outside the toll chargers system are cleared through the 'EasyGo' hub system, enabling users to receive a single invoice covering travel on all roads covered by EasyGo.

At the time of writing, the EasyGo system is being expanded to include Austria and potential other toll chargers to form a regional electronic toll service.

7.6 The future

7.6.1 Business drivers

With large numbers of ETC systems in operation since the early 1990s, and with national free-flow systems in operation since the mid-2000s, there is now considerable operational experience within road operators and highway authorities in Europe.

The systems are also 'mission-critical' as they are key to the collection of revenue from tolls which is, in most cases, the primary source of income for the highway operator.

7.6.2 Payment trends

ETC systems have many parallels with other electronic payment systems.

A trend in electronic fare collection systems for public transport (transit) systems is for operators to move customers from bespoke transit payment cards to either multipurpose city cards or to contactless debit and credit cards issued by banks. In London, Transport for London has recently introduced contactless bank card payment on all its buses, tube lines (metro system) and over-ground rail. Doing this allows operators to spread the operating costs of card payments with other vendors.

In the case of ETC systems, we could expect highway operators to follow a similar trend and to become less involved in running their own payment services. With on-board units that are able to be associated with a method of payment such as a user's bank account or debit card, the back office functions of the toll charger can be reduced.

7.6.3 Technology trends

The need for an account balance to be known by the on-board unit is now less important because drivers can be notified of their balance by means of mobile phone data connections. The need to issue pre-paid tags is also likely to reduce as bank accounts and credit/debit cards become more ubiquitous amongst Europe's driving population. Balances can be held centrally rather than on the on-board unit.

The function of the on-board unit becomes more and more focussed on being a secure means of identification of the user and a means of verifying the users entitlement to use the service or to use the toll road.

To achieve interoperability, the ETC on-board unit will, in the future, need to utilise several communications capabilities including DSRC and GSM. The unit will therefore either carry these capabilities itself or be able to communicate with the vehicles' own communication channels and services, for example, through the vehicle's telematics platform or the users' smart phone. However, the security of payment transactions required by toll chargers means the dedicated on-board unit will probably still be around for the foreseeable future.

Notes

1. Directive 2006/38/EC of the European Parliament and of the Council of 17 May 2006 amending Directive 1999/62/EC on the charging of heavy goods vehicles for the use of certain infrastructures.
2. ASFA (Mesqui J.) presentation to European Parliament TRAN committee 4 November 2014.
3. ASFA Key Figures (Chiffres Clés) 2014.
4. AISCAT Figures, REETS meeting Rome 29 April 2014.
5. ASETA (B. de la Fuente) January 2014.
6. Multilane free flow system White Paper (Q-Free 2005-64-009).
7. HVF EETS Domain Statement (Swiss Federal Customs Administration, 2012).
8. The German Satellite System (Steffan Koserski, Toll Collect, IBTTA Dallas 2008).

Bibliography

European Parliament. 'Pricing systems for road freight transport in EU Member States and Switzerland'. 2008

Q-Free 2005-64-009. 'Multilane free flow system White Paper'

RDW/EReg. 'Road pricing in Europe'

Schild R.R., on behalf of Národná diaľničná spoločnosť, a.s. 'Multi-lane tree-flow electronic tolling in the Slovak Republic'

Chapter 8

Business model and solutions for user-centric ITSs

Chi-Sheng Lin[1] and Jia-Chin Lin[2]

Next-generation telematics solutions are driven by the mature and recently employed intelligent transportation systems (ITSs), which are assisted by integration and rapid collaboration with information communication technology (ICT) markets as well as the automotive industry. ITSs are a rising technology and include many aspects. For ITSs, smart vehicles and wireless communications are promising technologies that may improve driving safety, reduce traffic congestion, and support information services in vehicles; further, green life for a blue planet is our ultimate goal. Despite the recent global economic downturn that has negatively affected the automobile industry, active research continues in these areas, and new technical challenges have emerged that demand research and development. Dedicated short-range communication (DSRC) exemplifies these challenges because wireless communication techniques have become significant and relevant for vehicular environments.

The direct link between clean mobility and ITSs includes less gasoline consumption and toxic gas emission. Thus, to maximize ITS efficiency, the population that adopts ITSs for daily transportation should increase as much as possible. To achieve this goal, the ITS design should not be limited by technique; the "user-centric" notion is the most important requirement and should be considered.

8.1 ITS: intelligent transportation system and intelligent total solution

8.1.1 The history of ITSs and vehicle communication as well as its future

A "smart city" is an ultimate goal for the modern world. The notion underlying a smart city includes the following: smart energy, smart water, smart transportation, smart building, and smart government. Smart transportation includes ITSs that play

[1]Institute for Information Industry, Taiwan, Republic of China
[2]National Central University, Taiwan, Republic of China

a fundamental role in all smart cities. Traffic is the first link between groups of people. Modern traffic is derived from more than 100 years of vehicle operation, since the first automobile was manufactured. People have continued to improve vehicle function and capabilities. Further, a flourishing car market yields many traffic problems and has become a global burden. Automobile popularity has complicated the sustainability burden for society and the natural environment, such as through safety threats to pedestrians, roadway congestion, fossil-fuel depletion, and toxic emission, which results in the greenhouse effect and pollutes the environment. We can perceive ITSs as a blood vessel in a smart city and a connector between smart cities. Considering the global socioeconomic importance of transportation, ITSs are envisioned as playing a significant role in the future and will yield safer, more efficient transportation. Enhanced efficiency is expected from better use of our limited capital assets, such as highways and railways.

Vehicle communication methods must improve to construct an ITS scenario. From a macroscopic perspective, vehicular communication simply entails communicating information between a vehicle and either another vehicle or the infrastructure. The old vehicular communication method includes applications, such as voice communication between vehicles and service provider, as well as entertainment, such as a radio broadcast to the public in their vehicles. A new era of vehicular technology is approaching, and it includes drivers, pedestrians, control centers, and law enforcement; this new era is driven by both (pre-competitive) public and private-sector funding. Not only are safety-related applications considered, but non-safety multimedia content providers also compose novel research topics. Key players in the industry, such as automobile companies and government agencies, are investing heavily in advanced research and development for many research projects. Recently, vehicle-based communications and the Vehicular Ad hoc NETworks (VANETs) have attracted research attention and development efforts in the United States, European Union (EU), Japan, and many other nations due to vehicle safety, climate change, and natural resource stewardship concerns. ITSs exhibit lower traffic accident incidence and social cost from transportation congestion; furthermore, these systems should improve economic growth, enhance national productivity, slow natural resource exhaustion, and reduce climate warming. The primary research focus has been on system development and telematics standardization; until recently, system testing was rarely investigated. The Vehicular Information Infrastructure (VII) Program clearly indicates that functionality should be tested and system performance should be assessed simultaneously when a DSRC system is developed to accurately verify the system operating stability and production reliability.

In October 1999, the US Federal Communications Commission (FCC) allocated a band at 5.9 GHz for DSRCs, which the ITS application used [1]. In addition, the European Telecommunications Standards Institute (ETSI) also allocated a band in the 5.9 GHz range for ITSs in the EU in August 2008. The bands were selected at approximately 5 GHz because the spectral environment and propagation characteristics are suitable for vehicular communications (e.g., the waves that propagate in these bands offer high-data-rate communications for long distances (up to 1 000 m) with negligible dependence on the weather). In addition, IEEE 802.11p, which is also referred to as wireless access in vehicular environments (WAVE) [2], is an extension of the IEEE 802.11 standard family. Furthermore, IEEE 802.11p can function in a similar manner as the original IEEE 802.11 standard to support ITS applications that operate in the DSRC band. In Japan, ARIB STD-T109 was formulated for ITS services [3]. This standard operates on the 700 MHz band, and the basic design is similar to IEEE 802.11p.

8.1.2 Related research and conventional technologies

Several globally notable vehicular communication-related projects are also described in References 4 and 5. It is impossible to list all of the conventional technologies. The major ITS research areas are Europe, Japan and United Sates; many research projects have been conducted in these areas. Further, many national-level projects are heavily pursued, such as the VII (2004–2011) Program in the United States, Advanced Safety Vehicle (ASV 1991–), Vehicle Information and Communication System (VICS 1990–), Smartway (2004–) in Japan, and Car2Car Communication Consortium (C2C-CC 2003–) projects in the EU. The conclusions from these projects are in References 2 and 5. Here, we only list certain major projects or projects related to our topic.

A. Communication device design and improvement

Most studies have focused on DSRC. The DSRC band is regulated by the US FCC to improve traveler safety, decrease traffic congestion, reduce air pollution, and aid in conserving vital fossil fuels [1,6]. The DSRC communication modes can be divided into three types: vehicle-to-vehicle (V2V), vehicle-to-infrastructure (V2I) communication, and vehicle-to-vehicle/infrastructure (V2X) communications. The entire DSRC spectrum is equally partitioned into seven channels; each channel includes a 10 MHz bandwidth; the channels include four service channels (SCHs), one control channel (CCH) and two special-use channels. DSRC can exchange data at a high rate between either a mobile or stationary vehicle with an on-board unit (OBU) and stationary road-side units (RSUs) mounted on structures along roadways or through hand-held devices. Therefore, DSRC applications are one- or two-way, short- to medium-range wireless communications; in particular, the applications are often designed for automotive use through protocol settings. In the near future, a large percentage of vehicles will be equipped with devices that include mobile computing, communication, and navigation abilities. The DSRC applications can be simply divided into two main categories: safety-related and non-safety-related applications.

The hardware is primarily in the transceiver structure. Although the transceiver structure follows the vehicular communications standard, the receiver can be improved, such as through channel estimation (CE) and synchronization as well as software-defined radio (SDR) and cognitive radio (CR), especially a relay network. Several problems have arisen from these new techniques that must be solved.

A protocol's concern is security, for which standards are necessary but insufficient. This issue is highly related to user privacy. Many ITS applications may be supplied or managed by the government. In this case, a particular security scheme can be standardized, but the operational rules must be coordinated with the certificate procedure; moreover, a certain legal status must be provided to a certificate authority.

Based on the aforementioned notion of an insufficient conventional standard, potential improvements to or variants of a telematics subsystem design have been investigated in this context. Such modifications often support efficiency enhancement or standard improvement. Examples include different guard interval values to overcome inter-symbol interference (ISI) and inter-carrier interference (ICI) [7]. Midambles have been proposed to resolve problems due to high mobility [8–10]. In a previous work [11], a pseudo-random-postfixed orthogonal frequency-division multiplexing (PRP-OFDM) could be used to manage problems caused by high frequency selectivity. PRP-OFDM is similar to China's digital terrestrial television broadcasting standard, GB 20600-2006, and the proposed technique may integrate into a multiple functional car TV set.

B. Improve communication quality and environment

Channel characterization and field trials have been considered by many researchers, such as in the COST 2100 project [12], which was conducted in Vienna, Austria, by the Vienna University of Technology. Several similar experiment reports have been released in the past 10 years, such as Reference 13; these data are from different transmission bands and scenarios. A paper that extensively considers this topic is given in Reference 14. The authors discuss many channel models that relate to vehicle communication.

Many vehicular communication studies have focused on the DSRC channel; however, due to certain DSRC limits, researchers have searched for an alternative, better performing radio channel to support or replace the DSRC. For example, in DSRC-based inter-vehicle communication, multi-hop transmission is necessary to process safety-related applications that require high reliability and low latency (i.e., 100 ms) within a wide service area (500–1 000 m) and high speed scenario (90 km/h–). However, multi-hop dissemination typically introduces severe contention and collision, which becomes more serious when many messages are generated by multiple sources [15]. Thus, in References 16–18 the researchers attempt to evaluate and model the UHF band for vehicular communication.

TV white space (TVWS) is a hot issue and supports better penetration, wherein the broad-area vehicular communication seems possible. Several studies have evaluated TVWS and used it for certain applications. In References 15 and 19, certain computer simulations and field trials are described.

At the Intelligent Transportation Systems World Congress 2013, Toyota, ITC demonstrated the benefits of TVWS for the Tokyo Big Sight. The scenario includes solving an impending disaster, wherein all telecommunication systems fail and people can use TVWS to connect Wi-Fi to work in infrastructures and to upload short messages to the Internet [20].

Improvements not only include hardware or physical layer (PHY), but the software or system processing flow may also require improvement. Primarily, the improvement entails adding new components. One simple example is a traffic jam.

Traffic jams are often due to a traffic accident or construction on the roadway. Other times, they are simply due to an inefficient route plan. In other words, too many cars operate on the same road simultaneously. Vehicular communication and networking can aid in using these resources better by collecting traffic information and through novel broadcasting means. By avoiding traffic accidents and better route suggestions for drivers, traffic resources can be used efficiently.

C. Developing services

ITS applications have generally been classified into three main categories in terms of functionality: *safety, efficiency, and comfort/value-added* applications; these applications are well-classified in Reference 21. More simply, the applications can be divided into safety and non-safety applications.

The first category includes vehicle safety requirements, for which operating scenario definitions and exploring many suitable system specifications are the key targets. In the final US VSC Project report [22], the DSRC equipment developed for the VII Program was implemented in various scenarios, and appropriate system specifications were investigated under all possible traffic conditions. For the requisite communication coverage and baud rate under different traffic conditions, suitable evaluation methods that measure transmitted energy have been extensively discussed elsewhere [23]. In accordance with the discussion in Reference 24, evaluation methods that involve system reliability can be classified as communication- and application-level reliability, for which packet delivery ratio and tolerance time window considerations depends on relevance.

The ASV project in Japan [25] noted the three major players in an ITS scenario: humans, roads, and vehicles. For the ASV project, the connection between these three players aided in harmonious operation of the system. The ASV project developed new technology or applications to aid drivers, but this project is also important because it created a safety scenario for people either in the car or on the road.

The second category includes non-safety applications, these applications can be separated to efficiency and comfort applications. The non-safety applications do not entail a strict time latency requirement. Most applications in this catalog are V2I scenarios; these applications include point of interest (POI), Electronic toll collection (ETC), route suggestion, and fleet management. Certain applications attempted to combine different communications systems; for example, Papadimitratos *et al.* [4] discussed developing a prototype that combines cellular communication and IEEE 802.11p.

D. Computer simulations or field trials and quality evaluation

Most computer simulation studies propose a new technique to improve conventional vehicular communications techniques or search for new research topics; these studies are classified in the second category. Certain computer simulations evaluate standard performance in a theoretical environment. In Reference 26, computer simulations and field trial experiments were conducted. The computer simulation experiments provide insight into the relationship between bit-error rate (BER) performance and signal-to-noise ratios (SNRs). In addition, the field trial experiments were carefully conducted on a vehicle operating in a well-controlled, highly safe, internationally standardized and accurately calibrated laboratory. A reasonable, practical equipment-testing and product-certification procedure was proposed. Using the systematic procedure, the field trial experiments evaluated packet loss and latency, which are thought to be the most critical parameters for a safety-oriented DSRC system. Based on these experiments, challenging telematics design issues involving ITSs were identified, and these issues will require relevant research attention and development efforts to conduct future research in this area.

The literature may be divided into two main groups. The first group employs current Wi-Fi technology as the wireless transmission application studied. These experiments are referred to as VANET. An access point (AP) is mounted along the road in an urban area, an OBU is equipped with a personal computer (PC) embedded in an automobile to record data transmission conditions, and the automobile is driven along predetermined routes. The main goals of this experiment are to verify network connectivity between the OBU and AP as well as to evaluate the IEEE 802.11a/b/g equipment transmission efficiency in this mobile environment. Often, the speed of the OBU movement is the dominant factor that effects communication effectiveness and efficiency [27]. Moreover, the concepts underlying wireline LAN may also be applied to test the platform for delay, timing jitter, and loss in streaming video conferencing [28]. The aforementioned parameters can be ranked as "good," "acceptable," and "poor." The audio and video transmissions are also characterized as either smooth or rough; thus, the aforementioned assessments are intuitive.

In accordance with the aforementioned literature survey, vehicle safety improvements due to wireless communication may be the most important concern. Most previous studies conducted experiments using actual automobiles on physical/real roads or lanes. Often, these studies analyze either comprehensive measurements or extensive recorded data to facilitate performance assessments and determine implicit problems that may be hidden within the current specifications or standardization processes. Although these comprehensive measurements and the extensive data collected from real automobiles on the roads are most similar to actual working scenarios, several disadvantages are apparent. (1) Experiments in a real-world environment often include too many factors or environmental parameters that are beyond the researchers' control. These factors may be interlinked in a complex manner, key factors may not always be easily recognized, and important parameters are rarely easily extracted for further research either through computer simulations or using statistical analyses. (2) The measurement/testing process is difficult to repeat, and reproducible results are difficult at other locations because measurements are not exhaustive. Incidentally, the experiments are not only time-consuming but also expensive. (3) To conduct the experiments, the researchers must drive at high speeds (e.g., 160 km/h). (4) To the best of our knowledge, previous studies were conducted using IEEE 802.11a WLAN equipment. Systems that rely on the 5.9 GHz DSRC specifications and IEEE 802.11a WLANs are significantly different.

Various scenarios and methods for field testing have been discussed for the second category; in particular, previous studies emphasize scenario classification, investigated platform settings, and operational test analyses [29,30]. The investigations were conducted using real, physical roads or lanes. These tests included certain extended-duration and large-scale field investigations conducted in urban canyons and highways; two cars transmit data to each other to measure the effects due to multipath interference/fades and relative motion. The investigations also considered situations where another car passed between the test cars and non-line-of-sight (NLOS) transmission influenced the received signal. Scenarios involving stops at a traffic light (red light) with zero relative velocity have also been studied. Local coverage, short-duration, and small-scale field trials were conducted to explore the relationship between relative velocity and communication performance. Researchers from the Georgia Institute of Technology, which was actively involved in IEEE 802.11p research, describe small-scale fading channel measurements in Reference 13. Wang *et al.* describe several representative V2V channel models based on the measurements in Reference 14.

8.2 Boundless, seamless, limitless

The spirit of ITS can be summarized in three words: boundless, seamless, and limitless; these three missions can be implemented to construct a ubiquitous ITS scenario. We can ensure that, if this mission is fulfilled, people can use ITS more often.

For clean mobility and ITS, we assume that an ITS can provide an easy and convenient method for daily life, wherein time, energy, and money are saved and a better quality of life is provided.

The title of this section is "Boundless, seamless, limitless" because, for ITS service, the coverage must be boundless so people can connect different types of applications; the transportation service must be seamless so all services feel like one service; and we hope that all services are limitless and can be upgraded to correspond with people's needs.

The user-centric notion has been previously discussed. In Reference 31, a reference model was proposed. Because an ITS will not satisfy each individual's needs, "I" is almost impossible, and we classified user's needs to simplify the question.

8.2.1 ITS scenario and player roles

Typically, researchers begin analyzing the ITS using certain techniques. In several studies, the thinking begins in a similar manner as the scenario figure from the ETSI [32]. Based on the idea and scenario proposed, many research projects were created; these projects may only include a part of the overall scenario, but they incrementally meet the overarching goals.

Certain useful and classic vehicular communication means related to traffic safety and efficiency applications support services such as collision warnings, speed and route management, fleet management, traffic flow optimization, street navigation, and several other similar services. Moreover, certain new and special applications that provide value-added services that include point of interest, travel information, and location-based services are also important. The operating environment for vehicular communication can be classified as V2V, V2I/I2V, V2X, and in-car.

In this scenario, almost all technologies have been developed. Although the performance is not sufficient to solve all needs or problems, the technologies can be classified by the service range. Certain applications require a short range to prevent over charges (i.e., ETC), and certain applications require long-range coverage (i.e., TMC broadcasting). Regardless of the type of application or service, each type can be perceived as aiding an individual's need. An individual can be a man or a specific functional device.

To design useful applications and services, we should understand what users/costumers want. Figure 8.1 shows the classifications for the roles and their needs in the ITS environment. We can classify the user in the ITS environment as a commuter, traveler, and driver (taxi or buses), among others. Based on this classification scheme, we can discern the ITS use cases, from private to public.

Not only are the private to public cases shown in Figure 8.1, this figure also illustrates user, operator, and regulator concerns. Because users may compose the core of the scenario, the operator and regulator are the service provider and supervisor, and they complete the scenario. What are the players' concerns? Users are primarily concerned with convenience, operators are concerned about benefits, and regulators are concerned about order.

The role of the user will change with different scenarios. For example, a user may be a driver when they are driving, and they may park their car and take subway or bus to the destination. Thus, the role and transport method change. The user's role may be a basis for classifying the routing solution.

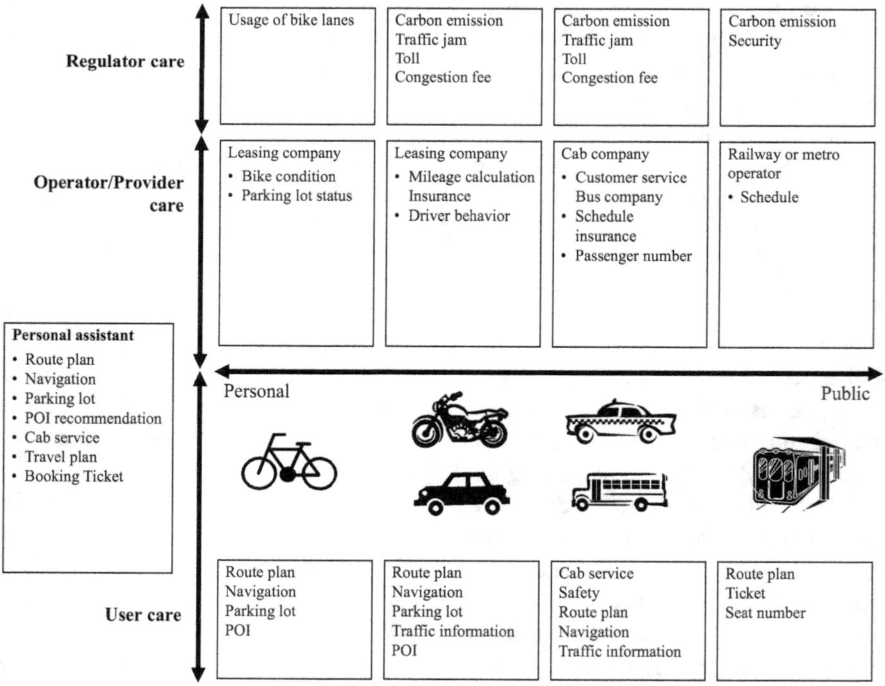

Figure 8.1 User-centric classification for ITS applications

The parameters in Figure 8.1 can be extended. For example, age can be used as a parameter and extend the third axis in Figure 8.1; thus, to consider the passenger on the bus, we may discuss children and senior passengers, among others. If we classify the user in more detail, we may generate a more specific business model for an ITS. However, for simplicity, we will not extent the user-centric classification chart herein. Throughout this chapter, we will only discuss the simple case.

Certain basic coefficients underlie routing solutions, and certain coefficients have not been previously considered. For example, when a service plans a subway transfer, it will not consider "is there any available seat on it?" The number of available seats will be changing with time. This, if the service is implemented, the system must monitor seating on the subway trucks.

We may also consider an emergency scenario, such as an earthquake or typhoon, wherein the government must evacuate many people. The system may provide all users with the same leaving strategy, but if all people use the same strategy, they will cause a traffic jam, which may be a significant consideration for next-generation navigation.

Does each scenario include these three players? Primarily, but an operator can also be a regulator or user. For example, a user may ride his own bike; he/she owns (not rents) the bike and, thus, is the operator. Alternatively, a user may drive a car through an ETC gate; if the ETC gate is operated by the government, the operator and regulator are the same.

8.2.2 Bridging all transportation methods

For analyzing all players' needs in an ITS scenario, the general needs can be concluded using the following terms: "*safer, faster, and easier.*"

Currently, most routing plans only support one or two solutions. However, exceptional routing plans may consider all possible transport means and provide a solution based on the users' needs; where all potential transportation means work together is "bridging."

A basic example is a smartphone application that suggests how a user navigates buses and subways. Certain such applications allow a user to construct conditions that a user prefers, but generally, the application constructs a route based on the shortest distance or shortest time. However, the shortest distance or most time-saving routing plan may not satisfy the user. For example, the shortest distance may require that the user pass a crowded street; or the most time-saving routing plan requires that the user change several subway trains. A suggestion may include the aforementioned three needs, which may not be easy to generate. Normal navigation systems will provide such suggestion, and improvements are the subject of current research efforts. One typical service that bridges different ITS services is a public transportation routing plan. A user can construct his or her own public transportation plan, such as "subway, bus, and rental bike" that combines "subway," "bus," and "rental bike." With successful bridging, users find that they may not need to own a bike or private car.

We realize that, with sufficient data, a suggestion may be more satisfactory to users. However, we occasionally require a revolutionary change to a conventional business style to render it appropriate for an entirely new system. Considering the above example, if the "subway, bus, and rental bike" plan must add the function "reserve a bike," then how is the function added? Alternatively, if the user desires to rent a car, can the rental company place a reserved car near the station, where the user does not need to travel to the rental agency, but they use an encrypted digital key for the car. Currently, rental businesses do not operate in this manner, but to link this operation with a new service, new technology must be developed.

Constructing a *ubiquitous service* is a difficult problem for every operator or regulator. A ubiquitous service typically requires combining two or three ITS services. Thus, the information between these systems must also be integrated and organized. Typically, integration requires multiple technologies. In addition, the infrastructure must also coordinate with the service. If the resources are insufficient, the services may satisfy a few people. Thus, every design may require a redundant plan to ensure the users are satisfied by the service.

8.3 Business model and solutions

An ITS is not a customized system, it must be designed for most users. In many countries, ITSs are constructed by the government, where the players only include the regulator and user. Typically, the service will become conservative and rigid. After an economic depression, certain governments may prefer a private company to construct and operate ITS services. However, whether the government or a private company runs the ITS service, the business model must be complete, and the benefits must be clearly defined.

8.3.1 Conventional ITS architecture

In Reference 33, an architecture was proposed by the Research and Innovative Technology Administration (RITA). This architecture includes three layers:

- The **Institutional Layer**
  This layer includes the institutions, policies, funding mechanisms, and processes required for effective ITS implementation, operation, and maintenance. These roles appear as a combination of regulator and service operators. This layer is a foundation for the entire ITS because all applications operate based on this layer.
- The **Transportation Layer**
  This layer composes the center of the ITS architecture. The communication layer composes the media that supports the transportation layer, and the institutional layer is the operator that manages the transportation layer. The transportation solutions are defined through the subsystems and interfaces as well as the underlying functionality and data definitions required for each transportation service. The transportation solutions are similar to the illustration in Figure 8.1.
- The **Communications Layer**
  The core component of ITS is communication system. We can say that ITS is built on information exchange. This layer provides for the accurate and efficient exchange of information between systems.

If we closely observe the transportation layer, we can group all units in the high-level architecture diagram shown in Reference 33, which depicts travelers, vehicles, fields, and the center. The traveler is characterized as the user, the vehicles are the scenario, the field is the operator, and the center is the regulator. Thus, the RITA architecture is similar to our proposition in Figure 8.1.

Reference 33 includes four large rectangles that represent classes filled with smaller rectangles, which represent subsystems.

- **Travelers**: this group includes remote traveler support and personal information access. The traveler support system is popular for smartphone public transportation system guides.
- **Centers**: this group includes information service provider, traffic management, emissions management, emergency management, transit management, payment administration, fleet and freight management, commercial vehicle administration, archived data management, and maintenance and construction management. Typically, cloud computing systems provide these services. The system must gather, classify, and analyze useful information.
- **Vehicles**: maintenance and construction vehicles, transit vehicles, commercial vehicles, emergency vehicles, and vehicles are listed. Although these vehicles may not include all types of vehicles, they represent the major types.
- **Field**: roadway, security monitoring, toll collection, parking management, and commercial vehicle checks. These fields may not fully express all ITS scenarios, but they include most applications.

These subsystems are interconnected by communication links that are represented by straight lines and extend through one of four items, which represent different types of communications:

- Wide-area wireless (mobile)
- Fixed-point to fixed-point (I2I)
- Vehicle-to-vehicle (V2V)
- Field-to-vehicle (V2I)

A model for RITA architecture is shown in Reference 5. This model was designed for the United States and may not be fully suitable for other countries. Thus, if we design a customized ITS service, we may refer to the model for generating a new model.

8.3.2 Business model generation

In this section, an overall business model is shown, and we discuss how to generate a business model that renders an ITS beneficial. This business model was established based on a report published by several research centers and the Market Intelligence & Consulting Institute.

This business model is not a complete solution, but it is a reference for a system designer to understand what they should consider and users' concerns. Moreover,

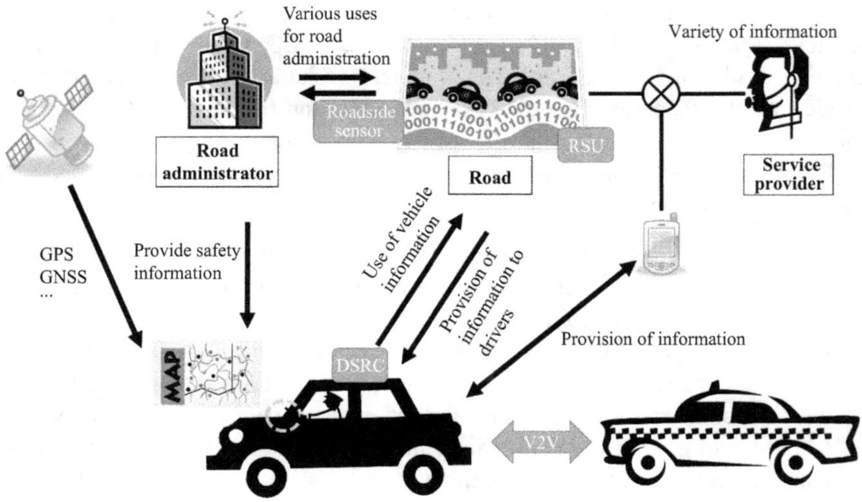

Figure 8.2 The business model proposed in this chapter

it is also a reference method for company survival. In fact, ITSs require local information, and each design must be adapted to local people.

Several business models and ideas relate to a smart city or ITS, such as in References 35–39; these models and ideas are useful for aiding planners.

We will use conventional architecture and the architecture proposed in the previous section to generate the business model.

The objective for the business that we address here is constructing a ubiquitous service. In this service, we wish that the users travel based on their preference and that the user can obtain necessary suggestions. However, *a business model may not always yield new services; it may eliminate a conventional service.* We also eliminate conventional services in our proposed model (Figure 8.2).

The business model adopts the method proposed in "Business model generation" [40] written by Alexander Osterwalder and Yves Pigneur. The model includes certain units, and we will briefly list the important items (it is impossible to list all items here).

First, we should assume an all-powerful ITS service provider. Second, the goal of this provider is to support everyone's needs regardless of whether the requirements are from the regulator, user, or provider.

We understand that an all-powerful service provider does not exist; thus, in the real world, the service provider in this business model may be a "service provider group," which means that two or more service providers may cooperate to realize the model.

In the following text, we discuss nine terms in the business model (Figure 8.3). Each term is defined in Reference 40.

Key partners

Typically, a key partner is also referred to as a key player. In this model, we list three key partners: the regulator, operator, and provider. This model does not include a user because users use the services produced by this model.

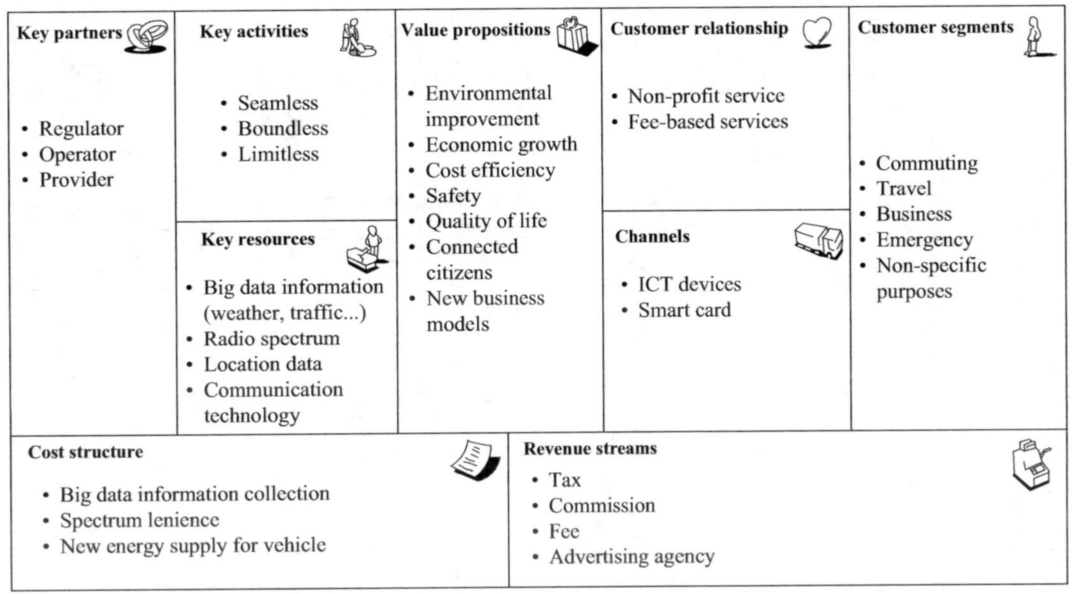

Figure 8.3 The business model proposed in this chapter

The difference between the operator and provider is that the operator manages the service, but the provider supplies the hardware or software that is required by the operator for managing the service or that allows users to enjoy the services.

In certain special cases, ITS users are also key partners. Because the users are a type of "observer" or "moving sensor" in the traffic; if they reply and update the status of a road to the traffic center, the users become one of the partners. However, to avoid mixing the characteristics, if the reply process is not the key activity, the designer should not include the users as a key partner.

Key activities

The key activities for a ubiquitous service are considered under the primary goals: *seamless, boundless, and limitless*. To satisfy this goal, the key activities should include conventional and developing services; certain conventional services may be reconsidered where novel activities are developed in the future.

Key resources

Key resources should be included in the hardware and software; even the operator quality is a concern.

Certain areas include suitable manufacturing abilities, which will lower the cost of the hardware. Further, good cash flow is an important resource.

In certain cases, flexible policies or legislation may aid a government in including private resources for ITS construction.

Value propositions

The main value proposition for a ubiquitous service is "safer, faster, and easier." We believe that this proposition includes basic values for ITS construction. In fact, value propositions can decide that an ITS service is based on pecuniary or social interests.

Consider the failure of the 700 MHz D Block auction in the United Sates, a few years ago; the auction failed to generate a substantial reserve price level, and thus, the subject band was relegated to square one. The government carrier is responsible for generating a dedicated public safety channel at the band [41].

How is the tangible public safety value evaluated? It is impossible to describe a rule for such an evaluation. Thus, the value propositions in this case only consider social interests.

Consider the following question. If the customer is the government, and the government's goal is to decrease CO_2 emissions, what is the value proposition?

First, we should realize that the value proposition underlying a user-centric ITS solution should closely mirror users' feelings. Decreasing CO_2 emissions may be highly beneficial for users (i.e., healthy body); thus, the benefits from decreasing CO_2 emissions may compose the value proposition, but decreasing the CO_2 emissions may only be "Key activities."

Customer relationship

Who is the customer? This question may not be a concern for typical products, but it is typically a concern for an ITS operator because a regulator may also occasionally be a customer. Occasionally, if a company operates a service that supports regulator in better organizing traffic, the customer is the regulator, not the drivers or pedestrians. However, how is good service evaluated? We consider good service based on the end users' feeling, such as the drivers or pedestrians.

Generating a good customer relationship with ITS users may differ from relationships involving normal products. Typically, the services are not discounted, and occasionally the government supplies ITS services; thus, the only way to build a good customer relationship is by providing a useful service that does not bother people.

Channel

We can consider the channel as the media that carries the ITS service. It may be a RFID tag, if the service is an ETC system. It may also be periodic beacon broadcasting.

Occasionally, a channel is also a means to promote or construct service center/station hardware/software and to allow users to begin using a service.

A novel challenge for constructing a channel is security. Occasionally, certain confidential and sensitive data may pass through the channel. The important data-protection issues include stealing or rewriting data.

Customer segments

The customer segments is the most important part of a user-centric ITS business model. The segments can be separated into the commuter, the traveler, the trip taker, an emergency, and special cases. The more each segment's needs are understood, the better a system be designed. Further discussion follows.

- Commuter: routine route. The user may not ask for the shortest path but a stable and risk-free route.
- Traveler: may stay for a short period of time. The user requires the easiest means for transportation.
- Trip taker: includes a long or short trip or business or normal trip. The user may ask for a time-saving transportation means.
- Emergency: includes evacuation. These routes must be safe, time saving, and without risk.
- Special case: these routes may be atypical and be a combination of the previous considerations.

Revenue streams

In many countries, ITS services must be compensated by the government. A good business model can aid a government in saving money and providing opportunities to find private company involved in the ITS ecosystem.

Revenue streams compose the most difficult consideration in an ITS business model; a policy maker and companies who are interested are required to facilitate the abstract model.

We suggest that the government should first prepare funds and operate the revenue streams efficiently.

Cost structure

The cost structure is rarely discussed in previous research because typically technical discussions do not mix with business concerns; however, for business models, these two considerations must be discussed simultaneously.

In most countries, the government constructs or invests in the ITS infrastructure. Occasionally, the operation is outsourced to private companies. However, if the government cannot afford the construction, cooperation with private financial support is necessary.

Typically, people only consider the hardware cost, such as the cost for constructing the infrastructure a cloud computing data center. Occasionally, the cost for developing software is considered, but typically, the training cost is neglected.

An efficient ITS service is necessary for humans to operate the entire system. If the operator can find a flaw in the system, the cost may decrease.

In addition, the government or regulator should also consider the cost to people's feelings. Certain ITS services may be linked to law enforcement, and occasionally, the users feel that the government constantly watches them.

These invisible costs may impact the success of an ITS service.

The model is the foundation, but we must determine the type of service considered.

Typically, we will determine a starting point. In Reference 40, several means for designing product are discussed. However, ITS applications or services are special; primarily, they include public works. Therefore, the ITS infrastructure typically requires government support, but a private company may operate applications or services. These applications or services may relate to public interest or an additional service that is unnecessary but may improve the quality of life. In Reference 39, a public business model grid was proposed. Although this model is designed for a smart city, the idea can be used for ITSs. We can use the case-mapping diagram to modify the business model that we generated and determine whether the entire service is disoriented.

If the ITS service is located in the third quadrant, the service will include limited government involvement and an indirect public value (Reference 39). Thus, we should reconsider the business model and decrease the weight of "Regulator" in "Key partners," and the "Value propositions" will focus on customer satisfaction, not all people that live in the city. Moreover, when a regulator becomes an advisor, not a partner, the "Cost structure" may need to include the regulator's policy.

After the modification, the designer can use a SWOT analysis or other analysis method to determine the feasibility and overcome the difficulties.

8.4 Key technology

In Reference 5, the authors discuss all technologies and use cases around the world. Based on this paper, we can depict a large ITS and smart city scenario. However, the solution may change when the requirements change.

Several projects and studies have been conducted to realize the ITS and smart city. Although the Smartway, the connected vehicle and smart routing are not novel issues, they may be combined with other issues to generate a novel issue or application.

It is too difficult to discuss each ITS application in this chapter; therefore, we refer to Figures 8.1–8.3 for two promising applications that may include the technology requirements listed in Table 8.1.

Table 8.1 The requirements for realizing the technologies

Technologies	Requirement
Connected automated vehicle	1. Communication technology • reduce mobility influence • interference compensation • multi-functional SDR transceiver (DSRC, LTE, TVWS,..., etc.) 2. Safety improvement vehicle (i.e., ASV) • Collision warning or prevent • Incident warning system • High sensitivity sensor • Object distinguish 3. Cloud transportation assistant system information • Data analysis • Data search • Artificial intelligence 4. Security • Encrypt transmission
Wireless communication	1. Wireless spectrum • Wider bandwidth • Better penetration ability 2. Multiple functional communication system
Clean energy and energy arrangement	1. New energy resource 2. Energy management

The two technologies highlighted in this section include a connected automated vehicle with wireless communication as well as clean energy and energy configuration. The technologies we address here include technologies that facilitate the business model or change a conventional ITS scenario and produce an evaluation.

8.4.1 Connected automated vehicle

A connected automated vehicle is a solution for user-centric ITSs because it is easy to construct a user-centric ITS scenario using an automated vehicle, especially a connected automated vehicle. Consider that you need a car one day because you must go to the supermarket or a warehouse and carry many items. What are the options? Do you call a taxi or borrow a car from a friend?

You can select both options. However, if you cannot drive, the remaining option is "call a taxi." We introduce a new option for the scenario wherein you cannot drive, but you know the destination. The new option is a "connected automotive vehicle." You are not required to drive; the car can drive itself.

Generally, automated vehicles are a type of private vehicle. However, for a rental business, renting an automated vehicle may become a type of taxi service.

This is a new type of business model, but as we stated in the previous section, "a business model may not always yield a new service; it may eliminate a conventional service." The final solution for addressing this problem may not include technology; legislation may be required.

If the law and ethics issues can be solved, the connected automated vehicle may be the "last mile" solution for an ITS. In other words, it can connect different transportation methods.

In the business model, we require a small unit to link different transportation systems; for example, people can borrow a bike from a certain area to ride to a subway station.

The connected automated vehicle will also become a useful transportation method. For example, if one must travel to an area without a subway or buses, the traffic center can deploy a connected automated car to enlarge or link the ITS service area; in particular, it can also help senior people. Connected automated vehicles include not only normal gas vehicles but also a micro-mobility/micro-vehicle. Occasionally, it will become another type of taxi or short-range public bicycle.

We observed a business opportunity for long-distance travel in the business model. For long-distance travel, one may only drive or use public transportation. However, certain people prefer to determine their own schedule without following a train schedule. A connected automated vehicle is a solution for this scenario.

A connected automated car may prevent an accident or energy consumption due to a driver's bad habits. Based on pre-programmed behavior, a vehicle may operate more efficiently or safely.

Normal vehicles may be a better routing solution based on time and energy efficiency through adding automated vehicles to ITSs. However, a normal automated vehicle that can only drive itself but not efficient is insufficient. A vehicle must know the details and the latest traffic information to save more energy.

In the following text, we discuss the definition of a connected automated vehicle and the type of connected automated vehicle we require.

First, let us consider the definition of an automated vehicle. The National Highway Traffic Safety Administration (NHTSA), 2013, categorized and defined four levels of automated vehicles [42]:

- Level 0 – No Automation
- Level 1 – Function-Specific Automation
- Level 2 – Combined-Function Automation
- Level 3 – Limited Self-Driving Automation
- Level 4 – Full Self-Driving Automation

Based on these levels, research on automated vehicles advances at various stages. The experience from developing advanced driver-assistance systems (ADAS) can support or accelerate research progress. Thus, this is a good example for demonstrating how a conventional product can become a new application. Although most of the vision-assistant equipment has been upgraded for the controller in the vehicles to read and recognize objects. The sensors must sense the following: (a) speed and acceleration of neighboring vehicles; (b) lanes; (c) traffic signs and speed limits; (d) pavement markings; (e) pedestrians; and (f) obstacles. These characteristics are clearly similar to conventional ADAS.

A major issue for automated vehicles is **decision**. Who makes the final decision while driving? For example, who decides the road used. Referring to the five levels defined by the NHTSA, beyond level 4, a passenger can make the final decision when necessary. However, at level 4, only the car decides. However, at each level, the decision is not limited to the car. If the car connects to the cloud server, it may use the server to aid in better and quicker decisions.

The simplified four steps for automated vehicle decision making are as follows:

- Step 1: Sense the environment – What is in the vicinity?
- Step 2: Decipher the location – Where am I?
- Step 3: Plan the next move – How should I proceed?
- Step 4: Execute the plan.

Step 1 depends on the vehicle sensors (Lidar, radar, IR, passive cameras, intelligent cameras, etc.) or data received from the traffic center. GPS can be used for Step 2. *The vehicle or the passenger can decide in Step 3.* Ultimately, what decisions is the vehicle responsible for? The decision can include the routing plan or reactions during an accident.

"Connected automated vehicle" is defined as a vehicle that combines an "automated vehicle" and a "connected vehicle." A pure connected vehicle is not constructed using automated equipment; thus, the automated level is 0. On the other hand, a pure automated vehicle operates in isolation from other vehicles; it operates using internal sensors. Upon combining these two types of vehicles, a connected automated vehicle can include automated and connected vehicle components [43].

A self-driving vehicle uses information from sensors and certain geo-information from the Internet. With only few vehicles on the street, the bandwidth can be used to easily transceive these data through wireless communication; however, if the number of vehicles increases, the wireless channel is limited to transmissions. Therefore, the next section will discuss the new dedicated radio band requirement.

A damage mitigation brake system, geo-location system (GPS/GNSS,...), lane-keep assist system, adaptive cruise-control system, and brake-control system are included in an ASV. A vehicle uses these technologies to sense the road and prevent dangerous activities. The above technology can operate independently, but if the information can be transferred to neighboring cars, the neighboring cars may generate predictions. Further, if the information is transmitted via a road-side infrastructure, the control center can also oversee all the traffic conditions and even guide drivers to fix problems on their cars.

The connected vehicle not only includes V2V and V2I, but it can also aid each car in obtaining information from surrounding vehicles, and the control center can "communicate" with the vehicle to determine the vehicle's health and suggest a routing plan. Toyota Motor Corporation has processed the field trial for their connected automated driving scenario [44].

Table 8.2 Wireless communication techniques adopted in a connected automated vehicle

Scenario	Application
Large-scale broadcasting	Extension of traffic sign Sudden danger
V2V communication	Location and GPS status Vehicle attribute Vehicle status Position and vehicle intention (on time) Collision avoidance Sensor calibration Extended (i.e., emergency vehicles, high priority vehicles, work status of a road work vehicle, etc.)
V2I communication	Emergency report (fleet, etc.) Vehicle status Position and vehicle intention (not real time) Traffic probe
I2V communication	Routing or speed suggestion Signal extension Emergency alert Local traffic information

In a connected automated vehicle, cloud service is important and can generate a routing plan, avoid traffic accidents, and monitor the moving vehicle. Further, consider a vertical crossing road; the location service must also estimate the vertical position. Otherwise, we can use weather information and emergency alerts, among other sources.

The following paragraphs discuss wireless communication techniques that will be adopted in connected automated vehicles. Table 8.2 shows the scenario, techniques, and data that will be transmitted.

The traffic probe in V2I communications is not used for regular processes. It is only used for improving the system. Considering the probing data may overload the traffic center, the car that will be used in probing the traffic will be random or only selected from cars at a location the traffic center is interested in.

Occasionally, the V2V connection can also be a way to enlarge the area available to the sensors. When a car senses potential danger in its environment, it can immediately warn the trailing cars. Using this path, the message can be directly transmitted without passing through the center; thus, the time consumed may be decreased.

Three different types of administration policies may be used to discuss connected automated vehicles.

- Centralized: all cars are controlled by the traffic center.
- Cooperative: cars decide their route via discussions with other cars. This scenario is most similar to the currently governing driver-centric scenario. However, this will result if computers are used instead of humans.
- Hybrid: A central body only provides suggestions; cars can decide, and then provide feedback on the decision to the center.

Administrative policy is also related to decision issues, which has been previously discussed. A central server can perceive a centralized policy as well as all decisions. A cooperative policy can be perceived as one in which the car makes all decisions. The above three types of policies can be perceived as decisions that are made by the car; thus, they each satisfy the NHTSA level 4 conditions.

Currently, most conventional automated vehicle research uses cooperative rule for operation (or only process one-car field trial). Thus, a vehicle will sense the surrounding area and avoid collision. In addition, because connection capabilities are unavailable, a vehicle cannot avoid traffic jams or construction. Google self-drive uses both sensors and cloud data, and we assume that it avoids heavy traffic.

Thus, after discussing the requirements for an automated vehicle and a connected automated vehicle, clearly, the major transmission between a vehicle to the infrastructure and V2V is sensor data. Classifying the ADAS system in References 43, 45, and 46 and the sensors that are typically constructed in an automated vehicle, the Google Car [47], the sensors are as follows:

1. Range: Laser range finder and Radar
2. Object: Video camera
3. Location: GPS receiver

If we compare the above sensors with sensors used in ADAS, we will find that both are similar. However, in an automated vehicle, the driver becomes a control processor; thus, all images, alerts, and numbers must be converted into a format that the processor can read and judge. Certain studies [7–9] focus on "how to drive better using sensor information."

An addition notion is also important; when humans drive a vehicle, many "feelings" and "intuition" as well as many "instincts" must be converted to a digital driving style. To enhance the reliability of each driving style, a control center is necessary. This control center can organize each vehicle and suggest or command the vehicle to drive harmoniously (without haste or careless lane changes).

To connect vehicles and to connect a vehicle with a control center, the communication techniques should operate at high mobility and transmit at a high data rate with low latency. DSRC and dynamic spectrum access (DSA) have been referenced in certain papers. In certain after-market products, mobile communication (3G, LTE) has also been adopted. However, the bandwidth or practicality of each technique is not solely based on operation. The next subsection details this notion.

8.4.2 Wireless communication

Because telematics manufacturing and the need for safety-related applications have significantly expanded in recent years, automotive electronics are becoming increasingly important. Advanced wireless communication technology as gained sufficient momentum to realize ITSs.

Wireless technology is necessary to connect all vehicles and monitor traffic. However, no specific wireless technology can satisfy all ITS applications.

Currently, certain conventional interactive ITS applications have been established for mobile communication (e.g., 3G technology). Under this scenario, a vehicle requires a built-in cellular module or to transmit data via a driver's cell phone. Currently, this is an appropriate solution, but more data or users will exhaust the overall bandwidth.

In addition, the conventional DSRC band is not satisfactory for V2R. Certain V2R applications must transmit over a long distance and through obstacles. The 5.9 GHz band cannot penetrate buildings or other objects (including other vehicle). Thus, certain researchers have searched for a good low-frequency (e.g., UHF) band to support the DSRC band.

Broadcasting data may not be considerable, but consider the area it must include, wherein LTE may not be appropriate. Whether the DSRC system should be used for broadcasting depends on the area. If the target area is approximately 1 km, DSRC may be appropriate. For larger areas, DSA is a better solution regardless of the ability to penetrate or the coverage. In Reference [10], a vehicular DSA scenario is discussed, wherein vehicles coexist with other wireless networks (802.11p, 802.11h) along the road. In this scenario, semi-auto driving has been tested using dynamic frequency selection (DFS).

We cannot ensure (or it may be impossible to ensure) that the mobile communication base station or DSRC RSU can be seamlessly employed. Under these conditions, the communication device must be designed as a multiple functional transceiver. Based on the connected vehicle, the communication device should include DSRC, Wi-Fi, mobile communication (3G, LTE), GPS, and TVWS. The advantages and disadvantages of several different communication technologies were analyzed in Reference 48.

A low-frequency ITS band is an international issue. Because almost every application must be transmitted via wireless communication, a specific band is necessary to transmit these data. In Table 8.2, V2V and V2I/I2V, but not a broadcasting signal, can be transmitted via a high-frequency band. These data can be transmitted via mobile communication (i.e., LTE), but mobile communications are too busy to allocate sufficient resources to an ITS service. The conventional band for ITS is DSRC, which operates at 5.8 or 5.9 GHz and has garnered over 10 years of research. Several V2V or V2R applications have already been designed using this band. However, DSRC is not sufficient to include all ITS applications. In References 15 and 19, the authors indicate the most important defects: coverage and penetration ability. When the coverage and penetration ability are insufficient to directly connect vehicles, in an inter-vehicle communication scenario, DSRC applications must be transmitted via multi-hop; however, multi-hop communication causes serious contention and collision. One concern is that, in an emergency, the emergency messages generated from the surrounding vehicles will not correctly transmit [15].

An emergency scenario is also necessary. In Reference 19, an emergency communication that can be employed by UHF TVWS was proposed. This scenario addresses one reason that wide-range ITS communication is necessary and is not supported by DSRC. Before we adopt TVWS for communication, we should allocate a dedicated ITS band at UHF.

The DSRC band may remain valuable for ITS, especially for V2V or local V2I applications. However, certain ITS band requirements have not been fulfilled: long range, better penetration performance, and high mobility support. A UHF band can support high penetration and long-distance broadcasting, which render UHF appropriate for constructing an ITS scenario.

Three UHF bands may be good candidates for the ITS band:

- 600 MHz:
    - In many countries, this band is used for TV broadcasting (analog/digital); thus, it is difficult to allocate an ITS channel.
- 700 MHz:
    - This band is also referred to as the "digital dividend" band. This band was originally adopted for analog TV broadcasting; while it was originally used for digital TV broadcasting, the government could rearrange the spectrum for other uses. The radio appearances for 700 MHz are suitable, but certain countries already allocate this band for LTE (i.e., APT 700).
- 800 MHz:
    - In certain countries, this band is the second choice. Because a GSM or CDMA system may be upgraded to LTE, such as through the 700 MHz band, the government may have an opportunity to rearrange the spectrum resources.

However, because the radio spectrum resources are limited, allocating a UHF band is difficult, especially bands allocated to LTE. As Figure 8.4 shows, only Japan has allocated a UHF band for ITS. In Japan, a 10 MHz ITS band is allocated at the center gap of FDD-LTE. This is an example and potential solution for arranging the ITS band. The field trial has also be proceeded in Japan [49].

Research on multiple functional communication systems is not novel; this research has been persistent since people have attempted to combine the DSRC module with mobile communication (e.g., 3G) or a Wi-Fi module [50,51]. The research will continue to integrate different communication systems, such as TVWS and UHF/VHF modules.

The research on multiple functional communication systems includes basebands, protocols, and RF. A multiple functional communication system does not mean placing different systems into one box. The inner portions must share resources and cooperate because these standards may have certain similar portions (i.e., all modulation techniques use OFDM), but the transceiver structure may be entirely different. Therefore, this problem does not have a universal solution.

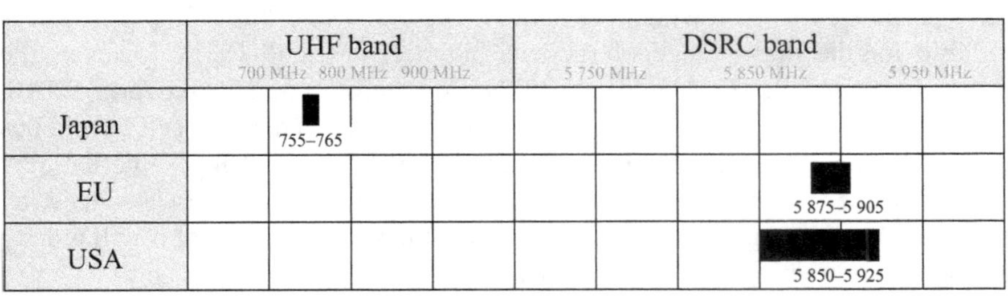

Figure 8.4 The ITS band in EU, Japan, and the USA

8.4.3 Clean energy and energy management

This section does not include energy consumption during ITS vehicle, infrastructure, and equipment manufacture. The only energy-saving topic discussed is the energy consumed during operation.

As mentioned previously, the link between clean mobility and ITS is lower gasoline consumption and toxic gas emission. After we discuss the users' needs, system design rule, and operation strategy, in this section, we conclude on these ideas as well as attempt to determine how much energy we can save. Generally, only people that drive a private car will be interested in how to save energy because saving energy saves money. However, if we consider the public transportation operator's perspective, saving energy also lowers the cost of business operation. When costs decrease, an operator can lower a ticket price or provide better service to the customers. An increase in a mass transportation system use rate can also yield similar results. When the passenger numbers increase, more people share the energy consumed.

Clean energy and energy consumption compose the "Value propositions" for the business model. Drivers can save money through low energy consumption. Further, the government will appreciate the decrease in pollution.

Typically, traffic consumes unnecessary energy due to poor routing plans and traffic jams. A better energy choice and means for use and configuration are discussed next.

Saving energy is one thing; the other is to develop a new energy source. Two novel energy sources are discussed in the following text.

EV

Electric vehicles (EVs) have been developed for many years. Although battery charging remains a major problem for EVs and certain electricity is generated through unclean means, for vehicles, EVs are the cleanest transportation.

For more convenient EVs, the charge problem must be solved. Certain studies have discussed means for more rapidly charging EVs; however, this topic is primarily unrelated to ITSs. Thus, we may discuss how to more easily find vehicle chargers or better consider charging in a routing plan.

Fuel cell

The CEO of Tesla, Elon Musk, commented regarding the usage of fuel cell for car's propulsion as following: "Oh god, a fuel cell is so bullsh*t. Hydrogen is suitable for the upper stage of rockets, but not for cars" [52]. However, the major car manufacturers have designed their own prototypes. Moreover, in 2010, Toyota Motor Corporation and Hino Motors, Ltd. produced a fuel-cell hybrid bus for a commercial bus route between central Tokyo and Tokyo International Airport (Haneda Airport), which is operated by Airport Transport Service Co., Ltd. [53]. In addition, the Japanese government has announced that it will aid the industry in promoting fuel cell vehicles [54].

The fuel cell is a means for using clean energy; however, it may be unstable and should be frequently monitored. In the future, automated fuel cell vehicles may be realized, and traffic centers must monitor all fuel cell vehicles. The monitoring function is also included in ITSs.

ITS technology may aid drivers and operators in saving energy.

Energy management
The basic energy configuration is the most efficient means for using energy. For ITSs, this configuration entails a means for a vehicle to use the least energy from A to B. To realize this goal, the routing plan must be more powerful. Most currently used navigation systems are offline and cannot obtain updated traffic information, such as traffic jams, accidents, or temporary traffic control. Thus, while the planned route is the shortest path, it requires more time on the road. However, energy-optimal routing for vehicles is currently more important. Create novel algorithms that save both energy and time is increasingly necessary; related studies are described in References 55 and 56.

In other cases, charging station needs may require connecting the charging station with power grid; thus, managing the grid must also be considered. In powerUP project [57] and Amsterdam Smart City (ASC) [58] the field trial on vehicle-to-grid idea are addressed and processed.

8.5 Conclusion

The materials in this chapter are based on previous studies. We cannot ensure that the materials in this chapter include the latest and final technologies that will be adopted in the future; several new developments are not included in this chapter.

However, following global research, we note that the basic human need and resources we must protect are usually the same. Thus, we can create a new service by placing a user in the middle of a circle and extending all possibilities outward; the users' needs are the most important consideration.

The business model is not a prediction or service proposal. The business model here is macroscopic and is a solution for the entire ITS. Therefore, similar to every product that has been designed, it must first include a goal that the product can aid or solve. The business model we discussed here is based on constructing a ubiquitous ITS scenario and seamless, boundless, and limitless service.

References

[1] "Federal Communications Commission. FCC 99-305." October 1999

[2] Uzcategui R. and Acosta-Marum G. "WAVE: A tutorial." *IEEE Commun. Mag.*, **47**(5):126–133, 2009

[3] "List of ARIB standards for radio." [online]. Available from http://www.arib.or.jp/english/html/overview/st_ej.html. [Accessed 8 Sep 2014]

[4] Papadimitratos P., La Fortelle A., Evenssen K., Brignolo R., and Cosenza S. "Vehicular communication systems: Enabling technologies, applications, and future outlook on intelligent transportation." *IEEE Commun. Mag.*, **47**(11):84–95, 2009

[5] Karagiannis G., Altintas O., Ekici E., *et al.* "Vehicular networking: a survey and tutorial on requirements, architectures, challenges, standards and solutions." *IEEE Commun. Surv. Tutorials*, **13**(4):584–616, 2011

[6] "FCC Report and Order 06-110: Amendment of the Commissions Rules Regarding Dedicated Short-Range Communication Services in the 5.850-5.925 GHz Band." 2006

[7] Kiokes G., Amditis A., and Uzunoglu N.K. "Simulation-based performance analysis and improvement of orthogonal frequency division multiplexing – 802.11p system for vehicular communications." *IET Intell. Transp. Syst.*, **3**(4):429–436, 2009

[8] Kim S.I., Oh S.H., and Choi H.K. "Mid-amble aided OFDM performance analysis in high mobility vehicular channel." *Proc. 2008 IEEE Intelligent Vehicles Symposium,* 2008, pp. 751–754

[9] Grunheid R., Rohling H., Ran J., Bolinth E., and Kern R. "Robust channel estimation in wireless LANs for mobile environments." *Proc. 2002 IEEE 56th Vehicular Technology Conference Fall (VTC-Fall)*, 2002, vol. 3, pp. 1545–1549

[10] Cho W., Kim S.I., Choi H.K., Oh H.S., and Kwak D.Y. "Performance evaluation of V2V/V2I communications: The effect of midamble insertion." *Proc. 2009 1st International Conference on Wireless Communication, Vehicular Technology, Information Theory and Aerospace Electronic Systems Technology (Wireless VITAE)*, 2009, pp. 793–797

[11] Lin C.-S., Sun C.-K., Lin J.-C., and Chen B.-C. "Performance evaluations of channel estimations in IEEE 802.11p environments." *Proc. 2009 International Conference on Ultra Modern Telecommunications Workshops (ICUMT),* 2009, pp. 1–5

[12] Shivaldova V., Maier G., Smely D., Czink N., Paier A., and Mecklenbrauker C.F. "Performance analysis of vehicle-to-vehicle tunnel measurements at 5.9 GHz." *Proc. 2011 General Assembly and Scientific Symposium (URSI)*, 2011, pp. 1–4

[13] Acosta-Marum G. and Ingram M.A. "Six time and frequency selective empirical channel models for vehicular wireless LANs." *IEEE Veh. Technol. Mag.,* **2**(4):4–11, 2007

[14] Wang C.-X., Cheng X., and Laurenson D. "Vehicle-to-vehicle channel modeling and measurements: recent advances and future challenges." *IEEE Commun. Mag.*, **47**(11):96–103, 2009

[15] Lim J.-H., Kim W., Naito K., and Gerla M. "Interplay between TVWS and DSRC: optimal strategy for QoS of safety message dissemination in VANET." *2013 International Conference on Computing, Networking and Communications (ICNC)*, 2013, pp. 1156–1161

[16] Maeshima O., Cai S., Honda T., and Urayama H. "A roadside-to-vehicle communication system for vehicle safety using dual frequency channels." *2007 IEEE Intelligent Transportation Systems Conference,* 2007, pp. 349–354

[17] Sai S., Niwa E., Mase K., *et al.* "Field evaluation of UHF radio propagation for an ITS safety system in an urban environment." *IEEE Commun. Mag.*, **47**(11):120–127, 2009

[18] Iigusa K., Li K., Sato K., and Harada H. "Proposal and design of Yagi-Uda antenna for UHF-ITS mobile terminal." *2009 IEEE Antennas and Propagation Society International Symposium*, 2009, pp. 1–4

[19] Villardi G., Thadeu Freitas de Abreu G., and Harada H. "TV white space technology: interference in portable cognitive emergency network." *IEEE Veh. Technol. Mag.*, **7**(2):47–53, 2012

[20] Altintas O., Seki K., Kremo H., Matsumoto M., Onishi R., and Tanaka H. "Vehicles as information hubs during disasters: glueing Wi-Fi to TV white space to cellular networks." *IEEE Intell. Transp. Syst. Mag.*, **6**(1):68–71, 2014

[21] Dar K., Bakhouya M., Gaber J., Wack M., and Lorenz P. "Wireless communication technologies for ITS applications." *IEEE Commun. Mag.*, **48**(5): 156–162, 2010

[22] National Highway Traffic Safety Administration (NHTSA). "Vehicle Safety Communications Project Final Report, Appendix G: Field Testing and Evaluation of WAVE/DSRC Communication Functionalities." April 2006

[23] Xu Q., Mak T., Ko J., and Sengupta R. "Vehicle-to-vehicle safety messaging in DSRC." *Proc. 2004 1st ACM Workshop on Vehicular Ad hoc Networks (VANET),* 2004, pp. 19–28

[24] Bai F. and Krishnan H. "Reliability analysis of DSRC wireless communication for vehicle safety applications." *Proc. 2006 IEEE Intelligent Transportation Systems Conference (ITSC)*, 2006, pp. 355–362

[25] ASV (Advanced Safety Vehicle) | automobile comprehensive safety information [online]. Available from http://www.mlit.go.jp/jidosha/anzen/01asv/ [Accessed 16 May 2015]

[26] Lin J.-C., Lin C.-S., Liang C.-N., and Chen B.-C. "Wireless communication performance based on IEEE 802.11p R2V field trials." *IEEE Commun. Mag.*, **50**(5):184–191, 2012

[27] Bychkovsky V., Hull B., Miu A., Balakrishnan H., and Madden S. "A measurement study of vehicular internet access using in situ Wi-Fi networks." *Proc. 2006 12th Annual International Conference on Mobile Computing and Networking (MobiCom)*, 2006, pp. 50–61

[28] Calyam P. and Lee C.-G. "Characterizing voice and video traffic behavior over the internet." *Proc. International Symposium on Computer and Information Sciences (ISCIS)*, 2005, pp. 1–12

[29] Jerbi M. and Senouci S.-M. "Characterizing multi-hop communication in vehicular networks." *Proc. 2008 IEEE Wireless Communications and Networking Conference (WCNC)*, 2008, pp. 3309–3313

[30] Jerbi M., Senouci S.-M., and Al Haj M. "Extensive experimental characterization of communications in vehicular ad hoc networks within different environments." *Proc. 2007 IEEE 65th Vehicular Technology Conference Spring (VTC-Spring)*, 2007, pp. 2590–2594
[31] Arbanowski S., Ballon P., David K., *et al.* "I-centric communications: personalization, ambient awareness, and adaptability for future mobile services." *IEEE Commun. Mag.*, **42**(9):63–69, 2004
[32] ETSI. "ETSI Technical Committee Intelligent Transportation System"
[33] "National ITS Architecture." [online]. Available from http://www.standards.its.dot.gov/LearnAboutStandards/NationalITSArchitecture [Accessed 4 Sep 2014]
[34] "Smartway Project Advisory committee." [online]. Available from http://www.mlit.go.jp/road/ITS/index/indexSmartWay.html [Accessed 4 Sep 2014]
[35] Mulligan C. E. A. and Olsson M. "Architectural implications of smart city business models: an evolutionary perspective." *IEEE Commun. Mag.*, **51**(6): 80–85, 2013
[36] Dimitrakopoulos G., Demestichas P., and Koutra V. "Intelligent management functionality for improving transportation efficiency by means of the car pooling concept." *IEEE Trans. Intell. Transp. Syst.*, **13**(2):424–436, 2012
[37] Walravens N. "Validating a business model framework for smart city services: the case of FixMyStreet." *2013 27th International Conference on Advanced Information Networking and Applications Workshops*, 2013, pp. 1355–1360
[38] Sheng-Chin Yu, Chia-Jen Ting, Hsing-Chuan Lu, and Fong-Ling Fu. "Older-users' acceptance of smartcard payment systems: An investigation of an old-street venders." *2012 8th International Conference on Information Science and Digital Content Technology (ICIDT)*, **3**:551–556, 2012
[39] Walravens N. and Ballon P. "Platform business models for smart cities: from control and value to governance and public value." *IEEE Commun. Mag.*, **51**(6):72–79, 2013
[40] Osterwalder A. and Pigneur Y. *Business model generation: a handbook for visionaries, game changers, and challengers*. 1st edn. John Wiley and Sons; 2010
[41] "700 MHz broadband for mission critical public safety data – a technical discussion." [online]. Available from http://www.pscr.gov/about_pscr/highlights/citig_2012/citig_pscr_broadband_122012.pdf [Accessed 5 Dec 2014]
[42] "U.S. Department of Transportation Releases Policy on automated vehicle development | National Highway Traffic Safety Administration (NHTSA)." [online]. Available from http://www.nhtsa.gov/About+NHTSA/Press+Releases/U.S.+Department+of+Transportation+Releases+Policy+on+Automated+Vehicle+Development [Accessed 4 Sep 2014]
[43] "Automated Vehicle Institute." [online]. Available from http://www.usfav.com/currentAV.html [Accessed 4 Sep 2014]
[44] Automatic driving technology development | Toyota Global News Room [online]. Available from http://newsroom.toyota.co.jp/jp/detail/4228358/ [Accessed 16 May 2015]

[45] "TDA2x SoC for advanced driver assistance systems (ADAS) – TI.com." [online]. Available from http://www.ti.com/ww/en/TDA2x-ADAS/?DCMP=pro-ap-tda2x-131016&HQS=TDA2x&DCMP=summeradv&HQS=ep-pro-aup-summeradv-btwheel-lp-20140610-en [Accessed 5 Sep 2014]
[46] "Advanced Driver Assistance Systems (ADAS) | Freescale." [online]. Available from http://www.freescale.com/webapp/sps/site/overview.jsp?code=APLADASYS [Accessed 5 Sep 2014]
[47] "Graphic: How Google's driverless car works – San Jose Mercury News." [online]. Available from http://www.mercurynews.com/ci_21627954/graphic-how-google-driverless-car-works-robot-jerry-brown-law [Accessed 5 Sep 2014]
[48] Lu N., Cheng N., Zhang N., Shen X., and Mark J.W. "Connected vehicles: solutions and challenges." *IEEE Internet Things J.*, **1**(4):289–299, August 2014
[49] Toyota Motor Corp., introduced the cooperative driving support system which will be operated in 2015 and utilised ITS dedicated frequency | Toyota Global News Room [online]. Available from http://newsroom.toyota.co.jp/jp/detail/4228240/ [Accessed 16 May 2015]
[50] Tsuboi T., Yamada J., Yamauchi N., and Yoshikawa N. "WAVE design for next DSRC applications." *2010 Wireless Telecommunications Symposium (WTS)*, 2010, pp. 1–6
[51] Matsumura T., Sugimoto S., Murakami S., Oda T., and Takahashi M. "ITS information delivery enhanced by combining DSRC and cellular mobile networks." *2012 International Conference on Connected Vehicles and Expo (ICCVE),* 2012, pp. 286–287
[52] "Elon Musk calls hydrogen fuel cell cars 'bullsh*t.' " [online]. Available from http://www.wired.com/2013/10/elon-musk-hydrogen/ [Accessed 4 Sep 2014]
[53] "TMC, Hino to provide fuel-cell bus for Tokyo Airport Routes | Toyota Motor Corporation Global Website." [online]. Available from http://www2.toyota.co.jp/en/news/10/12/1207_1.html [Accessed 4 Sep 2014]
[54] Boost the popularisation of fuel cell vehicles: Hydrogen station construction in earnest [online]. Available from http://www.asahi.com/articles/ASG915-FY2G91ULFA020.html [Accessed 16 May 2015]
[55] Wang Y., Jiang J., and Mu T. "Context-aware and energy-driven route optimization for fully electric vehicles via crowdsourcing." *IEEE Trans. Intell. Transp. Syst.*, **14**(3):1331–1345, 2013
[56] Cela A., Jurik T., Hamouche R., *et al.* "Energy optimal real-time navigation system." *IEEE Intell. Transp. Syst. Mag.*, **6**(3):66–79, 2014
[57] "PowerUp project homepage." [online]. Available from http://www.power-up.org/ [Accessed 4 Dec 2014]
[58] "Amsterdam Smart City – Vehicle2Grid." [online]. Available from http://amsterdamsmartcity.com/projects/detail/id/72/slug/vehicle2grid [Accessed 4 Dec 2014]

Chapter 9
Digitalisation for sustainable sea transports

Mikael Lind[1], Anders Brödje[2], Sandra Haraldson[1], Mikael Hägg[3] and Richard Watson[4]

9.1 Introduction

Traditional industries are reconfiguring themselves in a change towards digitalisation, thus relying on new opportunities for reaching goals beyond the needs of the single organisation [1]. With more and more devices being connected and the augmentation of the 'Internet of Things', business changes and new business opportunities develop. However, in order to function, this new information area involves co-production of information, requiring engaged co-producers to actively share the production of value for the recipients of their information. The purpose of such sharing would be to avoid suboptimisation and fragmented distribution of value to the recipients the information producers collectively serve. An example of this is in multi-modal transportation processes, where this phenomenon builds on various co-producers of information generating value in a coordinated and integrated fashion so as to meet the increasing demands of users within a holistic transport system.

Sea transport is deemed to be the most environmentally sustainable mode for mid- and long-term transport. Yet, as distinct from other transport modes, sea transport needs to be further integrated into the full transport chain. With such an integration into the multi-modal transport chain, sea transport will meet the requirements of seamless integration and integrated performance so as to meet the three pillars of sustainability (cf. e.g. [2]). However, for sea transport, with a legislative regime based upon hundreds of years of trading, where business traditionally was based around the idea that the earlier you came to port, the better the quality of goods you would be able to load (most notably in the tea trade), adopting new principles for information exchange requires changes to that legislative regime. Many of today's contracts (in shipping normally called 'charter parties') between cargo owner and cargo carrier build to a large degree on the traditions of

[1]Viktoria Swedish ICT, Gothenburg, Sweden
[2]Chalmers University of Technology/Swedish Maritime Administration, Gothenburg, Sweden
[3]Chalmers University of Technology, Gothenburg, Sweden
[4]University of Georgia, Georgia, USA

shipping, thus for instance in tanker shipping, vessels gather at anchorage outside a destination port while awaiting a berth for unloading operations. The logic behind this way of operation is based upon the idea that the sooner the cargo carrier is able to arrive at the port for unloading, the sooner the vessel will be able to take a new load. Thus, as the vessel reaches the port of unloading, the ship operator sends what is called a 'notice of readiness' stating that the vessel is ready to unload. From that very minute, the cargo owner is normally bound by the charter party to pay not only for the transport of the goods carried, but also extra (in shipping this is termed 'demurrage') for the time the vessel has to wait outside the port until it can unload in port. Similar examples can be found in other shipping sectors, although in the tanker business this is particularly obvious. Thus, in most of today's charter parties, there is an incentive for the cargo carrier to sail the vessel as rapidly as possible (without causing damage to the cargo) [3].

Further, studies have indicated that vessels often sail longer distances than necessary when sailing from a to b. A study by Markström and Holm [4] has indicated that there would be a possibility of reducing sailed distance for ships by approximately 12%. For verifying the optimisation tool, historic Automated Identification System (AIS) data was used for analysis. Also, the voyage optimisation tool used high resolution chart data as well as ship dimensions, current loading condition and so forth, in order to optimise the route on an individual basis for each vessel. The study used the Kattegatt, a sea region between the North Sea and the Baltic Straits, as the location for a case study, but further analysis has indicated that the results could be extrapolated to apply to the whole of the Baltic Sea Region. The reasons why vessels sail longer distances than necessary vary, although Markström and Holm indicate that tradition plays a large part, as well as the temporary exclusion of some areas from sea traffic.

From a sustainability point of view, the consequence of such modus operandi is that the environment is more polluted than necessary, since the vessel sails at a high speed only to lie idle for hours, days and in extreme cases weeks, waiting for a berth in order to unload. Had the vessel been able to sail at a reduced speed, less fuel (in shipping termed 'bunker fuel') would have been used, thus reducing emissions. Further, for each vessel operating in this way, revenue becomes lower than what should be. Also, a very modest cost benefit analysis using traffic on the Baltic Sea as a case study, focusing on the relationship between sailed distance and reduction of cost for shipowners and society as well as touching upon the possibility of increasing the flow of sea traffic so as to optimise sailing speed, has identified that an average reduction of 1% sailed distance per ship would save approximately €100 million per year. The savings can be split almost equally between society, as a result of reduced emissions, and shipowners, primarily as a result of reduced bunker fuel costs [5]. Considering that sea traffic on the Baltic Sea makes up approximately 10% of European sea traffic, although it transports about 15% of the global goods carried by sea [6], the findings from the cost-benefit analysis indicates the potential savings on a European or even global scale. The above is certainly a call in itself for environmentally friendly principles (noted as 'green') applied to steaming. Yet, in order to fully achieve the benefits of green steaming, information sharing in the whole field of sea transport needs to increase dramatically.

However, achieving better flow, thus increasing the environmental sustainability of sea transport, not only requires the business logic of shipping to change, but also requires other parties not directly involved in this part of the transport chain to serve users in the shipping industry with correct and up-to-the-minute data adding value to all services involved in the transport process. The latter, in particular, requires ports to make more precise information available regarding when a vessel is expected to berth. Yet such precise information is rarely available in ports and this affects the possibility of serving other parties involved in a multi-modal transport chain with precisely correct estimates. This could be achieved by promoting enabling agents in the transport chain to share their information, which to a large degree can be facilitated by more standardised digitalisation. Today, the information run within the shipping industry, yet there are no standards for such data streams, nor is there a central directory for locating such streams and associated documentation. Consequently information sharing and in particular sharing of real-time data is fragmented, resulting in ample opportunity for creating an infrastructure facilitating real-time information exchange between key actors.

> A digital data stream consists of digital elements that describe an event (e.g., a sale, the berthing of a ship), concerning both intentions or the actual occurrence of the event, or the current state of an entity (e.g., the level of humidity in a field, the current mood of a person) that are available real-time. The seven basic elements of a digital data stream are the description of when, where, who, what, how, why and outcome. [7]

The purpose of this chapter is to introduce the reader to the challenges facing the shipping industry in regard to the introduction of digitalisation, and to consider how this change could be a major contributor to the achievement of a higher degree of environmental sustainability. The chapter proposes a data-sharing infrastructure, which will be needed for achieving Sea Traffic Management (STM), aiming at providing seamless and automated information flow for the optimisation of goods through the sea transport stage of the multi-modal transport chain. The work presented is empirically based on the work carried out in the MONALISA 2.0 project.

We will start by giving a general introduction to sea transport, with an emphasis on information exchange in this sector. This will be followed by a theoretical framework, covering value chains in value networks governing integrated performance in multi-organisational business processes and episodic coupling. The latter will serve as a basis for the introduction of the concept of STM, a coordinative approach for just-in-time (JIT) operations in shipping and its interface to port operations. STM has the overall goal of achieving higher levels of safety, environmental sustainability and operational efficiency in sea transport. Using the potential of digitalisation for the purpose of sharing information enables STM. The same section also covers the role of digital data streams for enabling distribution of data throughout the transportation system, contributing to a higher level of integrated performance. The introduction of STM is followed by an introduction of its four sub-concepts: Strategic Voyage Management (SVM), Dynamic Voyage Management (DVM), Flow Management and Collaborative Decision Making (Port CDM). These sub-concepts are in turn facilitated by the concept of SWIM (System Wide Information Management), forming the enabling digital infrastructure necessary for the realisation of STM and its four sub-concepts. This will be followed by some concluding remarks.

9.2 Background

Sea transport is the key link in the global logistics chain, representing approximately 90% of all goods transported on a global basis. Estimates show that there are some 51,000 merchant vessels, and countless smaller vessels, responsible for carrying goods at sea [3]. Within European waters – one servicing one of the world's major consumer markets – approximately 17,000 vessels transit every day, rendering in some 29,000 calls to ports within the territorial waters of EU member states every year [8].

However, accident reports from the European Maritime Safety Administration (EMSA) have shown that there are approximately 500–600 maritime accidents occurring annually within EU waters [8–11]. Apart from results such as potential loss of life and other immediate effects, such as severe long-term problems for marine flora and fauna [12], loss of cargo and loss of or damage to vessels are very costly.

Although depending on trade, most sea voyages start and end in a port. Hence, information regarding various aspects of the port calls will be of the essence for the planning of a sea voyage of the vessel. Likewise, information from the vessel, for instance regarding the goods carried, or amount of bunker fuel on board, is crucial for other parties involved in the whole transport chain, as described in Figure 9.1.

Looking at a different destination of the transport chain, a consumer, when ordering something from the Internet, will normally receive an email confirming the order and possibly later another email confirming the dispatch and dispatch number of the package. From this information the consumer will be able to track when the goods left the Internet dealer's warehouse, when the package was loaded on to another means of transport, etc. As well as this flow of information, a prediction is made of the time of delivery of the package. The latter is normally updated during the course of events, so that the consumer will be able to plan ahead

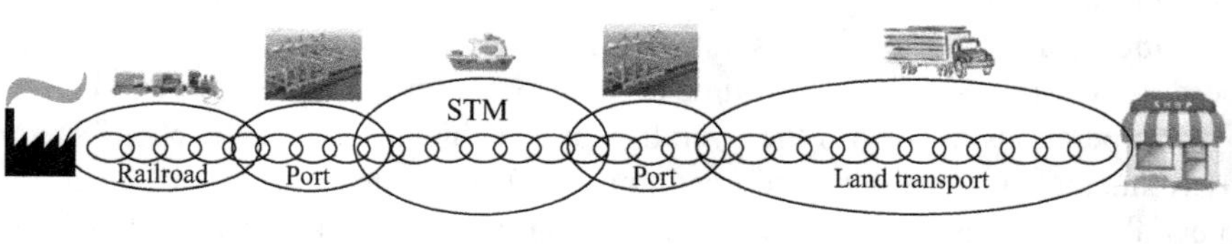

Figure 9.1 The interdependencies of maritime information among involved parties in the logistics chain

and be able to, if necessary, stay at home in order to take delivery of the package. To date, such information is not generally available in maritime transport.

Still, like all industries, shipping is supported by modern technology. Sophisticated equipment is in place on board vessels as well as with shore service providers, producing more and more information. However, much of the information produced cannot be easily transferred between equipment and information resources. So information gathered cannot be used on a wider scale or automated; it often needs some form of manual handling. Also, depending on the parties involved, both information requirements and procedures vary within regions and across the globe. An example of such a technically sophisticated, yet limited system is the AIS.

Over the past ten years, elements of the shipping transportation procedure have been included in the automation of information exchanges, through the adoption of AIS. One of the system's major purposes is to serve as an aid for navigators to increase their situational awareness through the automatic exchange of vessels' identification details (name, call sign, IMO-number [International Maritime Organization] and MMSI-number [Maritime Mobile Service Identity]) as well as certain information on vessels' static and dynamic data. The information is exchanged between ships sailing in the same limited geographic area [13]. Over time, AIS has come to be more integrated into various onboard as well as offboard applications. The information is fused with such information as the ship's position on an electronic chart, allowing the navigator to better understand the current surrounding situation through having data about nearby vessels.

Although a provider of data about the current situation, AIS does not provide information on a vessel's future movement. It does, however, have an embryonic possibility of specifying the intention of a vessel by allowing the navigator to specify the port of destination (which is a requirement of the International Maritime Organization). This information, however, is not sufficiently detailed, since it does not include any information about expected route and time of arrival. In practice, this means that very few users of such information know the intentions of the vessels in their area of interest.

An example of AIS as a shore-based application, is for instance the use of AIS information in Vessel Traffic Services (VTS) [14]. In this case, VTS personnel use the information provided by AIS, often fused with radar information, to monitor the current traffic situation. From this, although not provided by the system information, experienced VTS operators can predict how the traffic situation will look at some time in the future based on pattern recognition [15]. Still, the effects of a single vessel on other vessels' voyages or the flow of goods cannot be optimised by using the information obtainable from a single vessel. Nor can the information available for VTS be used to optimise the flow of sea traffic or goods as a whole, since VTS are only responsible for the traffic situation in limited geographical areas. Also, AIS information is now one of the major sources for retrieving ship positions in order to estimate a ship's possible arrival time at a port. However, even though the information from AIS is widely available on websites such as www.maritimetraffic.com, estimating a time of arrival using the information available from AIS requires considerable manual work.

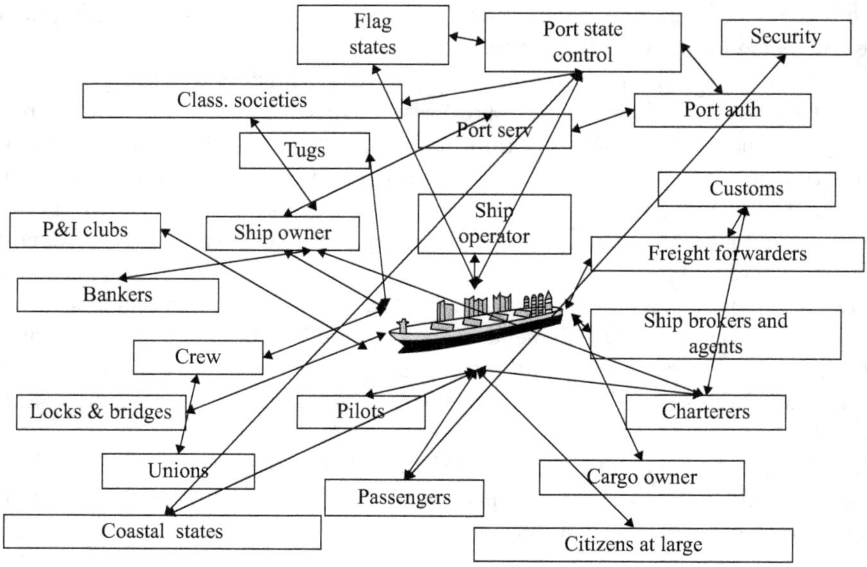

Figure 9.2 The information flow to/from a vessel

Vessel movements generate a flow of information through the various stages of each single movement. For instance, information is often required by Vessel Traffic Services, Custom Authorities, ship operators and cargo owners. Information also flows among vessels moving within the same geographical area, for instance via VHF radio [15]. At the centre of the flow of information is most commonly a ship's master. Managing this flow, which is best visualised in Figure 9.2 [16], is heavy and often time-consuming competing with more critical responsibilities, such as navigational safety.

As described in Figure 9.1, information regarding the voyage of a vessel is produced as well as held by a number of institutions, organisations and individuals. However, as described briefly above, these stakeholders all have different and sometimes diverging drivers for business in the shipping industry [17]. For instance, these stakeholders include linesmen at the quay preparing for the arrival of the ship, the loading master on an oil pier and tug operators. In all cases, multiple stakeholders typically have a need for information derived from other stakeholders involved in the successful and safe execution of a particular sea voyage. Yet, as a result of a lack of coordinated information flow, it becomes the role of a ship's master to facilitate that flow of information between the involved stakeholders. As the ship nears its port of destination, the role as facilitator of the information flow often moves somewhat towards a shore-based ship agent. Commonly, in both cases, most information is retrieved and distributed manually by the use of emails and phone calls, thus leaving ample room for information errors at the various stages of the gathering and spreading of data. Also, at sea, satellite communication is the major information carrier and because of its high cost, it is often used sparingly.

An important item of information needed many stakeholders is the Estimated Time of Arrival (ETA) of a vessel at its destination port. This is crucial information, for instance for maritime pilots or tug operators, as it is needed for planning the ready availability of a pilot or tugs to take the ship to port. Other stakeholders heavily dependent on the ETA of a vessel are the port as such, stevedores, VTS, and the ship and cargo owners. Other types of information may relate to the particular kind of goods on a vessel, which is of crucial interest for safety authorities, customs and the port. However, as explained above, the various stakeholders have different business drivers, thus there is not typically a common driver for the sharing of such information.

Originally introduced in the EU-funded EfficienSea project [18], 2009–2011, various aspects of the concept of route exchange have since been incorporated in three different EU-funded projects: MONALISA (2010–2013) [19], ACCSEAS (2012–2015) [20] and MONALISA 2.0 project (2013–2015). The three projects have shown how supplying vessels with the capability of seeing parts of each other's planned routes gives navigators a more complete understanding of how surrounding vessels are planning their onward voyage. Further, shore-based services are able to retrieve valuable information as well as supply vessels with advice on their routes. Such advice could be in the form of recommendations so as to avoid congested areas, advice on environmentally sensitive areas and Maritime Safety Information.

Thus, to add enhanced value to the logistics chain from maritime transport, information needs to be synchronised, making sure no piece of information is lost or incorrect. Lack of synchronisation will inevitably have a negative effect on the end users of the logistics chain. Thus, in its current form, the information flow of maritime logistics cannot be used to degree required for a broader information-sharing infrastructure. Yet, with the use of ICT solutions designed specifically for the maritime domain, such benefits may be achievable. The concept of STM, as implemented within the MONALISA 2.0 project, will affect the information flow within the maritime part of the logistics chain, allowing for the implementation of, for instance, general route optimisation. STM aims at integrating maritime transport into the whole chain from producer to retailer and early findings from the MONALISA 2.0 project have indicated that STM-based solutions will increase the efficiency of maritime transport, whilst at the same time increasing safety and reducing the environmental effects of shipping [5].

9.3 Theoretical framework

9.3.1 Value creation in multi-organisational business processes

Originally, a business process was conceived as a holistic concept, capturing value-adding activities transforming input to output that should be of value to the customer. This viewpoint is based on the classical definition of business processes given by Hammer and Champy [21, p. 35]: 'a collection of activities that takes one or more input and creates an output that is of value to the customer' (cf. e.g. [22]).

At the core of this traditional view, framed by Goldkuhl and Lind [23] as 'business process as sequential transformation', is a focus on ensuring activities performed for the customer create value. In the literature, value creation is often characterised and structured as value chains or value networks; and there exist contrasting views on value creation.

Within the management literature different ways of framing value creation have been proposed [24]. Initiated by Porter [25] the value chain model was the first step towards portraying the 'chained linkage of activities that exist in the physical world within traditional industries, particularly manufacturing'. Taking this point of departure, the different steps that a sea voyage could be divided into (including port operations) would be one natural conception of the physical transformation that is covered by moving goods and/or people from one location to another.

This metaphor has, however, been challenged by numerous scholars looking at networks and thereby introducing the notion of the value network concept: 'The focal of the value chain is the end product and the chain is designed around the activities required to produce it. The logic being that every company occupies a position in the chain; upstream suppliers provide inputs before passing them downstream to the next link in the chain, the customer. With the value network concept, value is co-created by a combination of players in the network' [24, p. 131]. The process of sharing information becomes an important enabler for such co-creation of value. In contrast to a focus on the role of the single company in a value chain, this shift from value chain to value network puts focus upon the value-creating system itself, in which several actors co-produce value.

In contrast to the value chain, a value network consists of specific roles and value interactions oriented towards the achievement of a particular task or outcome. The notion of relationship is the key in value networks. 'From a network perspective relationships are viewed as part of a larger whole – a network of interdependent relationships [...]. These relationships are "connected" since what happens in one relationship affects [sic] positively and negatively in others' [24, p. 133]. The value network perspective is promising, but a value chain perspective should be rejected, by, for example, asking, 'is it not of interest to focus upon actions performed in business processes'. Allee (2000, p. 439) claims, 'Value network analysis provides an opportunity to overcome the "split" in business management practices where human interactions and relationships reside in one world of models and practices and business processes and transactions reside in another'. A value network is therefore to be seen as 'any purposeful group of people or organizations creating social and economic good through complex dynamic exchanges of tangible and intangible value' (Allee, 2000, p. 429). Allee (2000, p. 439) further claims that 'reorienting toward networks means supporting people in wearing different "hats" and filling roles in multiple value creating networks'.

It would, however, be natural to combine these two, since both have their merits. The value chain perspective provides a structure for changes in the physical world, while the value network ensures essential relationships to meet the expectation of changes in the physical world. Building on both these complementary views of creating value, a multi-organisational perspective conceives value creation structured as value chains in value networks, meaning that value is created in actor relationships in interactions and in the actions performed. A multi-organisational perspective on value creation in business processes argues that all these value components and their interrelations are required to conceive value creation in multi-organisational settings. Multi-organisational business processes build upon the fact that different organisations, by undertaking different actor roles, co-produce value. A multi-organisational perspective thus has to build on a pragmatic conception of business processes, reflecting both transformative and coordinative dimensions of organisational work. The definition of a multi-organisational business process reads as follows. See Reference 27 for further elaboration on the concept of multi-organisational business processes.

One of the main purposes of business process orientation is to conceive structures for actions (cf. [23]). This means viewpoint realistic examination of business processes demands a pragmatic foundation emphasising different types of social actions (material and/or communicative) performed by actors acting on behalf of organisations and the business network. To address the weaknesses of the value network perspective, inasmuch as it remains unclear how value is created through interdependent relationships, an assignment view on business processes is adopted. This means that business processes are conceived as interactions between different roles in the creation of actor, role and action relationships. A multi-organisational perspective on business processes acknowledges business processes as assignments (cf. [28,29]) in which the establishment, fulfilment and evaluation of expectations are put at the core, and thereby constitute structures for actions. Expectations are covered by an assignment in which participating actors come to an agreement on what to realise. The ontological base for a view on business processes as multi-organisational does thus rely on business processes as assignment (cf. [28,29]). An assignment view on business processes has its foundation in the language-action tradition (cf. [30]) in which the viewpoint on business processes could be framed as 'business processes as coordination' (cf. [23]).

A multi-organisational perspective on business processes (cf. [28,29]), as noted, uses assignment structures as a basis for identifying interaction patterns. In action relationships, this means that expectations and commitments for future actions within assignments are created through the performance of actions based on certain role relationships. In multi-organisational business processes (MOBP) value is created through action relationships based on role relationships among several actor roles. Consequently value-adding activities are seen as parts of the establishment, fulfilment and conclusion of assignments. Assignment structures form the basis for coordinating and realising the (co-)production of value. Increased levels of digital connectedness and increased flow of data from assets and actions provide great possibilities for ensuring that the production of value becomes even more coordinated and optimised. In this way the transportation system's capability is expanded, which could enable a higher degree of competiveness, attracting beneficiaries and consequently more service providers.

9.3.2 Episodic coupling

As enumerated above, 'connected' relationships are essential. However, as the shipping industry, and many others, build upon the fact that each actor involved in producing value is an autonomous rational agent, it cannot be guaranteed that these relations remain connected over time. It has been claimed that the shipping industry is a complex adaptive system (cf. Cashiili and Medda, 2012) with a large number of self-organising autonomous agents and with variation in coupling (from loose to episodic tight coupling. We conceive relationships as contexts for expectation of future acts based upon historical interactions. In shipping the captain of the ship is the master and also responsible for the vessel propulsion. This means that different couplings would exist between the different actors over time. Watson and Boudreau [31] have formulated the episodic coupling principle as follows: 'The disparate systems and individuals within society can coordinate their actions through the sharing of information about those episodes when they want to interact, such as catching a bus, going to the cinema and finding a convenient flight. The two entities want to coordinate their actions for an episode or an event.' This means that the distribution of data about intentions, as for example when a certain waypoint is expected to be reached, becomes essential together with status updates, as for example when a waypoint is reached. Watson and Boudreau [31] further claim 'having the ability to learn what other entities are doing across a wide variety of timelines, from seconds to years away, enables people and enterprises in a highly decoupled society to coordinate their actions. Often this coordination is invisible to one of the parties, and there is no reason for it to be visible.' This would then mean a need for integration between the different actors, such as the vessel operator, the port of destination. This would however mean that agreements need to be made upon measures to share data about and that these measures have the same meaning for participating actors.

Watson and Boudreau [31] further claim that 'inter-organizational systems typify episodic coupling. By sharing data, firms can create tighter coupling of their activities. JIT is one of the well-known applications of episodic tight coupling. Firms in a supply chain share their input needs for a specific time so that suppliers can deliver as needed.'

9.4 The concept of STM

STM is a concept proposed for enabling a higher degree of integrated performance within the sea transport ecosystem. Its goal is to increase safety, environmental sustainability and operational efficiency in sea transport. STM relies upon involved actors sharing their short- and long-term intentions (e.g. estimates of when a state is to be reached) as well as information about reached states. STM is realised via four key concepts, (1) *SVM*, (2) *DVM*, (3) *Port CDM* and (4) *Flow Management*, all supported by a fifth concept, (5) a *SWIM* sharing of data in a common information environment and structure (e.g. Maritime Cloud [32], digital data streams [7] and

Open Bridge Platform (OBP) [33]) (see Figure 9.3). The content and development of these technologies have been inspired from the SESAR (Single European Sky ATM Research) programme and Airport CDM, which contribute to greener, safer and more efficient flight operations [34]. These concepts are further discussed in the sections 9.4.3 to 9.4.7.

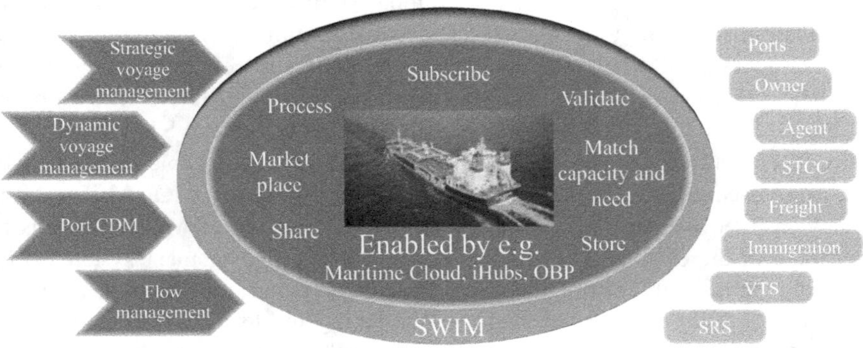

Figure 9.3 Sea traffic management within MONALISA 2.0

STM requires the engagement of many actors. Important enablers are an increased degree of connectivity, increased possibilities of digital collaboration, seamless interoperability between systems and highly distributed coordination (i.e. each actor taking responsibility for its actions) in sea transportation. This presents an opportunity to move away from a traditional approach to traffic management with a central governance unit. STM will involve and engage multiple actors on multiple levels and will require new procedures for information sharing in a distributed manner within each stakeholder's action scope.

Adopting such a modern approach to traffic management, as proposed by STM in MONALISA 2.0, enables and requires each involved actor to be engaged as a traffic management co-producer. Consequently STM will be performed on different actor levels, contributing to the overall performance of the transportation system. The co-production of STM will be designed to enable the involved actors to optimise their operations. Such optimisation, both for the performance of individual actors and for the integrated performance of the transportation system as such, requires stakeholders to share relevant information related to a shared common object of interest [1].

Transportation systems are ecosystems involving different actors performing different tasks based on episodic shared common objects and actions of interest. The common object of interest in STM is efficient, safe and sustainable sea transport. While the various stakeholders share this common interest, they interact episodically. Thus, a pilot might spend only an attempt guiding a ship into a harbour, and will then engage with another ship. The shipping industry operates as a series of episodically tightly coupled events when parties tightly coordinate their resources and then return to operating independently or tightly couple with another party.

The involved actors have to arrive at a consensus regarding the performance targets that govern the performance of the different focus areas. As mentioned previously, three areas of focus are safety, environmental sustainability and operational efficiency. Furthermore, performance targets within one area affect performance targets of other areas. For example, setting the right speed for reaching a destination according to an agreed ETA enables optimal consumption of bunker fuel (i.e. operational efficiency) and thereby minimises the effects on the environment (i.e. environmental sustainability). STM should support the orchestration of the efforts of all actors involved in sea transportation to achieve STM's goals.

Consequently STM, as conceptualised in MONALISA 2.0, explores alternatives to a centralised Sea Traffic Coordination Centre. Even though MONALISA 2.0 is largely inspired by Air Traffic Management, the objective of MONALISA is to contribute to the establishment of a distributed traffic management system rather than rely on a 'EuroControl' for Sea Operations. Such a distributed approach recognises that the culture and history of shipping results in the various parties having a high degree of autonomy, which is anathema to centralised control and command. A distributed data sharing design also gives room for new actors to enter the domain by providing new services building on data made available from the various stakeholders. Hence, STM favours a cooperative and coordinating model of data sharing, which fits the historical modus operandi and culture of shipping.

As inspired by Svallvåg 2013 [35], information sharing for STM in Intermodal Sea Transport could be expanded to include other transport means and thereby cover multimodal transport processes (cf. Figure 9.4). Hence, STM needs to become an integral part of the (distributed) management of the total chain of operations in multimodal transportation processes. The proposed distributed data sharing design can readily accommodate the inclusion of more stakeholders as higher levels of integration are sought.

The STM concept enables and requires that each involved actor is engaged as a traffic management co-producer. Consequently, STM will be performed on different actor levels, each contributing to the overall performance of the transportation system. The co-production of STM will be designed to enable the involved actors to optimise their operations. Such optimisation, both for the individual actor's performance and for the integrated performance of the transportation system

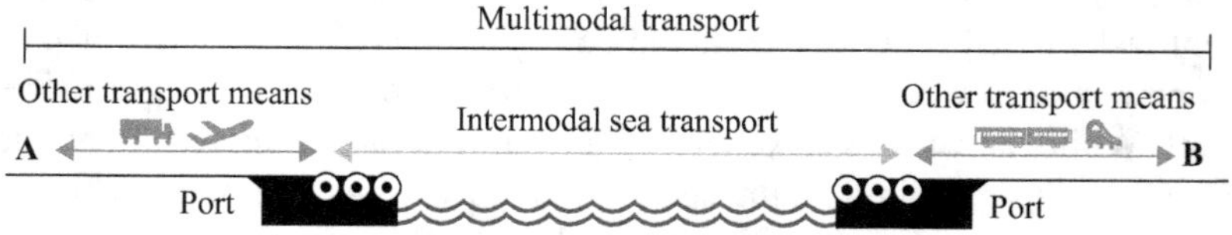

Figure 9.4 Intermodal sea transport as part of multimodal transport

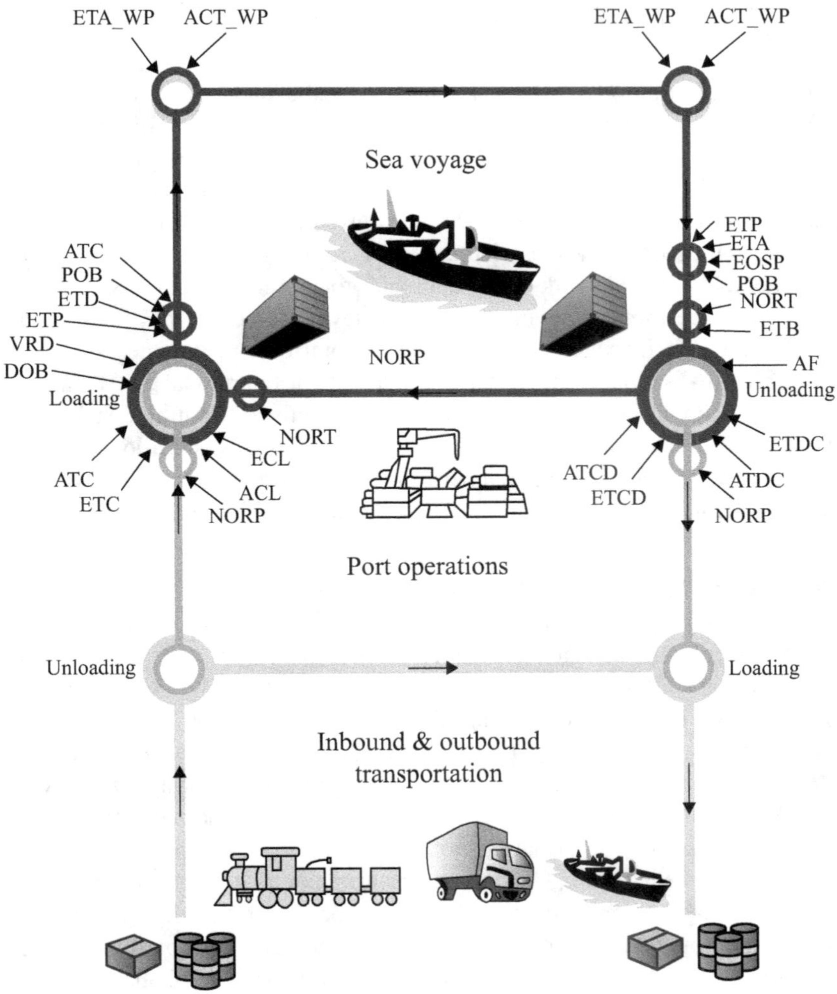

Figure 9.5 External control points enabling efficient 'turn around processes'

as such, requires that involved actors share relevant information related to a shared common object of interest.

Thus, STM will coordinate and orchestrate the efforts of all actors involved in sea transportation, leading to improved safety and punctuality with as low an impact on the environment as possible. It is of paramount importance that the performance targets within and between different areas are well aligned and create synergies in an optimal way. Figure 9.5 shows the relationships between key concepts in STM. It is important to notice that variants of the voyage and adjacent processes are to be determined based on different variables. The proposed control points are to be seen as generic (i.e. applicable for all types of voyages).

9.4.1 The goal of STM – increased collaboration and information sharing

Inspired by the definition of e-Navigation by IMO, the initial STM definition by the Swedish Maritime Administration, and the thoughts of modern traffic management as a distributed phenomenon, the following definition of STM [36] is proposed within MONALISA 2.0: *STM* is a concept encompassing all actors, actions and systems (infrastructure) assisting maritime transport from port to port. STM is a part of the multimodal logistics chain, encompassing sea- as well as shore-based operations. STM is a network-based approach for optimal Intermodal Sea Transport. STM is performed on multiple actor levels, where each engaged actor co-produces traffic management. These actors contribute to the integrated performance of the realisation of the performance targets of Intermodal Sea Transport as the shared common object of interest of the ecosystem constituting Sea Transport. STM puts an emphasis on interoperable and harmonised systems allowing a vessel to operate in a safe and efficient manner from port to port with a minimal impact on the environment. STM secures sea traffic flow and capacity optimisation.

9.4.1.1 Information sharing processes in STM

Various operational (acting) units, key actors, are engaged in sea transportation, where all operations strongly influence the performance of the ecosystem as a whole. Each operational unit is seen as a 'point of interest', collecting several actors acting on behalf of this 'point of interest'. Examples of 'points of interest' related to STM are vessels, ports, authorities and shipowners. These points of interest involve numerous actors that provide and utilise information to perform their tasks. Optimally, each operational unit manages data sharing between different actors by collecting information from providers and enabling information utilisers to access needed data. This means, for example, that ships would collect all relevant information from different sources on-board (e.g. the engine, the bridge), distribute the obtained information to different personnel on board the vessel as well as to other operational units on other vessels or onshore. Collaborative data sharing by all parties is an essential key success factor. An ecosystem, where the performance of an individual party relies on the integrated performance of different entities, requires that information is exchanged between entities efficiently and seamlessly (see Figure 9.6 [36]).

Within MONALISA schemas for information collaboration are being developed with the intention of enabling each unit to act in an optimal way to fulfil its goals and thus the goals of the ecosystem. The distributed nature of STM requires that each operational unit, and its subunits, acts in such way that the performance targets of the ecosystem are reached. Schemas of collaboration require the identification of (1) 'key' operational units, (2) key data to exchange/share (including standards), as well as conditions for the exchange of information.

The voyage order is the backbone of the process of information exchanges within STM (Figure 9.7). A particular voyage is initiated by a voyage order directed to the vessel from the operator (e.g. the shipping company or the

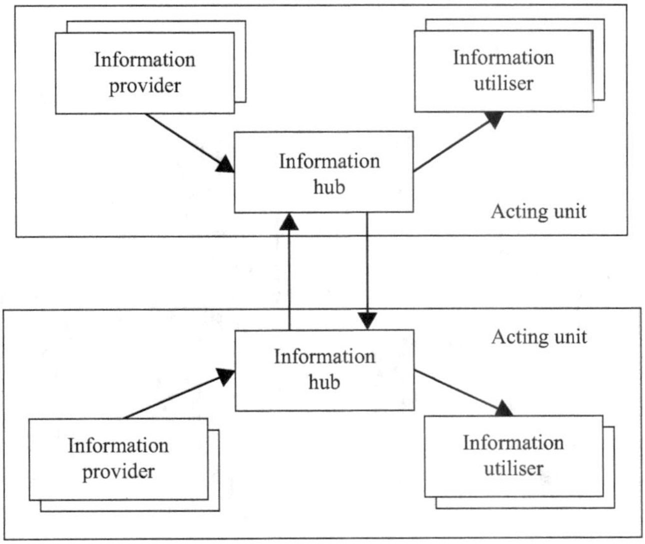

Figure 9.6 Distribution of information within and between acting units

charterer). This order includes a unique voyage number (Voyage-ID), which will be used to identify the voyage, and store information related to its performance, throughout its different steps. Access rights to the data associated with a voyage will be granted to different actors. This is done based on who is involved with the specific voyage (the relevant Sea Traffic Coordination Centre [STCC], VTS, ports etc.). This means, for example, that the relevant STCC will automatically have access to relevant data for vessels with a route passing through its area of operation.

For the vessel operator, the voyage order forms the basis for providing a dynamic voyage plan transferred through a Maritime Cloud (based on information in the ECDIS system). The design idea of STM is that this dynamic voyage plan is published in a maritime cloud (supported by SWIM). Additional information related to the voyage is continuously updated/confirmed through continuous automated position reports, leaving the necessity for noon and arrival reports obsolete. This dynamic voyage plan also includes several ETAs, as an ETA-table, for different waypoints, as well as the ETA for the destination.

In STM this (published) dynamic voyage plan (including the ETA-table) is used as basis for an STCC to propose an optimised route (including time-slot allocation with speed adjustments) for the vessel to take in order to reach its destination at the expected time. Proposed routes from the Dynamic Route Planning Process (managed by an STCC) and other influencing factors are the basis for updating the dynamic voyage plan (potentially this proposed route is also confirmed to an STCC by the vessel). Note that an STCC is one basis for providing optimisation of the voyage plan. The Captain and the shipping company might also use other suppliers of information for voyage optimisation. This means that the decision of changing a published voyage plan will be made by the Captain.

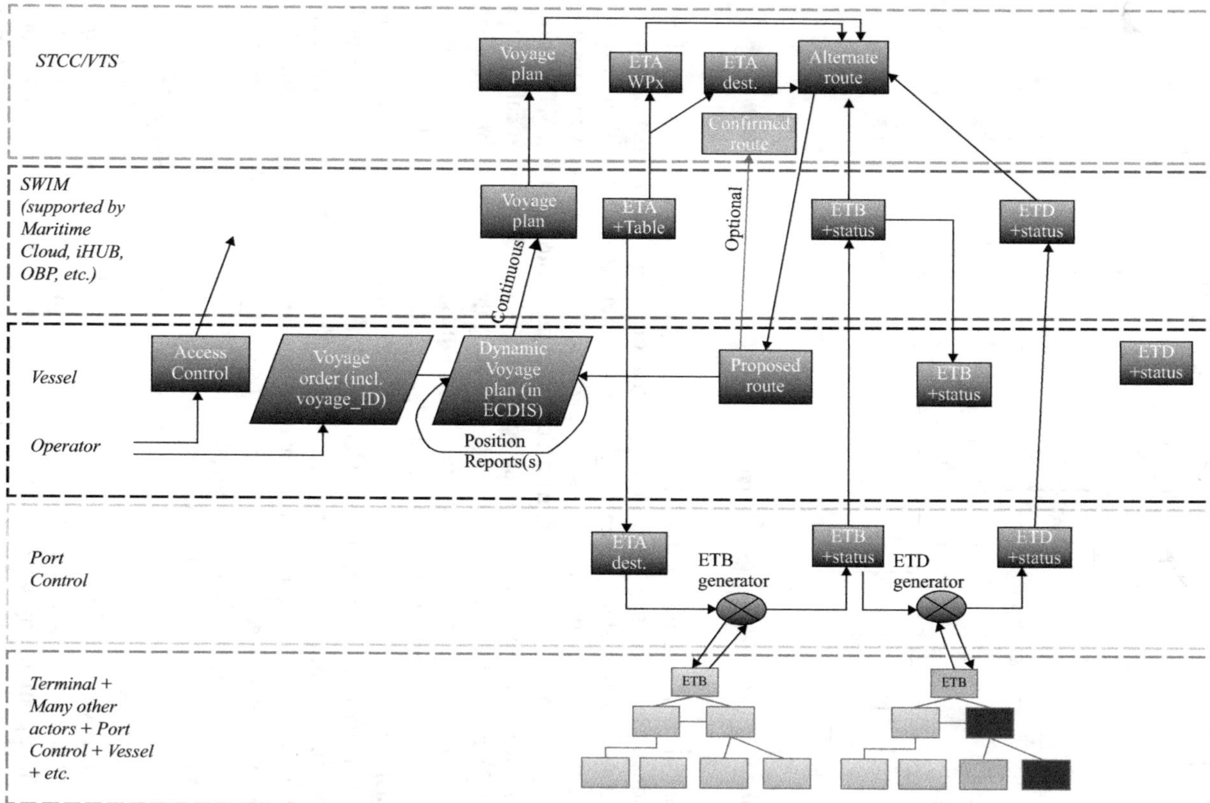

Figure 9.7 *Information sharing processes within STM (cf. Lind et al., 2014)*

The port of destination will subscribe to data that are related to the port. The time when the first ETA notification reaches the port triggers the 'ETB (Estimated Time of Berth (ALL FAST)) generator' to generate an instance of an ETB tree. In parallel, the 'ETD (Estimated Time of Departure) generator' generates an instance of an ETD tree. Actors subscribe to relevant measures enabling them to bid for, agree upon and plan for future operations.

These trees consist of numerous states requiring actions, performed by different actors, to be completed to enable berthing. This tree could/does (partly) form a marketplace for some actors to bid on the realisation of different actions to reach these states (such as towing the vessel to berth). During the sea voyage, we envisage a continuous bidding and agreement process for the different states to become confirmed. This approach would enable a marketplace and potentially encourage some competition among the different actors at ports where resources are under-utilised.

The ETB (and its status) will be continuously published in the maritime cloud as part of SWIM. A vessel will continually be informed of (updates of) the ETB and its status through its subscription to the maritime cloud. The closer in time that the ETB is to the actual time, the higher the precision of the ETB (= +/− × in time would be narrowed) is expected to be. The ETB and the ETD with its statuses also form the basis for the STCC to provide optimal (alternate) routes in its flow management procedures (see below).

A key enabler for STM is the notion of voyage numbers. A voyage, from port to port, can be referenced using a unique voyage number, where the voyage number serves as an information carrier for all parties involved. As soon as any new information with regard to the voyage, goods carried, etc., is known or modified, it is made available for others to use. By sharing information, all parties in the maritime transport chain can improve their operations. Ports will have up-to-date data on all ships and can plan accordingly. The captain will get information regarding port availability and can thus optimise his route. The key is to share information as early as possible, thus making work easier for other involved parties. Use of a voyage number can stretch from the planning of a cruise voyage years in advance, to updates on which tug boat will assist when leaving a berth. Even though information is shared, this does not imply that the information is publicly available. Only authorised parties will be able to access the information and the level of authorisation would vary depending on the specifics of the information.

For the notion of voyage numbers to come into effect there is a need for different external control points based on a combination of maritime/port standards (cf. table below) together with control points necessary for making ships' 'turn around processes' as efficient as possible. Some new measures (ETP, ECL, ETDC and ETCD) related to estimation have been proposed for the purpose of planning to enable successful realisation. Sustainable sea transport processes rely on high levels of accuracy (based on solid estimations) related to berthing, unloading, loading and departure. Solid conditions for the ultimate sea voyage can be established by enabling high levels of accuracy on the arrival, operations at port and the departure. Sea transportation is a multi-organisational business with numerous actors positioning and coordinating their performance in relation to different control points. One example is that the ETD is to be determined given that: (1) the loading of the cargo has been completed (cf. ATC); (2) the ship is ready to sail (cf. SRS (Ship Reporting System)); (3) the documents are on board (cf. DOB). In line with the ambitions of the MONALISA 2.0 project and amongst other factors, STM will be realised by sharing information about the status and values related to these control points for the particular voyage.

Abbreviation	Definition	Source (maritime standards)
ETA	Estimated Time of Arrival (ETA)	Standard abbreviation in shipping (cf. e.g. Baltic Exchange)
NORT	Notice Of Readiness Tendered	cf. e.g. GENCON
ETD	Estimated Time of Departure	Standard abbreviation in shipping (cf. e.g. Baltic Exchange)
ATD	Actual Time of Departure	Vessel sailed according to BIMCO, FONASBA
ETA_WP	Estimated Time of Arrival at Waypoint	Waypoint Route exchange protocol
ATA_WP	Actual Time of Arrival at Waypoint	Route exchange protocol
POB	Pilot On Board	Application of BIMCO, FONASBA in SOF
ETP	Estimated Time for Pilot (on Board)	*New measure* (when the pilot is booked)
NORP	Notice Of Readiness Port	Abbreviation used by port (various ports)
ECL	(Estimated) Time of Commenced Loading	Loading *New measure* (when loading is expected)
ACL	(Actual) Commenced Loading	Loading completed BIMCO, FONASBA
ETDC	Estimated Time of Discharging Commence	*New measure* (when discharging is expected to commence)
ATDC	Actual Time of Discharging Commenced	BIMCO, FONASBA
AF	All Fast (Vessel berthed)	BIMCO, FONASBA
ETC	Estimated Time of Completion	Standard abbreviation in shipping (cf. e.g. Baltic Exchange)
ATC	Actual Time of Completion	Loading completed according to BIMCO, FONASBA
ETCD	Estimated Time of Completed Discharging	*New measure* (when discharging is expected to be completed)
ATCD	Actual Time of Completed Discharging	Discharging completed according to BIMCO, FONASBA
ETB	Estimated Time of Berthing	cf. Baltic Exchange (when the Port advises vessel to be all fast)
DOB	Documents On Board	BIMCO, FONASBA
VRD	Vessel Ready to Depart	*New measure* (when vessel is ready to sail)
EOSP	End Of Sea Passage (Actual ETA)	Application of BIMCO, FONASBA in SOF

The consequences of this information sharing would enable:

1. Planning of operations to enable high berth efficiency (i.e. provide ETD and ETB as soon as the first voyage plan is submitted and published);
2. The status of the planning/agreement process with regard to ETB and ETD to be published continuously;
3. Actors enabling berthing (with the state of 'All Fast' as the end state of berth) and Departure (with the state of 'All Loose' as the end state of Departure) to be committed to established terms and conditions enabled by an open marketplace;
4. Green steaming (inspired by the concepts of slow and right steaming) to be enabled by route optimisation and highly accurate ETB and ETD;
5. The accuracy of ETB and ETD to be higher and thereby enable management of multiple instances of berthing and cargo operations seamlessly and smoothly.

9.4.2 Enabling integrated performance by digitalisation

STM is the concept of sharing and using all information from the maritime domain in real time. It facilitates seamless services in collaboration with all parties, involving both seaborne and shore-based functions. A key feature of the STM concept is that all parties involved in the maritime logistics chain *share* information.

STM is the dynamic, integrated management of sea traffic and maritime space, including sea traffic services, management of the maritime space and sea traffic flow management. STM is realised through the provision of facilities and seamless services in collaboration with all parties and involves seaborne as well as shore-based functions. The aim of STM is safe, environmentally friendly and efficient shipping. STM involves the aggregation of seaborne and shore-based functions required for ensuring the safe and efficient movement of vessels during all operational phases. The approach is based on experiences from Air Traffic Management (ATM) and the SESAR programme. Learning from ATM principles and experiences; structures, processes, methods and concepts will be adapted, scaled and adopted to the maritime domain.

9.4.3 Four concepts enabling STM

We now provide a short description of the four operational concepts of STM needing a digital infrastructure for their realisation.

STM builds on five defined sub-concepts: Flow Management, SVM, DVM, Port CDM and SWIM.

- Flow Management is primarily governed by such notions as single ship reporting areas, dynamic no-go-areas, arrival/departure management and capacity management.
- SVM builds on the JIT concept allowing non-stop voyages at the most economic speed from departure to arrival in port, and is enabled by Single voyage ID and Voyage ID assignment.

- DVM is made up of a Strategic Voyage Plan and Tactical Route Exchange, where the former is an iterative operation among involved parties and the latter is the exchange of a dynamic voyage plan in conning mode on board supporting the situational awareness of the navigator.
- Port CDM enables four collaborative arenas facilitating sustainable transports as a whole:
  - among actors operating within the port
  - between the port and actors realising sea voyages
  - between the port and actors realising inbound and outbound transportation (besides sea voyages)
  - between ports within each collaborative arena.

SWIM facilitates the sharing of information between different systems supporting the STM in making the right information available at the right place and time.

9.4.4 Strategic voyage management

In order to provide vessel operators with vital and necessary information to meet the needs of performing the most cost-efficient and environmentally sustainable passage at sea, it is important to create a dynamic flow of information. This can be updated in real time, immediately after changes of state, wherever in the transport chain a change occurs. Voyages making use of the advantages of STM are normally initiated by providing input to a voyage planner connected to the STM network. Based on the nature of the input made to the voyage planner the system will immediately make information available relevant to the plan for different actors. This could, for example, concern a vessel loaded with dangerous goods. As a result, the system will show information on restricting regulations along the route, or that the vessel is required to use a pilot when approaching the desired port. This requirement will immediately be known to the individual inputting data to the voyage planner.

When the voyage plan is completed, it is ready for nautical validation, service booking and confirmation and finally execution (the actual voyage taking place). When a voyage plan is 'released' by the planner, the different service providers such as port authorities, pilot services and VTS centres are made aware of the voyage and are requested to validate and confirm the part of the plan that relies on them. The port authority's administration system is connected to the STM and should confirm or decline the request, or even suggest changing the plan for arrival and departure (see the following Port CDM section). The planner will then be able to observe as the different service providers respond to the plan. Finally, through collaborative work the plan is ready for 'execution' and the dynamic phase of the voyage starts (see Section 9.4.5).

The voyage planner will also enable the establishment of an STM maritime marketplace. Such a marketplace is necessary for the planner, at the earliest possible stage, to make high-quality decisions and identify the maritime-service provider which best serves the ship's need. When a cargo owner needs his cargo transported by sea, there are normally several maritime transport companies able to take care of this need. The STM maritime marketplace is a way to connect a cargo owner's request directly to all relevant transport companies, to find the best suitable and affordable option. Transport companies, on the other hand, needs their cargo vessels fully loaded to achieve the best return on their voyage investment. The marketplace is available for them to make their free capacity known for all cargo owners.

In order to present all available options to the STM participant and optimise the use of capacity in every vessel, the design of the STM marketplace needs a digital infrastructure which will enable all concerned stakeholders to engage. When connected the individual planner, as well as the industry as a whole, will be able to reduce administration and increase both efficiency in operations and the utilisation of their available capacity.

9.4.5 Dynamic voyage management

To provide ship operators with vital and necessary information to compute the most cost-efficient and environmentally friendly passage at sea, it is important to create a dynamic flow of data that can be updated in real time, immediately after changes of state, wherever in the transport chain a change occurs. The dynamic voyage plan is an iterative phenomenon, with information shared in real time, voyage plans updated between involved parties in order to improve safety, efficiency and environmental performance by providing new possibilities for validation, optimisation, navigational assistance and situational awareness. The first three are services and the latter is a tool used in vessel-to-vessel communication provided by the technical enabler Route Exchange. SVM and DVM are parts of the STM overall voyage management processes. The relationships between these two voyage management processes are visualised in Figure 9.8 [37].

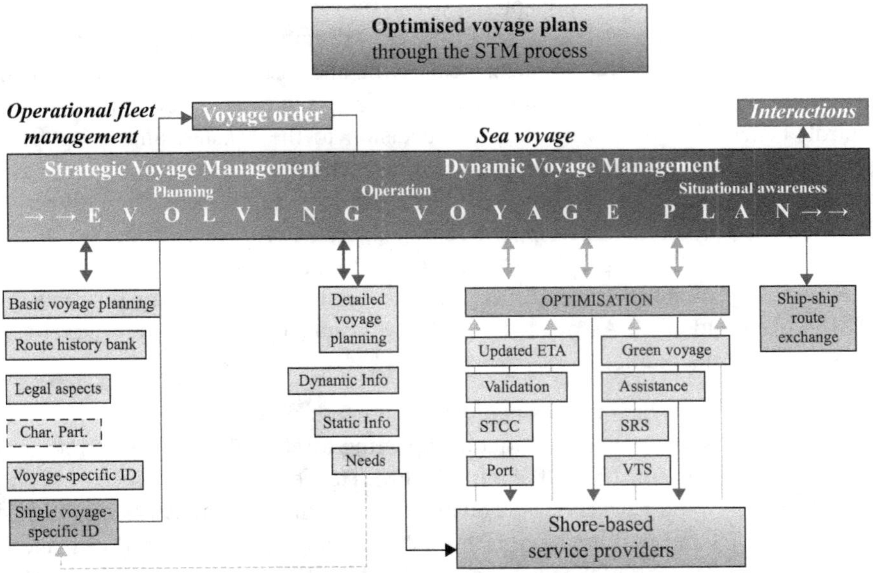

Figure 9.8 Linking strategic voyage management to dynamic voyage management

Associated with DVM are several services, including:

- Route Optimisation
  Public data stream services provided by different entities to support route optimisation, including data on weather, ice conditions, Maritime Safety Information (MSI), Maritime Spatially Planned areas (MSP), distance, speed, traffic congestion and bathymetric conditions. The route optimisation is an iterative process and can be performed both pre-departure and continuously en route as needs and conditions change. All plans can and must be changeable with short notice, and as soon as new orders and optimisations are completed a new agreement must be established and distributed.
- Route Validation (STCC)
  The intended voyage plan is sent to an STCC for validation, which can be done before a vessel's departure or before arrival at an STCC area. The validation control includes, but is not limited to, primarily an Under Keel Clearance (UKC), air draft check, no violation of MSP no-go areas, MSI and compliance with mandatory routing. No optimisation service as such is included in route validation.
- Assistance Services for route support
  In addition to monitoring, passive and automatic surveillance for detecting deviation from agreed routes, provided by Flow Management (see following), a voluntary Assistance Service for route support at various levels can be provided via the exchange of routes between a vessel and an STCC.

9.4.6 Flow management

Flow Management (FM) focuses on the whole traffic flow. The overall objective is to optimise and increase safety of sea traffic flow during all planning and executing phases. While Voyage Management deals with one instance, Flow Management deals with multiple instances. Optimising traffic is achieved by coordination, not control, hence always leaving the final decision to the Master, and using STM technical enablers. The FM concept should not contradict any regulation on navigational safety, such as COLREG (the International Regulations for Preventing Collisions at Sea). Authorities appointed by National Competent Authority (NCA) of Flag States will solely provide FM services. Port Control (existing organisation), VTS (existing organisation) and STCC (new organisation necessary for realising STM) will provide FM. Associated with Flow Management, there are several services provided, including:

- Traffic image for common situational awareness
  The establishment and continuous updating of a traffic image over geographical areas constitutes the base for FM. Hence, this is a necessary support feature for the other services. Today, real-time traffic images are established within VTS and port areas. In EU coastal waters, a near-time traffic image is established by the SafeSeaNet module STIRES. Still, in most cases, an STCC real-time traffic image will need to be established.

- Single ship reporting area
  In several earlier studies, seamless ships' reporting within the EU has been mentioned and proposed. An STCC area, as defined above, could potentially be implemented similar to a large-scale Ship Reporting System (SRS). The objective is to perform ship reporting at an entrance point, which is distributed to all stakeholders, such as VTS, involved along the route. Ship information is transmitted together with route information. Ship information will also be automatically handed over to the next STCC area.
- Enhanced monitoring in critical areas
  All ships participating in STM will have the ability to follow pre-planned routes that could be automatically or manually monitored and assisted from the STCC along the route. Deviations from an agreed route will be detected and measures taken when appropriate. Hence, the system will automatically detect if a ship is leaving its intended track or if a non-participating ship is manoeuvring in 'strange patterns'.
- Area management (dynamic no-go areas, MSI connected to specific areas)
  The exchange of route information and dynamic geo-locations opens up new possibilities for area management. A geographical area, which is considered to be sensitive during specific periods of time, can dynamically be classified as a restricted or no-go area. This area will be visible in Electronic Nautical Charts (ENC) and considered during route checking and route optimisation.
- Traffic synchronisation and capacity management
  Route optimisation could potentially consider traffic in congested waters. Hence, some kind of traffic synchronisation service will be needed in order to manage all already planned voyages and synchronise those with new voyages. This is mainly performed by using the concept of the ETA window, setting the safe haven in the long track direction and the dynamic separation, setting the safe haven in the cross track direction.
- Arrival/departure management at ports (ETA/ETD optimisation)
  Vessel traffic in port and quay planning are important input to the overall ETA/ETD optimisation process. This type of planning is conducted today either by a VTS centre offering a Traffic Organization Service or, more commonly, a Port Control function. This planning activity and information exchange is covered by the FM service arrival/departure management at ports. This FM service overlaps with the services of Port CDM (e.g. ETB and ETD generators).

9.4.7 Port CDM

Ports, serving as departure and arrival hubs for different means of transportation, require a coordinated transportation system addressing the goals of the transport system as a whole with smooth and seamless operations at sea, at port (reaching the port, departing from port, performing loading and unloading operations – and sometimes other maintenance and extraordinary administrative tasks) as well as connections to hinterland transportation. Seamless and sustainable transport enabled by STM requires a collaborative port. Inspired by airport CDM, Port CDM has been identified as a key enabler for reaching the full potential of STM. The purpose of port CDM is to provide a basis (processes, content, etc.) for the collaboration between key actors within the port and between the port and its surroundings. The overall goal of port CDM is to support JIT operations within ports and in relation to other actors coordinated by an efficient and collaborative port. Port CDM constitutes the interface between ports and STM. One driver for port CDM is to enable high accuracy in predictability leading to, among other effects, optimal berth productivity (as the number of cargo operations divided by the time at berth) [38]. Thus, essential boundary objects between sea and port are Estimated/Actual Time of Berth (ETB/ATB) and Estimated/Actual Time of Departure (ETD/ATD). ATB is defined as the time when the vessel is All Fast (at berth) and ATD as the time when the vessel is All Loose (from berth). Governance towards ETB and ETD give rise to green/slow steaming as well as reducing unnecessary waiting times, enabling substantial environmental and financial effects.

Port CDM functions for enabling four collaborative arenas enabling sustainable transports as a whole (cf. Figure 9.9). Within each collaborative arena, port CDM should support the development of efficient operations (e.g. integrating processes in the port (collaborative arena #1) so that the port is prepared for arriving ships, creating conditions so that a sea voyage will be as efficient as possible (JIT arrival) (collaborative arena #2)). Port CDM, as a common measurement, collaborative decision and information-sharing system, would support the integration of different processes and enable areas of collaboration to be completed with high efficiency resulting in JIT operations within and between the collaborative arenas.

In order for the port to optimise its operations, it is essential to receive real-time information of the status, together with updated estimates, of different transports that are affecting the operations at port. This means that the same measures function both as coordination mechanisms for optimising port operations (and creating readiness for managing necessary loading/unloading operations) and as boundary objects for other actors outside the port for their optimisation. These measures are at the core related to data being shared and the agreements being made within the collaborative arenas identified above. The measures are used for giving estimates (such as ETB and ETD as two essential boundary objects), giving actual times when certain stages have been reached and a basis for evaluation.

Port CDM builds upon different measures used as a basis for information sharing and making agreements around, such as ETA and ETD, where the overall goal is to arrive as close as possible to the provided ETA (ETx – ATx should be as close to zero as possible). The resulting deviation represents the predictability of the port as such (as a representation of an ecosystem of actors) and represents a measure of a port's performance in a synchronised transport chain (enabled by STM). This in turn would enable different actors to optimise their operations and their utilisation of physical infrastructure and variable resources. To reach the full effects of STM, and thereby enable sustainable sea transport processes, high accuracy (based on systematic estimation procedures) related to berthing, unloading, loading and departure, becomes necessary. Reliable estimates for a sea voyage can be established by enabling high accuracy of timing of the arrival, operations at port, and departure of a vessel. Different planning horizons are associated with

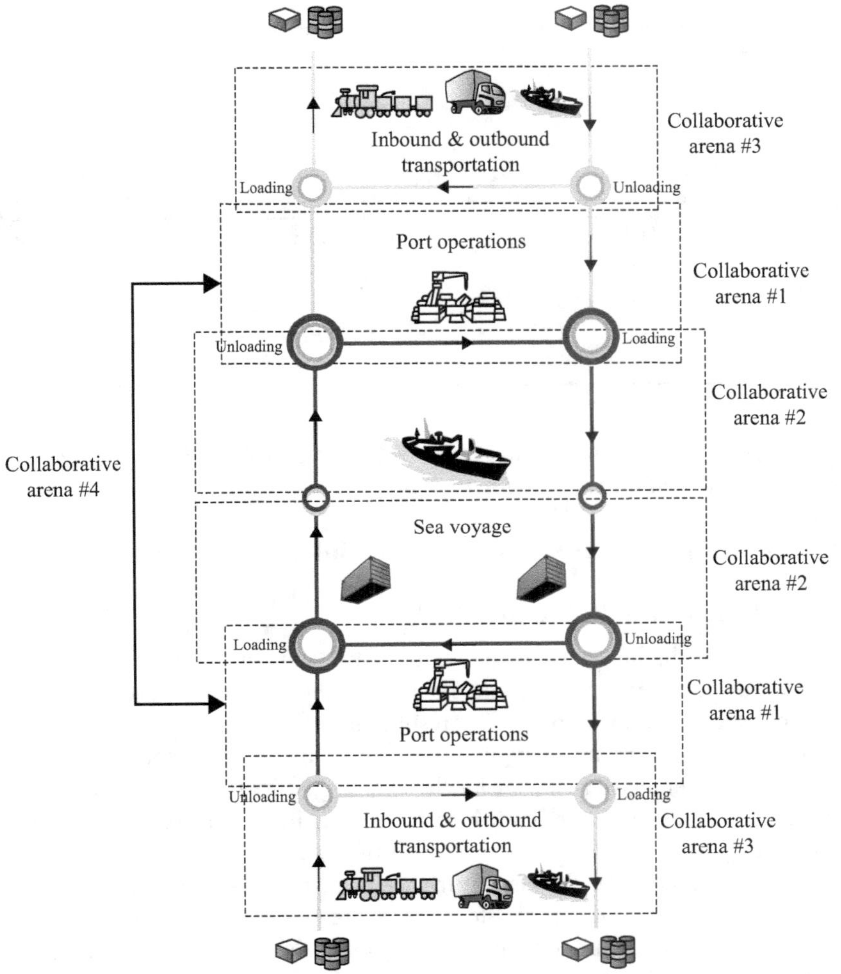

Figure 9.9 The identification of four collaborative arenas

different levels of tolerance for deviation between the estimated and actually reached state (the outcome) as depicted in Figure 9.10. The deviation should be diminishing with time; the closer to the Execution Phase the smaller the tolerance should be, until the actual moment of occurrence is reached for a certain state. This allows for the planning process, performed by the different actors, with different time horizons (i.e. long-term, mid-term and short-term planning) to be performed optimally, based on information about the interval of the outcome (e.g. a time range during which a certain state is reached). Sea transportation is a multi-organisational business with numerous actors positioning and coordinating their performance in relation to different control points. In line with the ambitions of the MONALISA 2.0 project and amongst other factors, STM will be realised by sharing information about the status and values related to identified control points for a particular voyage.

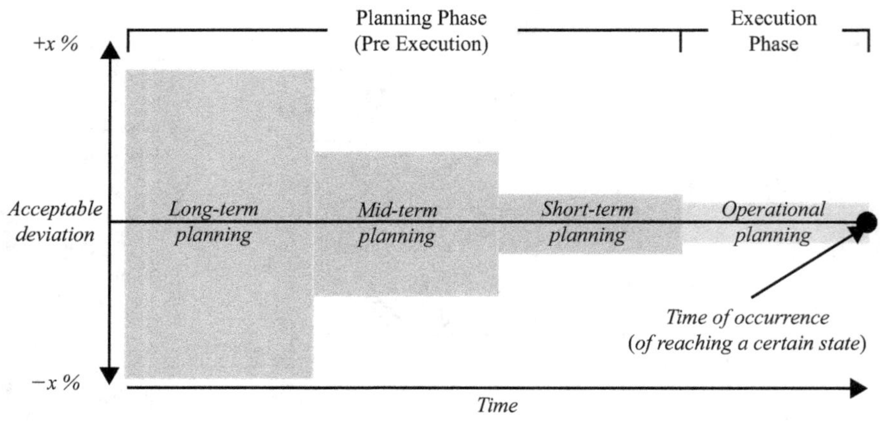

Figure 9.10 Acceptable deviation between estimate and actual occurrence (in different time slots)

9.5 Enabling digital streams and collaboration via SWIM

9.5.1 A distributed view on SWIM

Today, a lot of digital data streams exist in the shipping domain, yet there are no standards for these streams and no central directory for locating them and the associated documentation. As a result, the effective use of real-time data for shipping management is fragmented, uncoordinated and not very efficient. Consequently, an unused potential for the exchange of real-time data between key actors exists. Thus, to enable SWIM, an interoperability infrastructure needs to be established to facilitate the exchange of real-time data.

The current situation favours an incremental, market-driven approach to the development of SWIM. Step by step, these streams can be standardised, documented and made available to authorised accounts. As mentioned, data streams are the foundation of SWIM, no matter what its design. An incremental approach accelerates speed to operations and avoids trying to design a centralised system whose requirements will change as the value proposition of STM emerges from use and experience. Thus, SWIM becomes a central repository of data stream details.

9.5.1.1 Design principles for a distributed SWIM

The SWIM approach for distributed exchange/sharing of real-time data enabled by data streams relies on the following design principles:

- Digital data streams can be open, proprietary, or hybrid.
  SWIM cannot rely on a fully open data architecture. The intentions and performance of shipowners' operations can represent the competitive edge of the company; thus such data can be very sensitive. There are also service providers whose business models build on providing data in a condensed and quality-assured way (e.g. IHS-Fairplay, Q88.com, Purplefinder).

- Common standardised data format for all data streams in a SWIM repository.
  A data stream that is SWIM-enabled (i.e. connected to SWIM) must provide data according to the (SWIM-) defined data standard. Though data streams might exist for different sources, in a variety of formats, and serving different purposes, they should be translated into the SWIM standard for data exchange.
- Standardised API for accessing data streams to support interoperability.
  A data stream provider must provide a stream according to standardised definitions via a standardised SWIM API. A fairly comprehensive inventory of needed data services enabled by SWIM should be developed early in the implementation phase.
- Accessibility: Access to data should be governed by the data stream owner.
  The provider of the data assigns and manages access rights for a specific use of the data distributed in data streams, such as the voyage, to the requester. Subsequently, there is a need for a central repository of actors/actor-types/organisations which can be used by the data provider to assign access rights for the use of the data (i.e. by whom it should be used and under which conditions). In a decentralised model of access control, the data stream owner handles validation.
- Discoverability: The existence of a data stream should be informed by a discoverability mechanism.
  SWIM should contain mechanisms allowing a requester to learn about available data streams and access procedures. This means that a SWIM-enabled data stream should be requested to submit the existence of any data related to a certain voyage, vessel, etc. to this repository, but not the actual data. There is thus a need for a mechanism pointing at the existence of a particular data stream (e.g. data on a specific voyage, geographic area or port) and how it can be found. The discovery mechanism should automate connectivity to the desired data stream. For example, providing a pre-formatted message to the stream provider requesting access.

9.5.1.2 The advantages of an incremental approach to SWIM

An incremental approach can build upon existing data streams. When a core mechanism is in place, different actors can participate and 'SWIMify' their data streams at their pace or that dictated by competitive pressure. Interfaces for enabling access to data are not a revolutionary approach. If these kinds of interface are constructed the internal structures of the data providing organisations do not have to be changed. For enabling STM, these SWIMified interfaces do however need to be created by each data-providing organisation. The basic principle for creating SWIM-based interfaces from different systems is depicted in Figure 9.11.

The development of an information market should also be facilitated whereby entrepreneurs can aggregate data streams to create information and analytical services (different types of data streams could thus be created). The SWIM concept designs the standards, a metadata hub and a governance mechanism for the

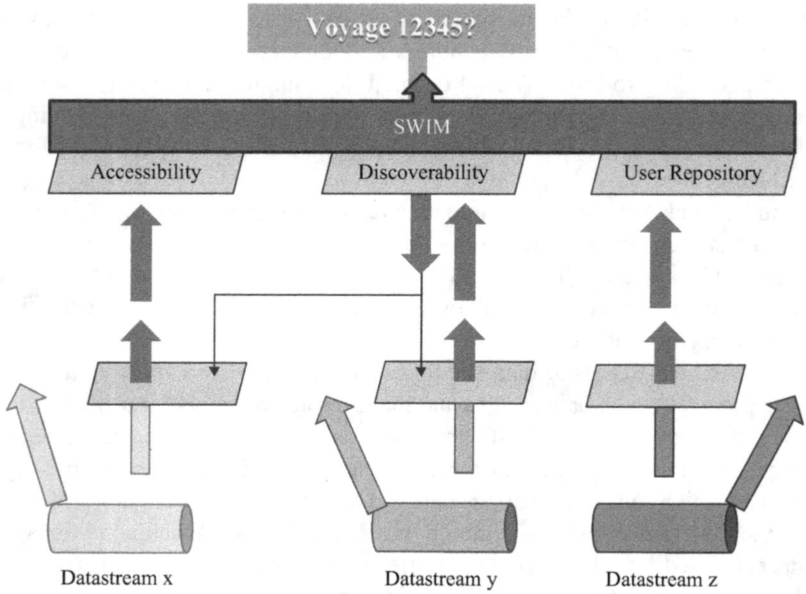

Figure 9.11 SWIMifying relevant data streams for STM

shipping industry data stream market. The provision of information is left to shipping industry participants and third parties.

9.6 Concluding reflections

The long history of shipping has created an industry with many autonomous players operating with a high degree of independence but with a need for episodic tight coupling to fulfil individual goals. For example, a ship cannot deliver its cargo without close cooperation with a pilot, tug and terminal operator. Despite this need for episodic tight coupling, the industry has not undergone the digital transformation that would enable data sharing to make coupling more efficient and reduce costs. Furthermore, we have presented a plan for realising STM that is highly cognisant of the nature of the industry. We advocate a digital transformation that preserves independence by allowing each party to manage access to its data and to decide when it will join the digital revolution. Our plan is based on creating an open market with complementary standards that facilitates coordination through data streaming and enables existing information intermediaries, such as shipping agents and new infomediaries provided by entrepreneurial information services, to adapt and emerge to meet the industry's information needs. In this chapter, we have described a series of digitally-based information services centred on STM that will accelerate digital transformation of the shipping industry and provide Europe with more efficient, safer and environmentally sustainable ship transportation.

As claimed by Watson and Boudreau [31] 'information enables episodes of self-managed episodic tight coupling in a society whose constituent elements are highly decoupled'. By adopting a multi-organisational perspective and structures for enabling *single reporting* and thus reducing the administrative burden, enabling *continuous updates on intentions and state changes, sharing real time data at the right time through data streams*, and meeting the challenges with *episodic connectivity* (because the vessel is not continually connected). A higher degree of openness and transparency, and supported by data streams, marketplaces for bidding and agreeing on assignments, is expected to become established. As explored in this chapter, an increased quantity of digitally shared information supported by STM looks at creating the following effects:

- High berth efficiency, by enabling ETD – ETB to be performed as fast as possible by assuring initiation as soon as the first voyage plan is submitted and published.
- The status of the planning/agreement process concerning ETB and ETD is published continuously.
- Actors enabling berthing (with the state of 'All Fast' as the end state of berth) and Departure (with the state of 'All Loose' as the end state of Departure) would be committed to established terms and conditions enabled by an open marketplace.
- Green steaming would be enabled by route optimisation and very accurate ETB and ETD.
- Higher accuracy of ETB and ETD, enabling management of multiple instances of berthing and cargo operations seamlessly and smoothly.

Based on the above it is to be acknowledged that the IS-community has substantial contributions to make in the development of a sustainable shipping industry empowered by digitalisation. The research discussed in this chapter is pointing at the necessity to empower multi-organisational collaboration enabled by increased systems interoperability, processes for episodic tight coupling and data streaming. As a basis for further research the notion of these concepts should be explored in a maritime informatics setting, where the application of these concepts would provide an additional empirical basis for scientific development within these discourses in the IS community.

References

[1] Adner R. (2006) Match your innovation strategy to your innovation ecosystem. Harvard Business Review [Internet]. 2014-07-20

[2] Elkington J. Cannibals with Forks: The Triple Bottom Line of 21st Century Business. New Society Publishers; 1998

[3] Stopford M. Maritime Economics. 3rd ed. New York, USA: Routledge; 2009

[4] Markström L., Holm H. Voyage optimisation on the shallow waters of the Baltic Sea. *SSPA Highlights*. 2013:4–5

[5] Andersson P., Ivehammar P. Economic Impacts – Cost Benefit Analysis of Implementing Dynamic Route Planning at Sea. Norrköping, Sweden: MONALISA Project Bureau; 2014
[6] Stankiewicz M., Backer H., Vlasov N. Maritime Activities in the Baltic Sea – An Integrated Thematic Assessment on Maritime Activities and Response to Pollution at Sea in the Baltic Sea Region. Helsinki, Finland: Helsinki Commission, 2010 BSEP-123 Contract No.: BSEP-123
[7] Watson R. Are You Ready to Profit from Real-Time Digital Data Streams? 2014
[8] EMSA. Maritime Accident Review 2010. Lisbon, Portugal: European Maritime Safety Agency; 2011
[9] EMSA. Maritime Accident Review 2007. Lisbon, Portugal: European Maritime Safety Agency; 2008
[10] EMSA. Maritime Accident Review 2008. Lisbon, Portugal: European Maritime Safety Agency; 2009
[11] EMSA. Maritime Accident Review 2009. Lisbon, Portugal: European Maritime Safety Agency; 2010
[12] Lindgren J.F. Assessing Sub-Lethal Impacts of Petroleum Compounds on Marine Benthic Systems. Gothenburg, Sweden: Chalmers University of Technology; 2012
[13] IMO. AIS transponders. Retrieved 2014-06-11; 2014
[14] IALA. IALA – Vessel Traffic Services Manual. Saint Germain en Laye, France: International Association of Marine Aids to Navigation and Lighthouse Authorities; 2008
[15] Brodje A., Lützhöft M., Dahlman J. (eds.). The Whats, Whens, Whys and Hows of VTS Operators use of sensor information. International Conference on Human Performance at Sea; Glasgow. Glasgow, UK: University of Strathclyde; 2010
[16] Svedberg U. (ed.) Sharing of Complex Maritime Information. e-Nav Underway 2013; Copenhagen, Denmark: Danish Maritime Authority; 2013
[17] Österman C. Developing a Value Proposition of Maritime Ergonomics. (PhD), Chalmers University of Technology, Gothenburg, Sweden; 2012
[18] EfficienSea Project Bureau. EfficienSea – Project Web Site http://www.efficiensea.org/default.asp: EfficienSea Project Bureau; 2011 [cited 2014-06-11].
[19] MONALISA Project Bureau. MONALISA 2.0 Web Site http://monalisaproject.eu/: MONALISA Project Bureau; 2013 [cited 2014-06-11]
[20] Accseas Project Bureau. Accseas Web Site http://www.accseas.eu/: Accseas Project Bureau; 2014 [cited 2014-06-11]
[21] Hammer M., Champy J. Reengineering the Corporation. A Manifesto for Business Revolution. London: Nicholas Brealey; 1993
[22] Davenport T.H. Process Innovation – Reengineering Work through Information Technology. Boston: Harvard Business School Press; 1993
[23] Goldkuhl G., Lind M. (2008) Coordination and transformation in business processes: towards an integrated view, *Business Process Management Journal*, 14(6)

[24] Peppard J., Rylander A. (2006) From value chain to value network: insights for mobile operators, *European Management Journal*, 24(2–3), 128–141

[25] Porter M.E. Competitive Advantage – Creating and Sustaining Superior Performance. NY: Macmillan; 1985

[26] Håkansson H., Snehota I. (2006) No business is an island: the network concept of business strategy. *Scandinavian Journal of Management*, 22(3), 256–270

[27] Haraldson S., Lind M. The Emergence of a Multi-Organizational View on Business Processes – Experiences from a Double-loop Action Research Approach. Lima, Peru: AMCIS; August 12–15, 2010

[28] Haraldson S., Lind M. Value Chains in Value Networks: A Multi-Organizational Business Process Definition, Australian Conference on Information Systems, Australia; 2011a

[29] Haraldson S., Lind M. Dividing Multi-Organizational Businesses into Processes: Capturing Value Creation in Assignment Structures, Australian Conference on Information Systems, Australia; 2011b

[30] Winograd T., Flores F. Understanding Computers and Cognition: A New Foundationfor Design. Norwood, NJ: Ablex; 1986

[31] Watson R.T., Boudreau M.-C. *Energy Informatics*. Athens, GA: Green ePress; 2011

[32] Brandt-Jensen E. The Maritime Cloud skal dække over skibsfarten http://www.soefart.dk/?art=3758&acc=LoginPopUp&typ=artikel: *Soefart*; 2013 [updated 2013-12-13; cited 2014-07-20]

[33] MARSEC-XL. Open Bridge Platform http://www.marsec-xl.no/MARSEC-XL_Norway/Products.html: MARSEC-XL; 2014 [cited 2014-07-20]

[34] EU. What is the SESAR project? Single European Sky ATM Research http://ec.europa.eu/transport/modes/air/sesar/: EU; 2014 [updated 2014-02-26; cited 2014-07-20]

[35] Södahl B., Hanning A., Hult C., Garme K., Hindrum K. Svallvåg 2013 – Svensk maritim forsknings-och innovationsagenda. Gothenburg, Sweden: Chalmers University of Technology; 2013

[36] Lind M., Brödje A., Watson R., Haraldson S., Holmberg P-E., Hägg M. Digital Infrastructures for Enabling Sea Traffic Management, The 10th International Symposium ISIS, Hamburg; 2014

[37] Svedberg U., Andreasson B. Concept: Dynamic Voyage Management; 2014

[38] Tirschwell R. Key Findings on Terminal Productivity Performance Across Ports, Countries and Regions. New Jersey, USA: JOC Inc.; 2013

Chapter 10

Team management and soft skills: case studies and cross-connections on the creation of an effective working team

Cristian De Mitri[1], *Raffaele Loscialpo*[2] *and Michele Fiorini*[3]

10.1 Introduction and purpose

Engineering systems, and transport systems particularly, although complicated and automatic, cannot work without human intervention. Aeroplanes are automatically driven by automatic pilots, high-speed trains are automatically driven by electronic high-tech equipment, even cars today can park alone and are starting to acquire more and more autonomy on the road. This fast evolution might lead us to think that soon there will be no need for humans to operate machines. Nevertheless, as automation develops in complexity, entire teams are put together to dominate the hi-tech complexity represented by machines. The question is: who dominates the organisational complexity of controlling other men's work? It is other men, not machines, not yet, at least. It is other men for a precise reason: because management is a complex art that so far no one has been able to translate into code for a programmable device.

Management has to do with people and teams. In the last century a new concept arose in the professional arena: the concept of team. No more artisans, then, as in the last century, rather groups of people called to work side by side, communicating and sharing. As society has evolved, so people's skills have had to evolve, too. And the 'new' skills are called 'soft' because they are difficult to structure: they make a big difference between a good and a bad team.

The chapter's aim is to present two management techniques and team selection models that would help everyone have a better understanding of how excellent teams can be selected, aggregated and then managed.

[1]Eggup, www.eggup.net
[2]Qhaosing®, www.qhaosing.com
[3]The IET Italy, www.theiet.org/italy

10.2 The Qhaosing® or cross-connections model

The vast majority of management schools used to teach control as the unique management technique. Today this concept is obsolete. Controlling a company, a society or even a small group of people means making things happen in exactly the way we decide; in other words, it means renouncing the enormous creative potential hidden in the group itself. Every day we have experience of startups that show the ability to change skin as soon as they realise a change in the market calls for this. Changing skin is a skill that the academic management theories do not support. New management tactics, based on complexity and chaos theory, help us to reinforce the skill of adapting, so important today to succeed in a world evolving faster than ever.

In order to illustrate the current management challenges, we will adopt a technique called *Qhaosing*® or 'cross-connection technique': to illustrate it, we will be using some performance models that only seem to have nothing to do with management; in fact, concealed in them are the key concepts of managing people (Figures 10.1 and 10.2).

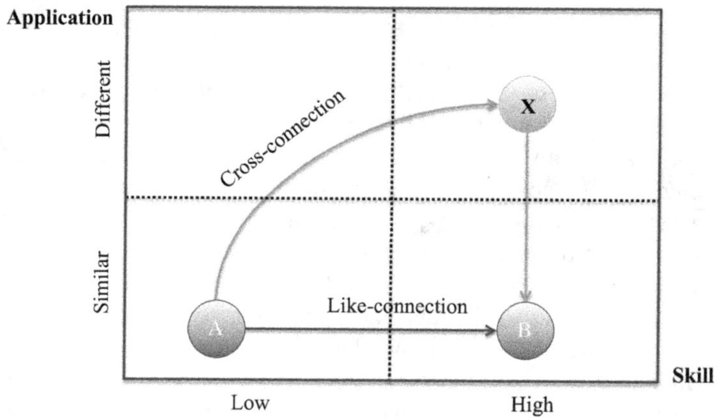

Figure 10.1 Qhaosing® or cross-connection model

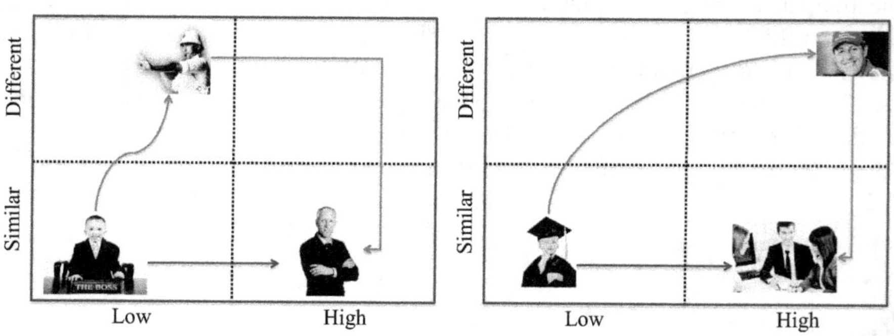

Figure 10.2 (a) 'Crossroads' model; (b) 'unpredictable driving' model

We will work principally with two models, one based on the 'crossroads' or 'chaotic management' concept, and the other on an 'unpredictable driving' model of the delegation mechanism. In order to derive maximum benefit from the Qhaosing® approach, it is necessary to identify oneself in the represented scenarios. To enable this, the techniques are always described in a narrative way. In this chapter, a synthesis of the two different scenarios will enable us to focus on the most significant teachings. Readers can study this approach in greater depth through the sources listed in the bibliography [1].

A key role in the narrations is played by Qhaosing® Ltd., a consulting company which aims to prod you out of your comfort zone. Year after year, we become more and more comfortable in our routines and behave as if on rail tracks, and when change comes, we are in trouble. Qhaosing® works in this area because we must pass through these feelings of discomfort if we want to change even small behaviours.

In both the tales we will look at, the protagonist is Manny, a young manager who has already, despite his youth, some conservative behaviour traits that make his new president doubt his potential ability to lead the company. This is the manager's last chance. That chance is an encounter with Qhaosing®: he is challenged with some tests which he finds very hard as they are so far from his daily routine. Only by showing he can adapt quickly to meet the different tests; can Manny convince his president that he is the right change agent and the right leader for the company?

10.3 Case 1: 'A crossroads to manage'

The crossroads illustrates perfectly the concept of complexity; and how to manage a crossroads can therefore be considered a typical complexity management skill (Figure 10.3).

Refining complexity management techniques is extremely valuable today, as we are surrounded more and more by complex systems which integrate hi-tech and standard operating procedures, machines and human beings. Controlling a machine is not properly an aspect of complexity management. As explained in the first story, we need to distinguish between complicated situations and complex ones, the latter being the ones we deal with when we manage a complex system. A system is complex only when two conditions occur at the same time:

- a *quantity*, meaning many interacting elements
- the behaviour of the quantity cannot be synthetised in simple cause–effect relationships.

Synthesis is not possible as the relationships among the elements are many and time-dependent. This model is powerful, as we will be expected to manage more complex situations in the near future and the hypotheses on which we have based our management behaviours will no longer apply.

In other words, a complicated thing can be understood, given more or less time, depending on the subject; whereas a complex thing can never be understood, even given infinite time.

Figure 10.3 A crossroads

Few of us really know in detail how a car's engine works; nevertheless more or less all of us drive a car. The car, though, is still a complicated system, and not a complex system. We can forecast the behaviour of each component, although it is difficult. We are able to use the sea to travel or to transport goods, without knowing anything of it. The sea is a complex system and it is hard to forecast with traditional mathematical theories; nevertheless, we are able to manage it, and transform it into a means facilitating our movements in space on the planet.

As indicated in the research, *Modellistica e Simulazione del Comportamento del Traffico in Intersezioni Stradali Urbane* [2], 'the dynamic problem of loading the road network depends on the complexity of reproducing the traffic dynamic status and is reinforced by the human behavior within the components'.

Two approaches are normally used to model the crossroads: the micro and the macro. In the micro approach, every single vehicle, every single pedestrian and every possible interaction are modelled with maths. It is easy to see the complexity within this approach and the high failure rate, too, mainly due to the low probability of modelling correctly a single entity or a single relationship. The macro model, on the contrary, is based on a fluid dynamics analogy. From a macro perspective, traffic can be seen as a flowing river; in fact, traffic flow is commonly likened to the movement of a fluid. Analogy with fluids, in the research work mentioned, leads us to focus on physical variables such as the flow, the concentration (spatial density) and the speed; the traffic is treated like a monodimensional incompressible fluid.

The difference from the micro approach is that the macro approach studies the traffic from the global perspective, neglecting each vehicle's dynamics.

A third model exists, the most futuristic model: it is described as *mesoscopic* and describes the statistical behaviour of the vehicles: they are aggregated in sets with similarities, in order to simplify the processing formulas. In this approach vehicles and pedestrians are not treated independently, but they are managed as a whole and their speed is calculated by a speed-density function defined for each road segment.

The bibliography treats a wide range of cases of crossroads with vehicles managed using the chaos theory: see [3–5].

The Qhaosing® model treats the case of crossroads management with hundreds of pedestrians. The difference is huge. As quantity is an essential characteristic of complex systems, this characteristic is even more evident in a crossroads busy with pedestrians, owing to their very great numbers. If a normal crossroads with vehicles on it is a good subject for statistical analysis as a parallel to complex systems, it looks as if crossroads with pedestrians is an even more appropriate subject.

In pursuit of the goal of sharing new management techniques for managing what is probably the most complex creature existing, the human being, we can find inspiration in observing Manny dealing with the crossroads: seeing him fail and succeed will give us valuable ideas about how to manage intensely engaged groups of people. The protagonist tries to use his own experience first, all based on controlling others. The first technique he tries is to use *authority symbols*.

10.3.1 Use of authority symbols

The manager changes from his normal clothes into a police uniform. He places himself in the middle of the crossroads, counting on the respect and fear produced by the uniform.

With his hands, he tries to redirect people flows first in one direction, then in the opposite one (Figure 10.4).

His open arms can physically block the flow on one side, but not on the other, so that travellers do not care and inevitably the two orthogonal flows conflict. The flow grows minute by minute and it becomes more difficult for him to change the direction of flow. After less than an hour, he gives up, stepping back into his 'considering corner'.

10.3.2 Use of the remote control

Having failed with his first attempt, he tries a second technique. This time, he decides to use a more effective method, using a remote control to change the lights. At first, it seems very effective: when the lights come on, people instinctively stop at the red traffic light, while a similar mass of people flows into the street perpendicular to the crossroads, as their traffic light is green.

He almost doesn't notice a few individuals who, after a furtive check, drive across the crossroads, despite the red light. There are just a few offenders at first, and the manager doesn't care. But their numbers increase as little by little the few are followed by others, who in turn are followed by others, and so on.

Figure 10.4 A traffic policeman

Within an hour, the crossroads is again overwhelmed by people coming from right to left and from left to right, as if the lights had disappeared.

The manager raises the level of control again: he combines the traffic lights (an automatic rules attribution system) with the authority symbol (himself dressed as a policeman) who could punish offenders.

He delegates the traffic light control and goes back to the middle of the crossroads. So now, there is a symbol of authority in the middle of the crossroads and the traffic lights are working by remote control.

As before, the manager/policeman gets some good results at first, but as soon as he stops the first offender, and punishes him, he realises how far the technique is from being efficient. In fact, even if he fines a couple of people, others break the rules, and still others see and follow their example. In a very short time, the situation reverts to chaos.

10.3.3 Use of a constraining control

Now the manager makes a third and final attempt: he further increases the level of control, using a technique of physical prevention. He uses a second remote control, which controls transparent barriers to physically stop the flow of people. This technique springs from the notion that people do not know what is right or wrong for themselves: so it is simply necessary not to let them be in situations in which they can break the rules.

Unfortunately, as the book reveals, even this attempt fails: this system resists the flow for longer than the other techniques, but in the end, under pressure from people who are excited and in a hurry, the barriers no longer work properly, the system collapses and the initial state of anarchy supervenes.

The lessons learnt by the manager in this process are several, all of them important:

- One authority figure alone cannot maintain control of a complex human system.
- The rules for managing complexity are not followed spontaneously.
- Even a few simple rules need many people following them to be respected.
- A rigid controlling organisation forbids people to break simple rules, even though the rules themselves do not fulfil long-term needs.

The most important of them is the lesson we will now unfold, which has to do with the chief law of chaos theory, the law of self-organisation. This is the key reason why the binding control fails.

The crossroads, in fact, acts like a complex system and reveals its attributes of complexity only after a certain amount of time.

The presence of the automatic barriers solves just part of the problem. It can be seen that the focus on problem-solving prevents the individual from seeing any side effects of the solution; so they will be revealed fully only at a later time. The use of barriers solves the problem of managing the crossroads only for a short time, but a few hours, or days, later, the complexity of the system will consume everything and the original situation will be restored.

'Strangeness' is a weird concept that has to do with the intrinsic characteristics of a complex system. Every system labelled 'complex', of course, by definition, cannot be described other than through complex mathematics, but even allowing for this, its prediction is impossible. Nevertheless, there are some kinds of slow movements in the system that can be predicted: and these movements are related to the intrinsic system properties that we cannot ignore if we want to change it.

In addition, the second essential aspect is that these changes should be gradually digested by the system. The transparent barriers evidently represent a solution that is not digestible, is too different from the intrinsic properties of the 'crossroad system'.

Basically, the movements are statistically subject to these factors:

1. The system is highly dynamic and fluctuating. It will oppose anything that would slow it down.
2. The degrees of freedom of the system components (walkers) are many (speed, acceleration, position) and able to adapt quickly to changing conditions.

The concept would not apply with the same impact to cars at a crossroads rather than pedestrians. It would apply partially; the first aspect, the dynamism, would definitely apply, whereas the second would not: a car has many fewer degrees of freedom than walking people and the space available for cars to run a crossroads would be reduced: free space would become a highly valuable resource, like all scarce resources, and as such, would be targeted by most cars at the same time. Without regulatory control, therefore, there would be too many conflicts. We need control, then, because of scarcity of conflicting resources.

Figure 10.5 People at a train station

A crossroads with people, instead, satisfies the two characteristics above, and is able to react, adapt and evolve more rapidly.

The problem is auto-solved. It is a natural property, as incredible as complex systems: this is called self-organisation.

Observing the system over time, we would notice the grouping of people with common attributes, for example, the slowest. The system could self-organise in a decentralised row made up of the slowest people. Similarly, the fastest will have plenty of space around them and this extra need for space will have been shared and accepted by all. In short, the system is in equilibrium, a constant dynamic equilibrium (Figure 10.5).

In order to improve the management of the crossroads, therefore, if the system is dynamic, for example, we can increase rather than reducing the degrees of freedom, and we might think for instance of creating some elevated roads. Moreover, for instance, if we really believe in the ability of this system to self-regulate and if the ability is more effective with more degrees of freedom, another way would be to extend them from a much higher viewpoint, so that the system could evolve more quickly. This could be achieved, for example, through mirrors placed in some specific places that could show the crossroads from many angles simultaneously.

Many situations, even those we do not expect, are characterised by complex behaviours that we often do not recognise because we are habituated to seeing them in the same way, a way that we think allows us to control them. Almost always, though, and always too late, this proves to be just an illusion.

And this leads us to the most important lesson:

- The only way to control a complex system is to help it to become self-organised.

The management lesson is clear: abandoning the temptations to control groups of people and basing the management style upon adaptation and listening.

10.4 Case 1: 'A strange car to drive'

If the first case was about managing a team from a working group, this case is about individual coaching, that is 1:1 management. Individual interaction is as important as team management, because one of the greatest gifts a leader or a manager should possess is the ability to make everyone feel unique.

In this scenario, the Qhaosing® approach, based on the use of a real experience to share and deepen key management principles, consists of driving a car that apparently is undrivable. The protagonist is Manny again, the manager going through the test week to learn how to manage change: this time the test consists of driving a car with many operational anomalies.

10.4.1 Knowing how to manage information

In the first scene, the manager is manoeuvring his car. It looks simple, but it soon turns out to be far more difficult: in fact, at the first curve, the car goes to the right instead of the left and accelerates instead of braking. After the first panicky moments, the manager responds to the data he receives from the vehicle. In fact, he has to deal with a schizophrenic speedometer, a deforming windscreen and accelerator and brake pedals that are totally unreliable.

What previously appeared to be reliable information (such as the speed shown on the speedometer, the road ahead seen through the windscreen, the resistance offered by the accelerator and brake pedals, the precision of the steering angle), now loses any logic and reliability. The manager looks for other information that can support him in driving the car safely home. In fact, just after the first contact with the pavement, he hears the right front wheel slamming repeatedly against the kerb. However, although a few minutes before, the collision with the pavement would have represented a danger for him, now it turns out to be a reliable reference point to maintain the curve trajectory. After two skids, first to the left and then to the right, the manager sees his destination and stops the car.

The manager's experience hides a great teaching: it is exactly what happens to all of us. Many people are in charge of a group of other people, some of a company, some of a working group, some of a football team, some of a family, some others of a group of boy scouts: whatever the walk of life, sooner or later it happens to all of us that we are in charge of a group of people, making decisions for them. And most, if not all of the time, we use information to take decisions or make choices. And most, if not all of us, do not wonder much about the correctness of the information. We just want more information, no matter whether it's practical or not, we just want more, rather than more reliable. And the risk is that constantly, we may run off the road. We rely on false information, and perform actions whose effects are nothing but unexpected. To use a not too strained metaphor, we steer too much, to follow a curve that is not in reality what we may believe from looking through the windscreen. And we end up, in fact, off the road.

There are other situations where information changes very quickly. This seems unreliable but this is because we do not pay attention for long enough: if we waited, we would be able to extract a sort of mean value. This is how a specific speedometer needle works: it is never still for a single moment, therefore it is instant by instant very unreliable, but observed over longer time, becomes kind of reliable.

The point is that judging whether information is more or less reliable is possible only when you know the outcome of the choice. So far, we have been leading a department in a steady and predictable market, in other words, we have been driving a fully working car along a curveless road. The market is changing and the unpredictability will be its most predictable attribute from now on: so we will have to drive an odd car like this, and we cannot trust the information that we have used to accelerate or steer so far. In other words, we cannot count on the traditional means we have used so far to continue to direct the car wherever we want. The same actions as those performed so far will have different results now. In order to succeed, we have to go on adapting by successive approximations to the information we receive.

Driving a car is a symbolic act. The car represents a company, the market is equivalent to the ever-changing context, accelerating and braking represent the actions taken by management, the speedometer represents the information received, and the windscreen represents our thought structures. Even the pavement is a symbol of something: a solution to demolish the thought structures that prevent us from adapting our responses.

We lead our teams, our companies, our football teams, even our families, looking through a windscreen which gets deformed with time and experience, and we cannot prevent that, it is part of human evolution. But as the manager's experience has shown, we can know our filters and their degree of deformation, counting on references external to the context, just like our friend, the pavement. The pavement can be anything that meets two requirements:

- It must be external to the context.
- It must be taken by a different mean (we use our eyes to see through the windscreen and we should use another mean, like our ears, to capture the vibrations coming from the basement).

In order to understand how this may work, let's imagine that we have a problem in the company and this problem has occurred several times before: we will see the problem through the distorting glass of our past experiences and if the context has changed, we will end up miserably off road. We should then look for an external reference to solve the problem: this has to be external to the context, and should not pass through the filter of our past experiences. We could, for example, ask a complete stranger for help. The ideal stranger will be one with a past experience very different from ours. He or she should be able to give us the best support, because it is independent of our experiences: so it represents what we need, an external reference to help us to hold our direction.

10.4.2 Knowing how to manage others

In the second scene, the manager is engaged to help a driver in another car very similar to the one he has driven before. The mistake he makes at the start is to take for granted that the two cars are identical. And his commands turn out to be all wrong, misleading the driver.

This is what happens to us every day, when we drive other people; we rely on our expertise to exert some strange form of power over others: we do not need actually to be experts, a few minutes of driving have been enough to make the manager feel expert and manage others with authority.

Most of us act in good faith for our own good, and the team's good in general, but this is not always enough to stop us going off the road. This happens often when we think we know a situation perfectly which, from the outside, seems very similar to another. We continue to issue commands using the same behaviours we have used successfully in the past but they are unsuccessful in new situations and we end up off the road.

Things start to work when we understand how to step out of the situation, putting ourselves into listening mode.

Stepping out of the contextual 'strange car' in this case means recognising that the car might behave unpredictably and that before taking decisive action in one direction rather than in another, we should study the situation through the feedback received after some micro input from it.

Things begin to take the right direction when we induce others to focus and pay attention to their car's behaviour, pushing them into making micro attempts to fully estimate the quality of the information received and the reliability of the commands that might be issued. It is also advisable to leave people to take the decisions they believe in (they are generally in the best position to take the best decision), and we feed back 'mirror' information on their performance regularly (every lap) so enabling them to improve with each lap.

Recapping, the second case and Manny's adventure reveal these important principles that every manager should remember:

- Too much information leads us to unconscious behaviours and actions.
- Often we adopt the same strategies for information management although contexts change.
- We guide our teams, our companies, our football teams, even our families, looking through a windscreen, which is deformed over time.
- We can measure the degree of deformation of our filters only by using references that are external to the context. Established thought patterns must be demolished.
- When we manage other people, we do overestimate our expertise and exercise a strange kind of power over others.
- Things start to work when we succeed in abstracting ourselves from the situation, shifting into a listening status.
- We should offer team leaders an external reference point with the right pace and continuous communication flow aimed always at continuous improvement.

10.5 The modern team

The successful team is the team which spends time on improving soft skills and knows exactly how to integrate the right person in the right role. The successful team is the team which has great 'group power' but knows also how to use the power of each individual member in specific situations.

We cannot plan everything in every project, especially team relations. Team relationships carry on in a natural way, continuously evolving during the whole of the team life cycle. The successful team environment is usually comfortable; it is a kind of environment where every team member can share ideas and knowledge, making a great contribution for the development of their work.

10.6 The talent

A talented person is one who can resolve, in the right context and at the right moment, problems which are not resolvable by other people.

Looking at the concept of talent in this way, one could say that talent is revealed just in certain specific situations. We must not think about a 'best performer' as an absolute. We must not imagine we can resolve every situation in the same way.

Talent is not inherent in a person from birth. Talent comes from a combination of person, context and situation. We can reach the talent, and work effectively with it. To find the talent we need to have excellent awareness of ourselves. The idea of talent that Eggup wants to introduce is that it exists between the person and the specific situation, so we might say that without that situation that person would not be able to use his or her talent.

10.7 A winning tactic

Successful teams have best practices and usually they achieve most of their goals. This is thanks to a combination of elements.

The combination of these elements makes it possible to reach the goal and be an effective team. Just as a tree extracts elements from the ground and exposes its leaves to the sun to produce energy, the successful team has the right skills to reach specific goals.

10.8 Team Role Theory

Team Role Theory is the brainchild of Meredith Belbin, who says that an effective team is the one which includes all the nine roles. Belbin describes eight specific roles in his book, *Management Teams*, and he added the ninth role subsequently.

Belbin says 'The team role is a pattern of behavior characteristic of the way in which one team member interacts with another where his performance serves to facilitate the progress of the team as a whole' [6].

Table 10.1 shows the nine Belbin roles [7].

Table 10.1 Belbin team roles

Team role	Contribution	Allowable weaknesses
Plant	Creative, imaginative, free-thinking. Generates ideas and solves difficult problems.	Ignores incidentals. Too preoccupied to communicate effectively.
Resource investigator	Outgoing, enthusiastic, communicative. Explores opportunities and develops contacts.	Over-optimistic. Loses interest once initial enthusiasm has passed.
Coordinator	Mature, confident, identifies talent. Clarifies goals. Delegates effectively.	Can be seen as manipulative. Offloads own share of the work.
Shaper	Challenging, dynamic, thrives on pressure. Has the drive and courage to overcome obstacles.	Prone to provocation. Offends people's feelings.
Monitor evaluator	Sober, strategic and discerning. Sees all options and judges accurately.	Lacks drive and ability to inspire others. Can be over critical.
Teamworker	Co-operative, perceptive and diplomatic. Listens and averts friction.	Indecisive in crunch situations. Avoids confrontation.
Implementer	Practical, reliable, efficient. Turns ideas into actions and organises work that needs to be done.	Somewhat inflexible. Slow to respond to new possibilities.
Completer finisher	Painstaking, conscientious, anxious. Searches out errors. Polishes and perfects.	Inclined to worry unduly. Reluctant to delegate.
Specialist	Single-minded, self-starting, dedicated. Provides knowledge and skills in rare supply.	Contributes only on a narrow front. Dwells on technicalities.

10.9 Stages of team development

In 1965 Bruce Tuckman proposed his 'stages of group development' model. According to this theory, there are four plus one stages in the development of a team [8–11] (Figure 10.6):

- *Forming*: this is the first step, when people are getting to know each other, sharing ideas and expectations.
- *Storming*: everyone tries to understand what they should be building and how it should be done. At this stage, usually, there is a lot of discussion.
- *Norming*: the team is beginning to establish information, and to build and define tasks and roles.
- *Performing*: every team member must know how to do his or her job effectively and efficiently.
- *Adjourning*: the team finishes the job; and all the tasks have been done.

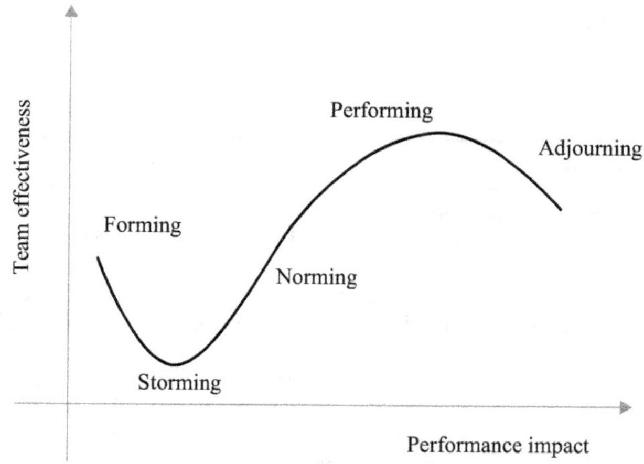

Figure 10.6 Tuckman's stages of group development

In this model, the most complicated stage is the *Storming* phase, in which the team may encounter problems and lose time.

There are two complicated aspects of the *Storming* phase: human relations and tasks. Human relations fraught with discussions and tasks are performed more slowly than usual. If this stage is managed well, it can be helpful in improving creativity and developing new ideas.

Today teams change much more quickly than they did in the sixties. Often people are forced to change goals and elements so as to maintain high performance standards. Whenever you make a change in the objectives, or even worse, in the elements that make up a team, the Tuckman model falls and you have to start again from the *Forming* phase. That is why the fate of any production depends on the quality of the *Forming* phase.

10.10 Build high-potential teams

There are a lot of case studies about group dynamics. The quality of relationships in a team is the key factor. All of these relationships are based on soft skills.

The results of many research studies tell us something about the relationship between the personality traits of each member and team performance.

Eggup has made a specific mathematical model to support the *Forming* phase. The model is based on a psychographic analysis of each team member with the 'International Personality Item Pool (IPIP) Big-Five Factor Markers' test.

The process is: if we know the psychographic characteristics of each team member we can assess the team potential, and then we are able to predict the level of performance of a team.

Using the 'IPIP Big-Five Factor Markers' model, each team member is assessed on five personality traits:

- *Extraversion*: a tendency to be energetic, to be enthusiastic, to seek stimulation and to enjoy interacting with people.
- *Agreeableness*: a tendency to be kind, compassionate, cooperative, generous and to avoid conflicts.
- *Conscientiousness*: a tendency to show aim for achievement, self-discipline, to have a focus and to act dutifully.
- *Emotional Stability*: a tendency to be calm, to experience pleasant emotions, and to be unlikely to feel tense or rattled.
- *Openness to Experience*: a tendency to appreciate art, emotion, adventure and unusual ideas, to be creative, imaginative and curious.

The Eggup model assesses the team and predicts its potential effectiveness.

10.11 Project life cycle and team roles

As defined in an article by Fisher, Hunter and Macrosson [12], every kind of project has a life cycle based on these five stages:

1. Planning & defining resources needed
2. Scheduling
3. Implementing
4. Measuring progress & reporting
5. Maintaining human relations

The five stages, combined with the 'IPIP Big-Five Factor Markers' model, enable Eggup to establish team role analysis. Eggup gives more attention to the soft skills of each team member and introduces the concept of Team Potential. Each of these roles contributes to each phase of the project life cycle (Figure 10.7).

The map of the five Eggup roles and the Team Potential are important during the 'forming' phase of a team.

Team Potential is primarily based on two aspects:

- Performance, that is, the ability to reach goals.
- Satisfaction, that is, the well-being of each member in having reached a certain result.

To have an effective team we need to have team *satisfaction* and team *performance*.

In a study carried out by Lester, Meglino and Korsgaard [13], an important relation is defined between job satisfaction for each team member and team effort. Further, they confirm a direct proportion between effort and the team performance.

During team formation, we need to give attention to the psychographic characteristics of each member. In this way we maximise the opportunity to create satisfied teams.

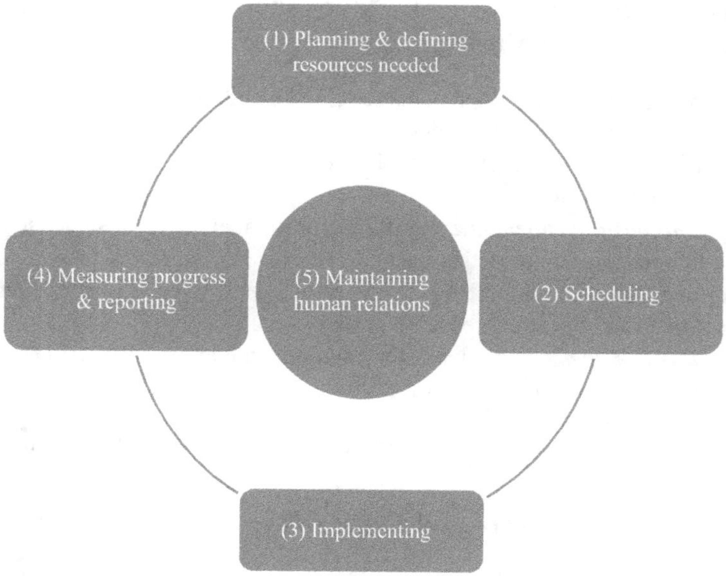

Figure 10.7 Project life cycle and team roles

During team formation we need to consider:

- Technical skills required to conduct a project in the right way.
- Soft skills useful for the establishment of effective teamwork.

One of Eggup's most successful case studies was a project developed for 152 students on the marketing course, taught by Professor Carlo Alberto Pratesi, of Roma TRE University, Rome, Italy.

The aim of this case study is to suggest, considering only students' psychographic characteristics, the best fit, in terms of people combination, to set up high-performing teams with equal individual participation levels to work on a group project.

Eggup, by considering only the personality of each student, was able to produce 31 teams with an average size of 5 people, meeting the requirement of high-performance with equal individual performance levels.

The results were:

- No clashes among team members during the execution of a project.
- Increasing performance levels and quality of output for each project team.
- Eighty percent of the entire student population were very satisfied by the composition of their teams and suggested Eggup also be used in other classes and/or in other projects.

Here is a spontaneous comment from a student about the use of Eggup:

> It was very interesting to work with people who I did not know before. We got in tune immediately and I enjoyed working with them. In the end it was a good project and I felt comfortable with them.

This kind of experimentation demonstrates how important it is to consider the distribution of behavioural characteristics in the team-*Forming* phase. With equal technical skills, what makes the difference is soft skills. The right mix of soft skills, from the start, allows the difficult *Storming* phase to be accomplished with no particular problems.

10.12 Conclusions

When building a team, certain key elements must be considered in order to transform a group of people into an excellent high-performance team. The innovative Eggup platform can be used to facilitate such a successful aggregation, as it includes all the proven algorithms, products of decades of research. Once the team is established, it's a matter of managing it, and in order to perform the fine art of team management, two cross-connected models can help us: the cross-roads model, which can remind us how important it is to deploy the art of listening to the 'system' to be managed, whether a small or large group of people; and the 'crazy car driving' model, which reminds us how important it is to let go of all our established thought patterns based on past experience, and to approach the team members with an open mind, giving non-judgmental feedback aiming at continuous improvement.

References

[1] *Qhaosing*® Ed Lulu 2009, by Raffaele Loscialpo

[2] Simona Saccone. *Modellistica e Simulazione del Comportamento del Traffico in Intersezioni Stradali Urbane*, University of Genova, Savona, Italy, 2007

[3] Abbas Mahmoudabadi. *The Assessment of Applying Chaos Theory for Daily Traffic Estimation*, Proceedings of the 2014 International Conference on Industrial Engineering and Operations Management Bali, Indonesia, January 7 – 9, 2014

[4] *The Application of Chaos Theory to Manage Congestion in Signalised Urban Networks*, TORG Seminar Series, Newcastle University, March 26, 2014

[5] Yinping Fu. *Research on the Short-Term Traffic Flow Forecasting Method Based on Chaos Theory*, Institute of Transportation and Economics, East China Jiaotong University, 2013

[6] Meredith Belbin R. *Management Teams: Why They Succeed or Fail*, ISBN 0 7506 2676 3, Butterworth-Heinemann, an imprint of Elsevier Science, Oxford U.K., 2002

[7] Belbin sample report, http://www.belbin.com/content/page/5639/Team%20report.pdf, p. 2

[8] Smith M.K. (2005). Bruce W. Tuckman – forming, storming, norming and performing in groups, the encyclopaedia of informal education.

http://infed.org/mobi/bruce-w-tuckman-forming-storming-norming-and-performing-in-groups/. Retrieved: 2014/12/18

[9] Gina Abudi. The five stages of project team development, 2009. Retrieved: 2010/05/18

[10] Tuckman B.W. 'Developmental sequence in small groups', *Psychological Bulletin*, 1965

[11] Tuckman B.W., Jensen M.A. Stages of small-group development revisited, 1977

[12] Fisher S.G., Hunter T.A., Macrosson W.D.K. 'A validation study of Belbin's team roles', *European Journal of Work and Organizational Psychology*, 2001

[13] Lester S.W., Meglino B.M., Korsgaard M.A. 'The antecedents and consequences of group potency: a longitudinal investigation of newly formed work groups', *Academy Management Journal*, 2002

Chapter 11

Integrated visual information for maritime surveillance

Domenico D. Bloisi[1], Luca Iocchi[1], Daniele Nardi[1] and Michele Fiorini[2]

11.1 Introduction

The protection of coastal areas is becoming a necessary requirement due to the increasing threats such as illegal smuggling, immigration, illegal fishing, oil spills, and, in some parts of the world, piracy. Moreover, the control of vessel traffic is often correlated to environment protection issues, since vessels carrying dangerous goods (e.g., oil-tankers) can cause huge environmental disasters.

There exist various surveillance systems for the maritime domain, relaying on different types of sensors, including Vessel Monitoring Systems (VMSs), Automatic Identification System (AIS) [1], ship- and land-based radars, air- and space-born SAR systems, harbor-based visual surveillance, and Vessel Traffic Services (VTS) systems [2].

These technologies integrate several information sources, for example, systems combining AIS data with SAR-imagery [3], buoy-mounted sensors with land radars [4,5], visual- with radar-based surveillance [6], and multiple ship-based sensors [7].

In particular, VTS systems combine radar and AIS data and are often equipped with long-range surveillance cameras. However, using radar and AIS data only is not sufficient to ensure a complete solution to the maritime surveillance problem, due to the following limitations:

- The AIS signal may be not available (AIS device not activated or malfunctioning).
- The recognition task for non-cooperative (non-AIS) targets can only be addressed visually by human operators.
- Radar-based systems are not suitable for vessel traffic monitoring in populated areas, due to high electromagnetic radiation emissions.

Replacing the radar with cameras is a feasible solution to address the maritime surveillance problem, without the need to place a radar antenna in a populated area.

[1]Department of Computer, Control, and Management Engineering, Sapienza University of Rome, Rome, Italy
[2]Selex ES S.p.A. – A Finmeccanica Company, Rome, Italy

An example is ARGOS system [8], designed to monitor the naval traffic in the center of the city of Venice (Italy) by using only cameras.

The aim of this chapter is to illustrate a framework for extending the functionalities of the currently used VTS systems, by adding a visual dimension to the usual radar-like view together with the possibility of automatically recognizing non-cooperative targets.

The framework is designed to:

1. detect boats by means of a classifier-based approach, using electro-optical (EO) images
2. find the horizon line in the observed scene
3. track multiple boats even in the presence of occlusions
4. fuse information from existing VTS systems with visual data
5. provide feedback information to the VTS system in order to adjust the radar sensitivity
6. be deployable in populated areas.

The main contribution of this chapter is to provide a data fusion (DF) scheme for combining in a unique view, radar and visual data. The experimental evaluation of the performance for the modules included in the framework has been carried out using publicly available data from the VOC dataset [9] and the MarDT – Maritime Detection and Tracking (MarDT) [10] data set, containing data coming from different real VTS systems, with ground truth information. Moreover, an operative scenario where traditional VTS systems can benefit from the proposed approach is presented.

The remainder of the chapter is organized as follows. After analyzing related work in Section 11.2, the proposed framework is detailed in Section 11.3, presenting also a possible operative scenario. A quantitative evaluation of the vessel detection module is provided in Section 11.4, while an application example in the air traffic domain is presented in Section 11.5. Conclusions are drawn in Section 11.6.

11.2 Related work

The following aspects made maritime automatic surveillance a very challenging task (see Figure 11.1):

- wide areas to be monitored
- highly dynamic background (due to waves and boat wakes)
- weather issues (such as heavy rain or fog)
- night-time monitoring.

In addition, the objects to be monitored can have very different size (ranging from few to hundreds of meters in length) and their number can be very high: 23 million tons of cargo and 55,000 passengers travel by ship every day worldwide [11]. Thus, in order to perform an accurate and effective monitoring of coastal areas, it is necessary to manage a huge amount of data coming from multiple (e.g., multiple cameras) and heterogeneous information sources (e.g., daylight and night cameras, radar, and AIS).

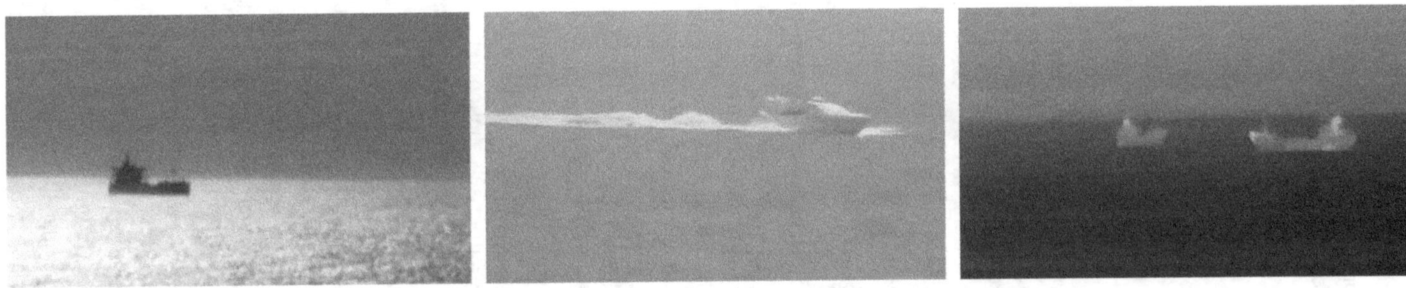

(a) (b) (c)

Figure 11.1 *Challenges in maritime automatic surveillance. (a) Reflections on the water surface; (b) boat wakes; (c) night-time monitoring*

Vision is becoming one essential component in recent maritime surveillance systems and this trend is supported by the deployment of a number of working systems [5]. Several approaches concerning information fusion solutions for the maritime domain are discussed in the rest of this section (see also a recent survey in [12]).

11.2.1 Systems using radars

Raytheon Marine Small Target Tracker [13] is a system deployed for surveillance of the Straits of Gibraltar and waterways near New York airports. The system fuses information from multiple radars, detecting and tracking small boats at a distance of 10 nautical miles (NM). The use of microwave radars limits the surveillance within the horizon.

Accipiter Radar developed a radar surveillance network solution [14], which is currently deployed for surveillance of portions of Lake Erie and Lake Ontario between USA and Canada. The network is built using off-the-shelf radars, which can be placed on rooftops, water towers, mobile vehicles, aerostats, and towers. Since only radar sensors are considered, the output of the system can provide only a 2D view of the observed situation.

11.2.2 Satellite-based systems

Automatic Sea Vision (ASV) [15] is an automatic optical system for maritime safety by using infrared (IR), GPS, and AIS data. To detect relevant objects, the sea area captured by the sensor is segmented and its statistical distribution is calculated. Any irregularities from this distribution are supposed to correspond to objects of interest. However, such an approach can produce false positives if there are wakes on the water.

COMMANDER [16] is a system developed by Thales Canada in which the Command, Control, and Communications (C3) nodes are connected through a satellite communication network to provide Canada-wide coverage. It is a system for the sharing of a Common Operating Picture (COP), including contact data, messages, and geo-referenced map overlays, between vessels, helicopters, and shore installations. However, the management of non-cooperative targets remains problematic.

11.2.3 Systems using sonars

Harbor Surveillance System [17] uses multiple sensors, including radar, sonar, and EO-IR devices to detect divers, swimmers, and small boats. The system is developed by DSIT Solutions and uses an autonomous underwater vehicle (AUV) that performs underwater surveys using forward-looking and side scan sonar systems. The declared detection ranges for the underwater surveillance unit is 500 m.

Data coming from radar, EO-IR, and sonar are used in the system called HarborGuard2 [18] to provide over and underwater surveillance. The system, developed by L-3 Klein, is currently deployed by the US Navy for protection of base facilities; local governments for bridge, port/harbor, and critical infrastructure security; and commercial companies for oil drilling rig and critical asset protection. However, the detection is based only on the radar sensors and the human operator is responsible to assign a tracked target to the CCTV/Thermal Imaging camera controller.

11.2.4 Camera-based systems

SeeCoast system [19] detects, classifies, and tracks vessels by fusing EO and IR video data with radar and AIS information and provides decision support. It has been deployed in Coast Guard sites in Virginia, USA. The detection is carried out by estimating the motion of the background and segmenting it into components. However, motion-based vessel detection can experience difficulties when a boat is moving directly towards the Maximum Average Correlation Height (MACH) filters are employed for vessel classification in Reference 20. Vessel detections are cross-referenced with ship pre-arrival notices in order to verify the access of vessels to the port. As reported by the authors, such an approach tends to misclassify small vessels (e.g., fishing boats).

An object detection system for finding ships in maritime video is presented in Reference 21. The used approach is based on the Histogram of Oriented Gradients (HOG) [22]. Since the calculation of the detection features involves a significant amount of computational resources, real-time performance can be obtained only by means of hardware acceleration with programmable components such as FPGAs.

A method for visual surveillance in maritime domain with non-stationary camera installed on an untethered buoy is described in Reference 23. After the detection of the horizon line, a color gradient filter is applied to obtain a grayscale image with intensities corresponding to the magnitude of color changes. Detection of the objects of interest is performed through thresholding of such grayscale image into a binary map. The algorithm is limited by the assumption that all marine targets are located above the horizon line.

11.2.5 Discussion

As argued in Reference 12, none of the above-presented systems addresses the requirements for performing in highly cluttered conditions, for tracking targets in ambiguous situations, and for reporting suspicious activity. Moreover, most of the cited systems use the radar as the main sensor and such a solution may not be applicable to populated areas. As an example, just consider the maritime traffic in the Venetian Laguna or in Venice down-town itself [8] and, more in general, in a touristic or residential environment where it is not acceptable to live near a radar antenna.

In this chapter, we propose a camera-based framework in which cameras are used in combination with radar data in order to obtain a vessel traffic monitoring system with a high accuracy. In our approach, cameras are the main information sources, thus a system that adopts the presented framework can be deployed both in a site where radar is already existing (augmenting its performance) and in a site where it is not possible or not convenient to install a radar.

Furthermore, we propose solutions to deal with the challenges related to automatic video surveillance in the maritime domain [24]:

- non-stationary cameras (e.g., Pan-Tilt-Zoom (PTZ) cameras)
- different classes of targets (ranging from huge oil-tankers to small phishing boats)
- reflections and boat wakes on the water surface
- boats anchored off the coast, which can appear motionless.

In Section 11.3, the architecture of the framework is presented together with the approaches and methods that can be used to deal with the above-listed issues.

11.3 Architecture of the framework

The functional architecture of the proposed framework is shown in Figure 11.2. An EO-IR visual device is the main sensor and it can be moved through a control module by a human operator. The control module provides the orientation and the field-of-view (FOV) of the visual sensor to the Video Processing Unit (VPU). The VPU is responsible to detect and track the vessels by using visual information. It is worth noting that the detection task is made complex by the camera movements, since it is not possible to create a model of the observed scene.

The DF module receives data from both the VPU and the VTS system. It aims at associating the visual tracks coming from the video analysis with the tracks generated from radar and AIS data. In this way, it is possible to provide the user a new visual dimension, in addition to the traditional geo-referenced, radar-like VTS view. Moreover, the DF module sends feedback information to the VTS system, which can be useful to improve the detection accuracy of the radar by adjusting its parameters.

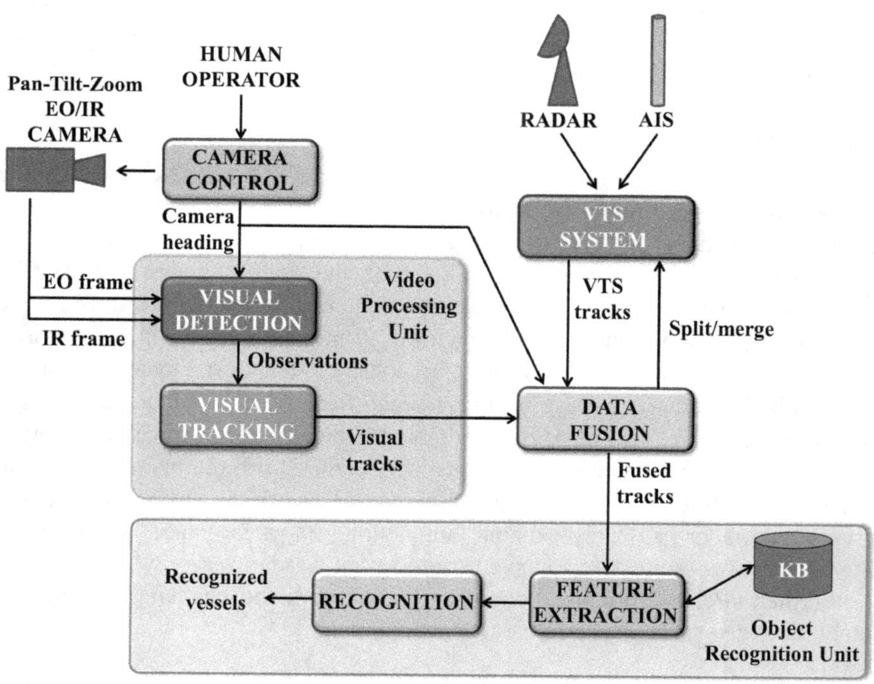

Figure 11.2 Functional architecture of the proposed framework. The DF module manages information coming from the Visual Processing Unit and the VTS system and sends validated tracks to the Object Recognition Unit

The tracks generated by the DF module are sent to the Object Recognition Unit, which classifies them according to specific visual features. In the remainder of this section, each module of the framework is described in detail.

11.3.1 Visual detection

The visual detection module is part of the VPU and it aims at processing the current video frame by searching for the targets of interest. Visual detection is a crucial task, since detection accuracy affects all the stages in the VPU pipeline and it must be as high as possible, while maintaining an acceptable computational load. The visual detection module is made of three main blocks: (1) the Haar-based boat classifier, (2) the horizon detection function, and (3) a filter for removing possible noise (see Figure 11.3).

11.3.1.1 Haar-based classifier

Since the camera can be moved by the user, a foreground/background modeling approach to detect vessels (e.g., background subtraction) could be not effective (due to the changes in the background produced by the movements of the camera by the user). A possible solution consists in adopting a classification-based method. Classification is used to learn a model from a set of labeled data instances (training phase) and then

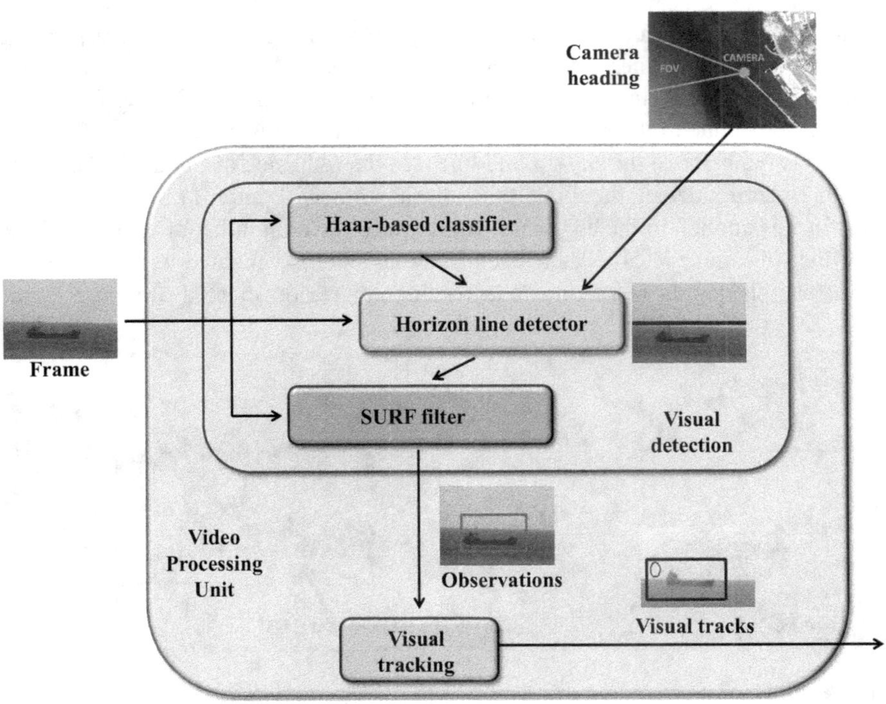

Figure 11.3 Video Processing Unit. Video frames coming from the camera are analyzed in order to find boats and the horizon line. False positives are filtered out exploiting a SURF-based filter

to classify a test instance into one of the pre-defined classes by using the learned model (testing phase) [25]. Classification-based detection can operate on still images, thus avoiding the creation of a model of the background.

Different methods can be used to create the classifier [20,9]. In order to obtain real-time performance, we discarded computationally expensive testing methods [22,26] and adopted an approach based on Haar-like features [27]. It is worth noting that such an approach has been originally designed for face detection, thus we suitably adapted it for boat detection [28,24]. Moreover, the chosen approach gives the possibility of combining different weak classifiers in a cascade, in order to obtain a strong classifier.

A set of 4,000 images not containing boats (negative images) and a set of 1,500 images taken from the Internet, depicting various types of vessels with different view angle (positive images), have been used as input for the offline training phase, obtaining a 24-level classifier (the training stage was stopped when the false alarm rate reached $5 \times 10^{-6} = 0.000005$. The OpenCV [29] HaarTraining functions have been used to create the classifier with a 60×30 search window. The size of search window has been experimentally found. The used set of positive and negative images can be downloaded from the MarDT [10].

11.3.1.2 Horizon line detector

Along with boats, it can be useful to detect the limit of the sea surface also. Depending on the heading of the camera, the module called Horizon Line Detector is responsible to find the horizon line, differentiating between sky-water and sky-ground line (Figure 11.4). Indeed, in presence of the coast, the probability of finding false positives increases due to the rich texture of the scene, thus it is convenient to filter out detections lying above the sky-ground line (see the right side of Figure 11.4).

In order to detect the limit of the sea surface (Figure 11.5a), the Hough transform is applied to the edge map of the frame, creating a list of candidate lines (red lines in Figure 11.5b). Then, each candidate line is validated with respect to a set of sample points belonging to a rectangular region created around the line

Figure 11.4 Sky-water and sky-ground line. The probability of finding false positives increases in presence of the coast, due to the rich texture of the scene (right side of this figure)
Source: The images are from the VOC dataset [9].

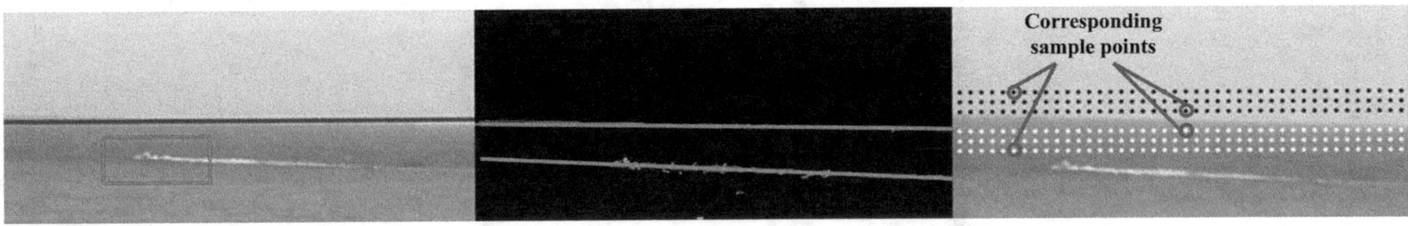

(a) (b) (c)

Figure 11.5 *Horizon line detection. (a) The horizon line correctly detected; (b) the edge map with the Hough line in red; (c) the set of sample points used for validating the line (two corresponding pairs are highlighted)*

(Figure 11.5c). The points above the candidate line are compared with the corresponding points under the line. If the percentage of corresponding pairs with different intensity values is over a predefined threshold then the line is considered valid. We set the threshold at 95%, in order to filter out false positive detection that can be generated by long wakes. As an example, in Figure 11.5b two candidate lines are individuated. The lower line is discarded because the sample points over and under the line have similar intensity values, while the upper line is considered as a valid horizon line.

11.3.1.3 SURF filter

The acronym SURF (Speeded-Up Robust Features) indicates a scale- and rotation-invariant detector and descriptor [30]. SURF can be used to filter out possible false detections made by the classifier. Indeed, in the maritime domain, false detections can be caused by waves, boat wakes, and reflections on the water surface. When the classifier detects an object of interest, its bounding box is analyzed in order to extract the SURF key points in it. If a negligible number of key points are present in the bounding box, then the observation is rejected.

In Figure 11.6, the above presented algorithm can be used to filter out a false positive detection caused by wakes (black bounding box). In Figure 11.7, the same algorithm allows to filter out two false positive detection caused by reflections on the water surface (the two black bounding box on the bottom of the image).

11.3.1.4 Observations

The output of the detection module is a list of observations, each of them being a bounding box representing a detected boat. Our classifier [28] is able to detect targets of different size, with blurring noise, in presence of boat wakes and reflections on the water surface (see Figure 11.8).

Figure 11.6 False detections caused by wakes can be filtered out analyzing the SURF key points in the bounding box. If a negligible number of key points are found, then the observation is rejected

Figure 11.7 *In case of reflections, SURF can be used for rejecting false positive detections, since we experimentally noted that the number of key points generated by reflections on the water surface is limited*

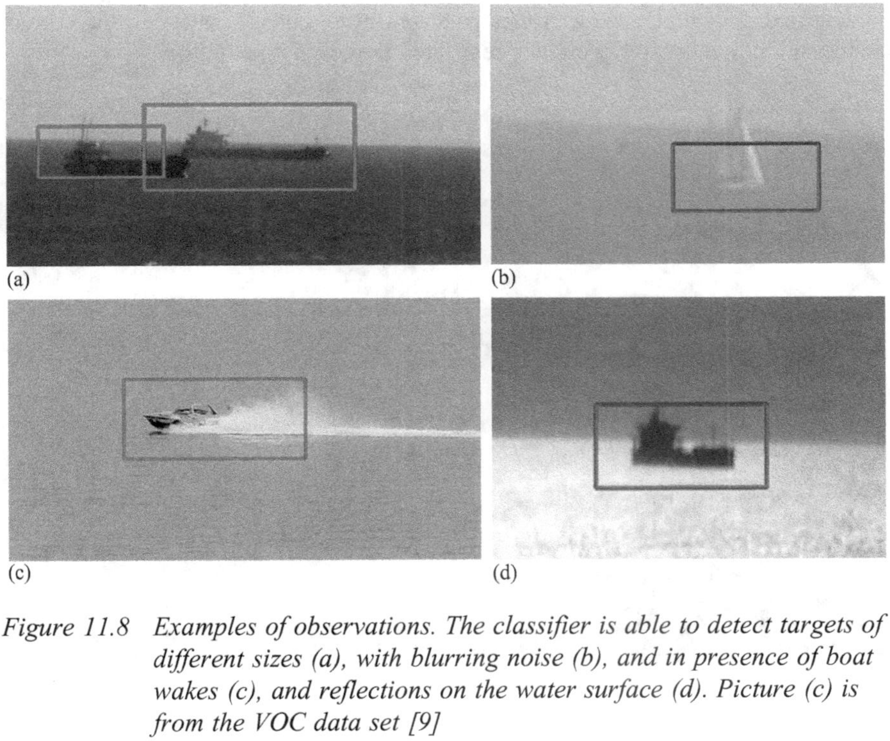

Figure 11.8 *Examples of observations. The classifier is able to detect targets of different sizes (a), with blurring noise (b), and in presence of boat wakes (c), and reflections on the water surface (d). Picture (c) is from the VOC data set [9]*

11.3.2 Visual tracking

The visual tracking module takes as input both the image flow and the observations generated by the detection module. It returns a set of visual tracks, that is, bounding boxes with an identification number, and has the role of discarding false positives by temporal filtering the received observations. Tracks and observations are associated by measuring the Bhattacharyya distance between the HSV color histogram of the track (updated over time) and the one of the observation. Only tracks that present a sufficient number of associated observations are considered of interest (the association threshold has been set to 10): such a policy allows to filter out a high number of false positive detections.

11.3.2.1 Group tracking

Sometimes occlusions can occur when boats are aligned with respect to the camera view or when boats are close to each other. Occlusion situations cause considerable difficulties for visual tracking. A possible solution is to consider partially overlapping tracks to form a group instead of tracking them separately (see Figure 11.9). When two or more tracks have their bounding boxes moving closer to each other (Figure 11.9a), the tracker starts considering them to form a group (Figure 11.9b) and it stores the HSV histograms of the involved tracks – the histograms will be used as models to re-identify the tracks when the occlusion phase is over. To this end, when only one track is completely visible, the tracker considers the occluded track as frozen (Figure 11.9c). When the occluded object becomes visible again, a new track is identified, but the tracker can use the saved histograms to re-assign the correct identification number to the already registered target (Figure 11.9d).

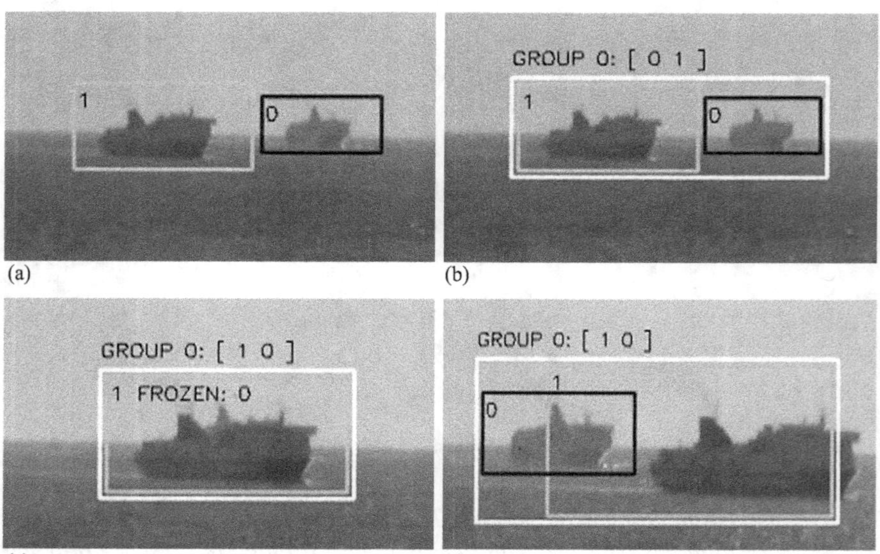

(a) (b) (c) (d)

Figure 11.9 Group tracking

11.3.3 VTS system

A Coastal Surveillance Site (CSS) is responsible for detecting and tracking vessels in a range up to 12 NM, the defined "Territorial Waters" [31]. A VTS system is a CSS designed to merge information coming from radars and AIS devices. The user is provided with a 2D geographical view, showing all the boats navigating near the site. For AIS-equipped vessels, the VTS system can also show additional information such as the name of the boat and its route.

11.3.4 Data fusion

The DF module aims at creating a combined view for the monitored situation, which is not available in today's VTS systems (see Figure 11.10). The following steps are performed by the DF module:

1. *Data synchronization*: VTS and video data are synchronized to perform a consistent data association. Time synchronization is required since VTS data have a refresh period of about 3 s, while the video stream is transmitted at 25 frames per second (fps).
2. *VTS tracks selection*: VTS tracks in the FOV of the camera are selected (see right side of Figure 11.10) by considering the current heading of the camera, which is received from the camera control module.
3. *Rotation*: VTS tracks are aligned with respect to the camera heading, since the 2D geographical view can be rotated with respect to the frames acquired by the camera.
4. *Common Space Projection*: Visual and VTS tracks are projected in a common, normalized 2D space (see Figure 11.11). This step can be performed since all the boats are moving on a plane (the water surface).
5. *Association*: A probabilistic association between VTS and visual tracks is carried out in order to build the final integrated view (see Figure 11.12).

DF between video and VTS data is performed on a probabilistic basis. VTS and visual tracks are projected onto a 2D common reference space in order to perform the association. For visual data, the first dimension (x) in the common

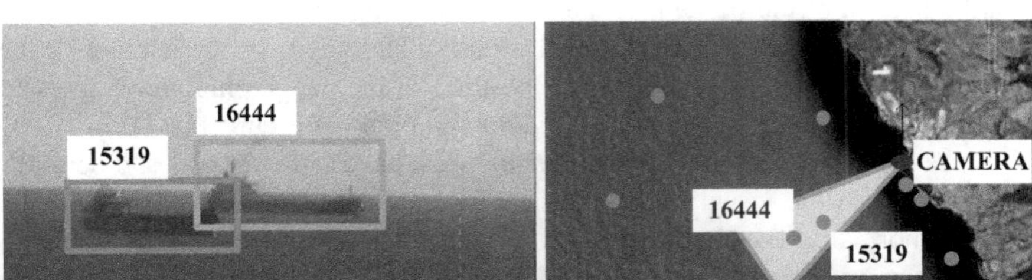

Figure 11.10 Combined view. The state-of-the-art VTS systems can be improved adding a visual dimension (left side of this figure) to the currently available radar-like view (right side of this figure)

Figure 11.11 Visual and VTS data projection. The values for the x *and* y *coordinates are extracted from the video frame (left part of the figure) and the geographical view (right part of this figure)*

Figure 11.12 Probabilistic association. The common space is shown on the left, with VTS data marked in red and visual tracks in black. The observed scene is reported on the right, with the AIS labels associated with the images of the boats. The current headings of the ships are also calculated

reference space is the distance (in pixels) of the bounding box from the left margin of the frame, while the second dimension (y) is the distance of the bounding box from the bottom of the frame (see the left side of Figure 11.11).

For VTS data, the distance (in pixel) of the VTS track from the left side of the FOV and the distance from the camera position in the geographic view represent the x and y dimensions, respectively (see the right part of Figure 11.11).

Since the video frame and the geographic view have different scales, the x and y values related to the visual track and the x and y coordinates related to the VTS track are normalized with respect to the width and the height of the common reference space window. The projected visual and VTS tracks are associated on the basis of a nearest-neighbor policy (see Figure 11.12), by using a predefined threshold. The proposed DF scheme is justified by the fact that the boats are moving on a planar surface, thus a boat that is closer to the camera than another one will appear in a lower position in the image [24].

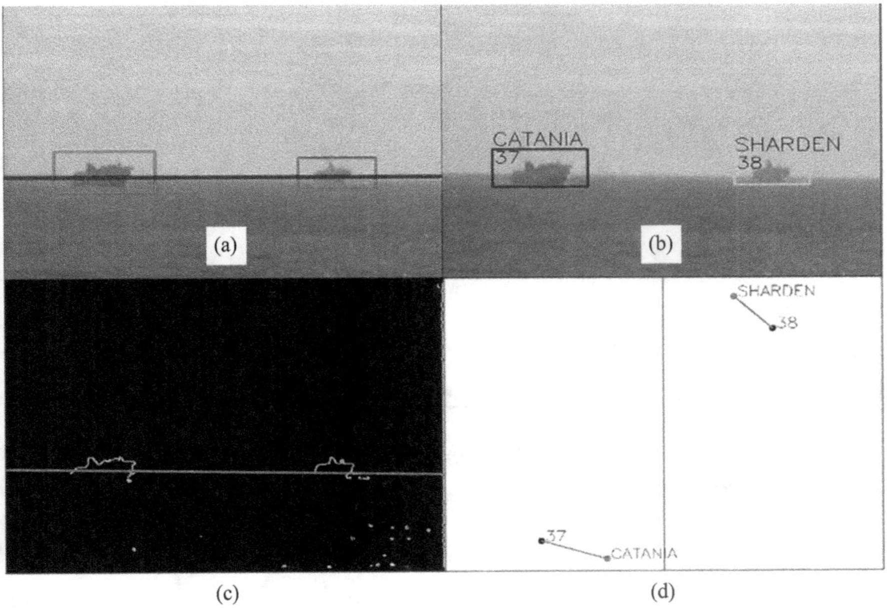

Figure 11.13 *Global view of the observed situation. (a) Visual detection; (b) visual tracking and VTS data fused in the same view; (c) edge map for sky-water line detection; (d) data association in the common space*

The final output presented to the user is a global view of the observed situation, as reported in Figure 11.13.

11.3.4.1 Feedback to VTS system

The DF module can help in correcting erroneous detection made by the radar. In presence of targets having a relevant size (e.g., oil-tankers) it is possible to have multiple VTS tracks for the same target (split case). On the other hand, in the case of a small boat navigating near a bigger one, it is also possible to have a single VTS track instead of two (merge case).

Figure 11.14 shows an example of a split situation: the system detects two visual observations (Figure 11.14a) and two visual tracks (Figure 11.14b), but multiple VTS tracks (projected in Figure 11.14c as red dots) also. The multiple radar detections can be clustered by exploiting the visual tracks (the black dots in Figure 11.14c referring to the visual tracks in Figure 11.14b) as centroids.

11.3.5 Object recognition

The Object Recognition Unit is responsible for recognizing the boats in the FOV of the camera. In order to deal with such a complex problem, a novel algorithm, called "barcode matching," is proposed. The method is designed to work in a well-defined operative scenario and to manage non-closed curves and low information images.

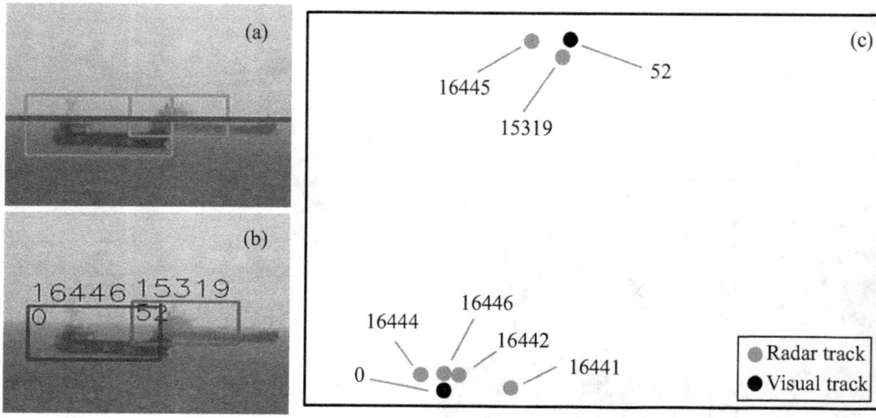

Figure 11.14 Feedback information can be sent to adjust the radar parameters in case of misdetection

The "barcode matching" requires a database containing a set of records of known vessels, each record containing the silhouette of a boat and a heading. Multiple silhouettes of the same boat can be stored, corresponding to different headings. All the stored silhouette images are normalized to the same size, in order to allow a comparison between them.

The barcode matching method consists of the following steps (see Figure 11.15):

1. The bounding box around the object to recognize is obtained by the list of valid tracks (Figure 11.15a).
2. The object's silhouette inside the bounding box is extracted and the edges belonging to the horizon line (sky-water line), if present, are filtered out. A histogram is computed starting from the object's silhouette. The histogram represents the distance of the object's silhouette from the top of the image (Figure 11.15b).
3. A model histogram is chosen in the database according to the heading of the object (Figure 11.15c).
4. The chosen model histogram is compared with the object's histogram by calculating the Bhattacharrya distance between them. Then, the barcode for the object's histogram is computed and it is compared with the barcode from the model silhouette (Figure 11.15d).

The more similar the two histograms are, the lower the Bhattacharrya distance value is. However, even if such a distance grows in proportion to the amount of differences in the two histograms, it is not possible to completely rely on it, in order to recognize an object. Indeed, it is very difficult to determine a threshold that can be used to recognize an object given a model's histogram. Thus, in order to improve the discriminative power of the method, an additional set of features has

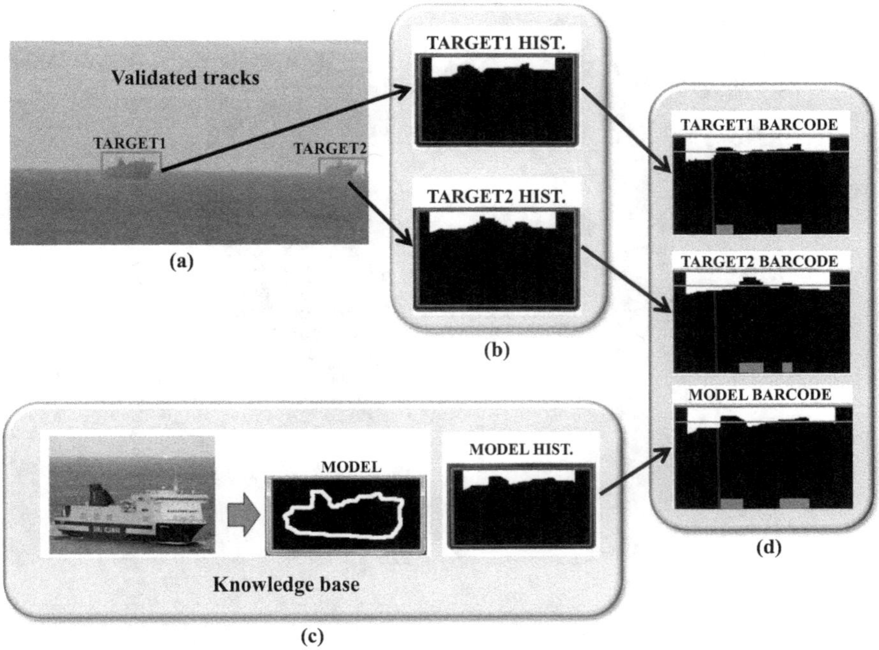

Figure 11.15 Target silhouettes extraction and comparison. (a) Validated tracks are analyzed in order to extract silhouettes; (b) targets' silhouettes are used to generate histograms; (c) the knowledge base in the Object Recognition Unit is made of histograms ordered by heading values; (d) barcode matching

been used. By computing the mean value μ of the histogram, it is possible to identify the bins in the histogram, whose values are below μ.

As an example, the silhouette produced by a chimney or by an antenna of a boat can be identified by analyzing the number and the position of the bins that are above the mean μ. Indeed, the barcode for each histogram highlights particular features of the silhouette of the vessel, which can be used to distinguish between two possible matches (see Figure 11.15).

It is worth noting that, even if the described recognition algorithm is strongly affected by the angle with which the camera captures the boat, in real scenarios the heading of the boats near the coast is rather stable and predictable. Indeed, usually vessels approaching or leaving ports have to navigate in specific corridors. Figure 11.16 shows an example where two boats are navigating in opposite directions, inside a corridor that is established by nautical charts. The example demonstrates that, in operative scenarios, the angle between the boats and the camera is predictable, since all the boats are forced to follow almost the same path and the camera can be deployed accordingly, in order to obtain the best possible view.

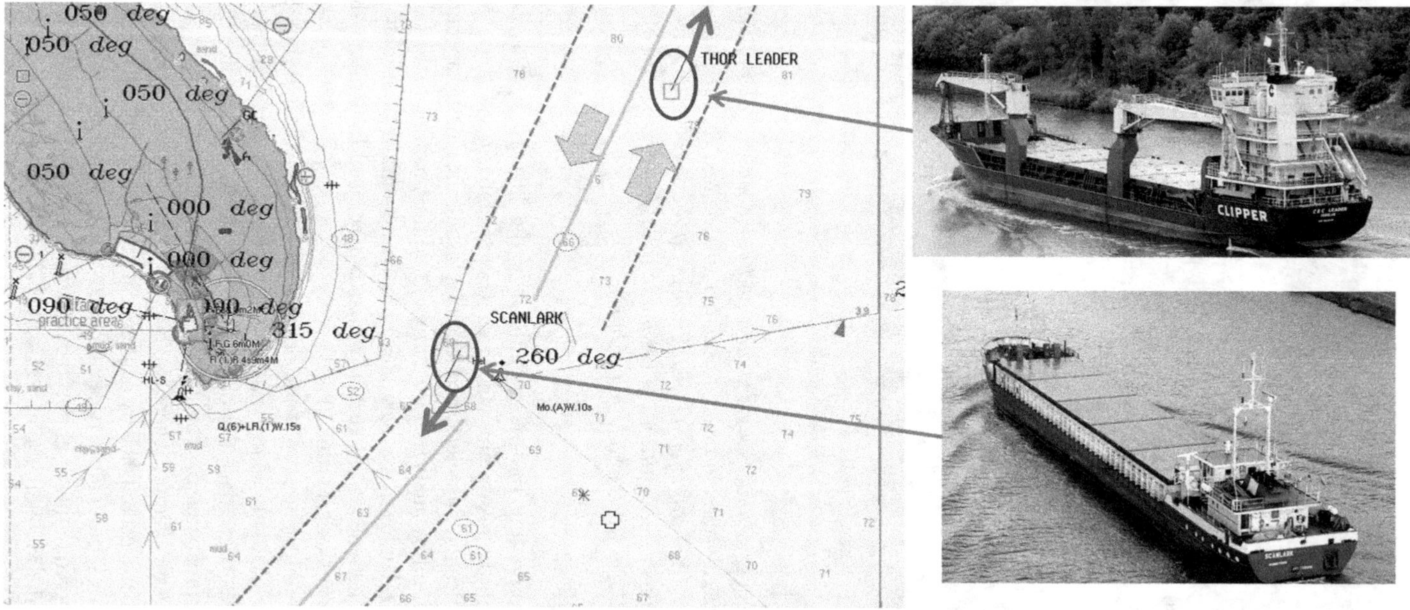

Figure 11.16 *Operative scenario. Two boats are navigating in a corridor to enter/leave a port, maintaining a predictable heading*
Source: Images are from the website www.marinetraffic.com.

11.4 Experimental results

The boat detection approach has been tested to evaluate both the detection accuracy and the required computational load. Images from two different publicly available data sets have been used:

1. MarDT, containing video sequences coming from two real VTS sites [10]
2. VOC, containing images coming from the publicly available VOC database [9].

11.4.1 MarDT data

In order to quantitatively evaluate the accuracy of the detection method, a data set of maritime video streams and images, called MarDT, containing ground truth data that can be used for evaluating the performance of automatic video surveillance systems and of computer vision techniques [10], has been built. In particular, for each video in the data set the following details are provided:

- sensor type (EO or IR)
- camera type (static or moving)
- location
- light conditions (sunny or cloudy)
- foreground masks to evaluate the foreground segmentation results
- annotations with the bounding box vertices to evaluate the detection results
- identification numbers to evaluate the data association and tracking results.

In addition, the videos have been recorded with varying observing angles and weather conditions. MarDT data set contains:

1. real visual data recorded in a VTS center in Italy;
2. real visual data recorded in a VTS center in the North of Europe;
3. real visual data from the ARGOS system [8] that monitors the naval traffic in the Grand Canal in the city of Venice, Italy;
4. visual data recorded in a port in Asia.

A set of 100 randomly chosen images taken from 11 videos recorded with different light conditions and camera positions has been used to test the accuracy of the boat detection. To filter out false positives due to wakes (see Figure 11.6) and reflections (e.g., in Figure 11.8), an additional weak-classifier has been created by means of a negative set made of 4,000 images of wakes and other false positive detections obtained by the original classifier. The results are reported in Table 11.1, in terms of detection rate (DR) and false alarm rate (FAR)

$$\mathrm{DR} = \frac{\mathrm{TP}}{\mathrm{TP}+\mathrm{FN}} \mathrm{FAR} = \frac{\mathrm{FP}}{\mathrm{TP}+\mathrm{FP}}$$

where TPs are the correctly detected boats, FN is the number of not detected boats, and FPs are the incorrect detections.

Table 11.1 Visual detection results on data coming from a real site

Classifier	Coastline detection	Detection rate (DR)	False alarm rate (FAR)
Without wake examples	NO	0.892	0.475
Without wake examples	YES	0.892	0.265
With wake examples	YES	0.928	0.251

Table 11.2 Visual detection results on data from VOC [9]

Coastline detection	Detection rate (DR)	False alarm rate (FAR)
NO	0.872	0.332
YES	0.872	0.198

11.4.2 VOC data

The VOC database [9] is a publicly available data set containing several images of interest for maritime surveillance applications. We considered important to include in the evaluation such database, because it contains a large variety of boats and it is thus very adequate to test in particular the detection phase of the system. Since the VOC database contains images belonging to different classes of objects, we selected a subset of the images containing boats, discarding images that are not suitable for a VTS application. The results of the detection using our Haar-based classifier are summarized in Table 11.2.

Results show good performance of our system and, in particular, the improvement that can be obtained in reducing false alarms (from 0.332 to 0.198) when the horizon detection is activated.

11.4.3 Computational speed

The detection method has been tested with two different CPUs:

1. Intel Core 2 U7300 1.30 GHz 4 GB RAM (2 cores);
2. Intel Core i7 3770 3.40 GHz 16 GB RAM (8 cores).

The computational speed has been measured using recorded data, live data coming from a commercial Real Time Streaming Protocol (RTSP) webcam, and live data coming from an EO camera on site. The tests demonstrated the capability of the system architecture of exploiting all the computational power available. The results are shown in Table 11.3.

11.4.4 Discussion

The analysis of the results demonstrates that:

1. The Haar-based approach is an effective solution to boat detection with moving cameras in the maritime domain. An elevated DR can be obtained on real

Table 11.3 Computational load for the detection phase

Frame size	2 cores	8 cores
319 × 261	10	65
352 × 288	9	60
414 × 338	6	43
586 × 479	4	25
704 × 576	3	20

data and on benchmark images. It is worth noting that our detector has been built using images coming from real data recorded in a real site. Those data differ significantly from the VOC data set data, thus a detection rate of 0.872 can be considered an acceptable performance.

2. The horizon line detection is crucial in obtaining a lower FAR. Indeed, an approach based on Haar-like features for building the classifier inherently produces a high DR, with an elevated FAR. Thus, the chance to lower it through the horizon line analysis allows improving considerably the detection performance.
3. The approach is suitable for a real-time application. With a commercial CPU it is possible to achieve a real time (25 fps) processing speed for 586 × 479 images.

Moreover, the observations obtained using the boat detector are sent to the visual tracking module and then to the DF process. In such a way, it is possible to reduce drastically the FAR, as well as to improve the DR, thanks to the temporal filtering carried out by the tracker and to the context-based information coming from the VTS system. However, it is essential that frame-by-frame detection provides reliable results, since the presented framework is also conceived to be used in place of radar based systems at the occurrence.

11.5 Enhancing air traffic control with visual data

The use of the presented framework is not limited to the maritime scenario. Indeed, the same framework can also be successfully applied in the case of an air traffic monitoring system. Advanced surface movement guidance and control systems (A-SMGCS) aim at providing routing, guidance, and surveillance for the control of aircraft and vehicles in order to maintain the declared surface movement rate under all weather conditions within the aerodrome visibility operational level (AVOL), while maintaining the required level of safety [32]. The necessary condition for all the algorithms (routing, guidance, and conflict avoidance) to work correctly is the provision of reliable surveillance data in all airport areas.

Usually, Surface Movement Radars (SMRs) are used as the main source of information for estimating aircraft positions. However, in complex airport layouts (see Figure 11.17), traditional radar-based surveillance systems are affected by limitations in their coverage due to reflections or shadows that are caused by buildings, equipment, or other reflecting objects on the airport surface.

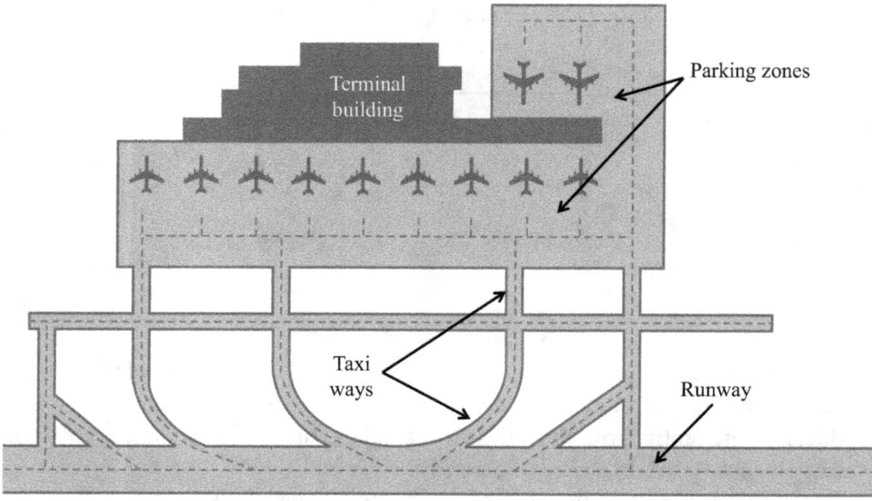

Figure 11.17 A general airport layout. Shadows and blind spots can occur in the parking zones due to the presence of buildings and reflecting objects on the airport surface

Cooperative systems for airport ground traffic surveillance, such as the Automatic Dependent Surveillance-Broadcast (ADS-B) system [33], can be involved for mitigating the problems. An ADS-B equipped aircraft determines its own position using a global navigation satellite system (GNSS) and periodically broadcasts its position as well as other relevant information to potential ground stations and to other aircrafts with ADS-B equipment. Nevertheless, the adoption of the ADS-B system does not provide reliable data, since the limitations mentioned earlier can also arise in the GPS signal propagation used by the aircrafts (or vehicles) to compute their position. Furthermore, not all the aircrafts are equipped with ADS-B transponders.

When the GPS signal is not available, it is possible to adopt a Multilateration (MLAT) approach. MLAT uses a grid of ground receivers to record ADS-B reports and to perform a time-difference-of-arrival analysis to estimate the aircraft position. However, MLAT can only detect cooperative targets since the aircraft transponder has to be switched on. In case of malfunction or if the transponder is switched off, the system cannot detect the aircraft [34].

For all these reasons, it is very difficult to provide a reliable coverage in all the airport areas without increasing the number of surveillance sensors (SMR radars or MLAT sensors). Moreover, electromagnetic pollution caused by radars is something that must be reduced to a minimum everywhere; therefore, a "green" technology should be used to get the exact position of all parked aircrafts reliably in order to provide the airport controllers a complete view of the situation at hand.

A possible solution to increase the overall system coverage as well as for reaching higher levels of the A-SMGCS specification [32] is to introduce a system able to act as a "gap-filler" for the airport zones that traditional sensors fail to cover reliably. Such a system should be:

1. less expensive than the electromagnetic sensors used in the state-of-the-art systems
2. radiation-free.

The framework discussed in this chapter, where the main sensors are electro-optical cameras, can be used to identify the presence of aircrafts in the parking zones (see Figure 11.17). The same assumption of using an uncalibrated PTZ camera (that can be moved freely by a human user) made in the maritime scenario is valid also when the goal is to detect and track aircrafts in parking areas. The information concerning the movements in the monitored area, for example, the departure of an aircraft can be merged with the data coming from the traditional radar-based system, thus increasing the overall system coverage.

11.5.1 A framework for ground traffic surveillance in airports

Parking zones in airports (Figure 11.17) represent a challenging scenario for automatic video surveillance [35], due to:

- quite fast changes of illumination (e.g., due to cloud movement)
- reduced visibility caused by adverse meteorological conditions (e.g., rain, fog)
- presence of not moving objects of interest: relevant objects should be detected both while moving and while they are steady (e.g., parked aircrafts).

Moreover, in order to limit the number of sensors, usually parking zones are monitored by PTZ cameras. Since PTZ cameras can be moved (automatically or manually), the conventional foreground/background modeling approach is ineffective being based on the assumptions that all the objects of interest are moving and that the camera is static.

To overcome these limitations, as in the above considered maritime scenario, the aircraft detection module has to be based on a classifier (see Figure 11.18).

The VPU is designed to detect and track the aircrafts. It takes as input a single image coming from the video stream (current frame). The DF module aims at associating the tracks coming from the VPU (visual tracks) and the tracks elaborated by the Multi-Sensor Data Fusion (MSDF) unit (system tracks). Indeed, the MSDF module collects and computes all the data coming from the existing electromagnetic sensors (ADS-B, SMR, MLAT).

11.5.2 Aircraft detection

The same Haar-like feature-based approach used to detect boats can be applied to detect aircrafts. To verify the applicability of the classification-based method for aircraft detection, we created a binary aircraft classifier by using the OpenCV HaarTraining functions.

A set of 1,000 images not containing aircrafts and a set of 550 images including different types of aircrafts (with different views) taken from the Internet have been used as input for the offline training stage obtaining a 15-level classifier

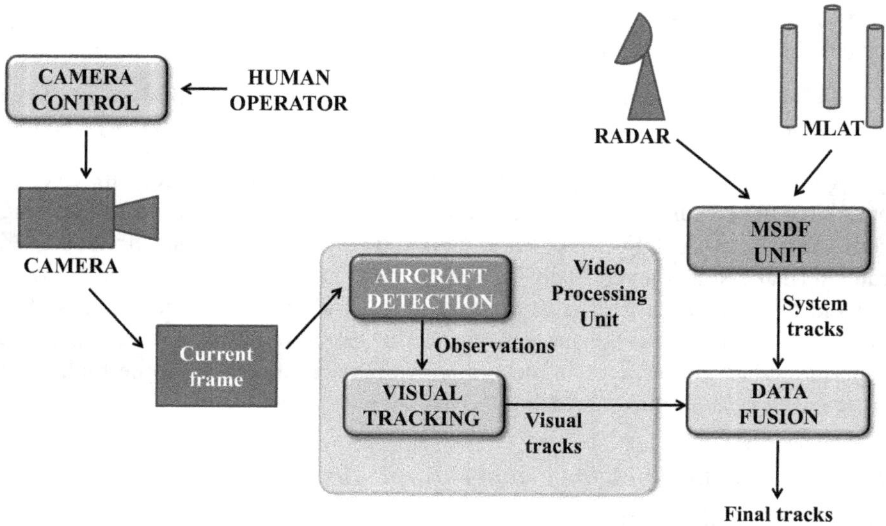

Figure 11.18 The proposed framework. Visual information can be used in combination with existing radar systems to increase the overall detection accuracy

(the training was stopped when the false alarm rate reached $5\times10^{-6}=0.000005$). We experimentally found that good detection results can be obtained using a search window of 90 × 30 pixels. It is worth noting that since the aircrafts can be viewed by different angles, the dimension of the search window is a crucial parameter for creating an accurate classifier.

The output of the aircraft detection module is a list of observations. Each observation is a bounding box representing a detected aircraft. In Figure 11.19 some examples of detection of different types of aircrafts are shown. The examples demonstrate that the proposed framework can be a flexible and cost-effective solution to improve existing systems performance.

11.6 Summary and conclusions

A framework for integrating visual information in an existing VTS system has been described. A major advantage of the proposed framework is that an enhanced, global view (VTS information plus visual data) of the observed situation can be built, by adding a visual dimension to the traditional view, which combines radar and AIS data only. Such an enhanced view is very effective for the user that has an immediate visual evidence of the situation at hand. The framework is designed to provide a feedback to the VTS system also, in order to improve the radar detection by adjusting the parameters of the sensors.

The results on multiple images and video sequences from real operational scenarios demonstrate the robustness of the approach with respect to the presence of changing light conditions, reflections on the water surface, boat wakes, and

Figure 11.19 Detection examples. The classifier is able to detect aircrafts given a video frame. The output is a list of bounding boxes (observations) containing the objects of interest
Source: Images are from YouTube videos.

targets that appear motionless (e.g., boats anchored off the coast). Moreover, the approach is robust to user-controlled motion of the camera.

The results on two different data sets (including the publicly available data coming from the VOC data set) show that the proposed detection approach provides accurate results, while maintaining a real-time computational speed.

Furthermore, in this chapter, a publicly available data set of videos and images called MarDT, containing data recorded in different scenarios, using both EO and IR cameras, with varying light and weather conditions, has been presented. MarDT contains the data used to validate the proposed framework and it will be improved in the near future, by extending the data sets and acquisition scenarios and by adding vessel silhouettes for object recognition.

Overall, we believe that our framework substantially improves the effectiveness of surveillance in maritime domains. The underlying enabling technology is provided by exploiting a proper fusion of the information coming from different sensors. The key feature of the proposed approach is the use of cameras as the main sensors, differently from traditional VTS systems that use radars and AIS data. Moreover, an application example of the presented framework on in an Air Traffic Control use-case shows the flexibility of the described functional architecture and its applicability in multiple contexts.

References

[1] I. Harre, "AIS adding new quality to VTS systems," *The Journal of Navigation* 3 (2000) 527–539

[2] Canadian Coast Guard, Vessel Trac Services (VTS) update study, vol. 159, 1991

[3] G. Saur, S. Estable, K. Zielinski, S. Knabe, M. Teutsch, and M. Gabel, Detection and classification of man-made offshore objects in terrasar-x and rapid eye imagery: Selected results of the demarine-deko project, in: Proceedings of IEEE Oceans, Santander, 2011
[4] W. Kruger and Z. Orlov, Robust layer-based boat detection and multi-target-tracking in maritime environments, in: Waterside Security Conference (WSS), 2010 International, pp. 1–7
[5] S. Fefilatyev, D. Goldgof, M. Shreve, and C. Lembke, "Detection and tracking of ships in open sea with rapidly moving buoy-mounted camera system," *Ocean Engineering* 54 (2012) 1–12
[6] M.D. Rodriguez Sullivan and M. Shah, Visual surveillance in maritime port facilities 6978 (2008) 11–19
[7] H. Wei, H. Nguyen, P. Ramu, C. Raju, X. Liu, and J. Yadegar, Automated intelligent video surveillance system for ships 73061 (2009) 1–12
[8] D. Bloisi and L. Iocchi, "ARGOS – a video surveillance system for boat traffic monitoring in Venice," *International Journal of Pattern Recognition and Artificial Intelligence* 23 (2009) 1477–1502
[9] M. Everingham, L. Van Gool, C.K.I. Williams, J. Winn, and A. Zisserman, "The pascal visual object classes (VOC) challenge," *International Journal of Computer Vision* 88 (2010) 303–338
[10] D. Bloisi, L. Iocchi, and A. Pennisi, Maritime activity recognition (MAR) dataset, http://www.dis.uniroma1.it/bloisi/mar, 2012
[11] Allianz Global Corporate & Specialty AG, Safety and Shipping 1912–2012: From Titanic to Costa Concordia, 2012
[12] J. Leggat, T. Litvak, I. Parker, A. Sinha, S. Vidalis, and A. Wong, Study on Persistent Monitoring of Maritime, Great Lakes and St. Lawrence Seaway Border Regions, Technical Report, DRDC CSS CR 2011-28, 2011
[13] Raytheon Canada, "Surface based sensors for maritime domain awareness," *FrontLine Canada – Industry Special* 2 (2012) 37
[14] T. Nohara, A commercial approach to successful persistent radar surveillance of sea, air and land along the northern border, in: IEEE International Conference on Technologies for Homeland Security, pp. 276–282
[15] N. Pires, J. Guinet, and E. Dusch, "ASV: an innovative automatic system for maritime surveillance," *Navigation* 58 (2010) 1–20
[16] Thales Group, COMMANDER C3, http://www.thalesgroup.com, 2012
[17] DSIT Solutions, Harbor surveillance system, http://www.dsit.co.il, 2012
[18] L-3 Klein, HarborGuard Command and Control, http://www.l-3klein.com, 2012
[19] B.J. Rhodes, N.A. Bomberger, M. Seibert, and A.M. Waxman, SeeCoast: automated port scene understanding facilitated by normalcy learning, in: IEEE Conference on Military Communications, pp. 3656–3662, 2006
[20] M.D. Rodriguez Sullivan and M. Shah, Visual surveillance in maritime port facilities, in: Proceedings of SPIE, vol. 6978, pp. 886–893, 2008

[21] R.G. Wijnhoven, K. van Rens, E.G. Jaspers, and P.H. de With, Tracking ships from fast moving camera through image registration, in: 31st Symposium on Information Theory in the Benelux, pp. 73–80, 2010

[22] N. Dalal and B. Triggs, Histogram of oriented gradients for human detection, in: IEEE Conference on Computer Vision and Pattern Recognition, vol. 1, pp. 886–893, 2005

[23] S. Fefilatyev, D.B. Goldgof, and C. Lembke, Tracking ships from fast moving camera through image registration, in: International Conference on Pattern Recognition, vol. 4861, pp. 3500–3503, 2010

[24] D. Bloisi, L. Iocchi, M. Fiorini, and G. Graziano, Camera based target recognition for maritime awareness, in: Proceedings of the Fifteenth International Conference on Information Fusion (FUSION), Singapore, pp. 1982–1987, 2012

[25] V. Chandola, A. Banerjee, and V. Kumar, "Anomaly detection: a survey," *ACM Computing Survey* 41 (2009) 1–58

[26] O. Tuzel, F. Porikli, and P. Meer, "Pedestrian detection via classification on Riemannian manifolds," *Patterns Analysis and Machine Intelligence* 30 (2008) 1713–1727

[27] P. Viola and M.J. Jones, "Robust real-time face detection," *International Journal of Computer Vision* 57 (2004) 137–154

[28] D. Bloisi, L. Iocchi, M. Fiorini, and G. Graziano, Automatic maritime surveillance with visual target detection, in: Proceedings of the International Defense and Homeland Security Simulation Workshop (DHSS), Rome, Italy, pp. 141–145, 2011

[29] OpenCV, http://opencv.org, 2012

[30] H. Bay, T. Tuytelaars, and L.V. Gool, Surf: speeded up robust features, in: Proceedings of the Ninth European Conference on Computer Vision, 2006, pp. 404–417

[31] United Nation (UN), Convention on the law of the sea: territorial sea and contiguous zone, 1982

[32] International Civil Aviation Organization (ICAO): Advanced surface movement guidance and control systems (A-SMGCS) manual, 2004

[33] S.D. Thompson and K.A. Sinclair, "Automatic Dependent Surveillance-Broadcast in the Gulf of Mexico," *Lincoln Laboratory Journal* 17 (2007) 1–15

[34] K. Dimitropoulos, T. Semertzidis, and N. Grammalidis, Video and signal based surveillance for airport applications, in: *Advanced Video and Signal Based Surveillance*, 2009, pp. 170–175

[35] J.A. Besada, J. Portillo, J. Garcia, and J.M. Molina, Image-based automatic surveillance for airport surface, in: FUSION, 2001

Chapter 12

AIS signal radiolocation, tracking and verification

Pier Francesco Sammartino[1], Michele Vespe[1], Dario Tarchi[1], Franco Oliveri[1], Francesco Papi[2], Francesco Borghese[3], Giuseppe Aulicino[4] and Antonio Vollero[4]

12.1 Introduction

Maritime surveillance aims at achieving the best possible awareness of what is occurring at sea by obtaining reliable and updated information on all 'objects' in a given area. The identification and tracking of all the vessels in the area of interest is today an essential element for dealing with a variety of security threats in the maritime domain. Objects at sea can be broadly divided in two categories, those sending information about themselves (cooperative), usually because requested to do so by a specific norm, and those that do not provide such information (not cooperative). The second type of objects can be detected using several types of sensors such as radar, sonar, EO cameras, which operate on a variety of platforms (ground, space and sea). Both classes of objects can be of interest for maritime surveillance purposes and often the above mentioned sensors are used to verify or enhance the information voluntarily provided by collaborative objects. Regrettably, these sensors may not be available in certain areas, thus the improved exploitation of information obtained by the analysis of cooperative data/communications can be an inexpensive alternative source of valuable knowledge: there are actually several operational examples of the possibility to locate a 'object' by exploiting the RF signals that are emitted by the target itself for different primary purposes.

In the framework of cooperative sensors the Automatic Identification System (AIS) has an important role as AIS transponders are operational on majority large number of vessels. AIS has been originally conceived for collision avoidance and is

[1]European Commission, Joint Research Centre (JRC), Institute for the Protection and Security of the Citizen (IPSC), Maritime Affairs Unit, Via Enrico Fermi 2749, 21027, Ispra (VA), Italy
[2]Department of Electrical and Computer Engineering, Curtin University, Kent St, Bentley WA 6102, Australia
[3]Elman s.r.l., Via di valle Caia km 4.7, Pomezia (Rome), Italy
[4]Italian Coast Guard Headquarters, Viale dell'Arte 16, Rome, Italy

a system whereby ships broadcast their presence, identification and location. Differently than other operational coastal active systems for maritime surveillance, AIS is characterised by a considerable terrestrial coverage (VHF propagation) together with a relatively accurate positioning (GNSS) performance [1]. The recent increase of terrestrial networks and satellite constellations of receivers is providing global tracking data that enable a wide spectrum of applications beyond collision avoidance. The central role of AIS has been recognised by international regulations. Ships of 300 gross tons and upwards in international voyages, 500 tons and upwards for cargoes not in international waters and passenger vessels are obliged to be fitted with AIS equipment as regulated by the IMO Safety of life and sea (SOLAS) [2]. Furthermore, all EU fishing vessels of overall length exceeding 15 m are also required to be fitted with AIS from May 2014 [3].

In more detail, AIS messages encode state vector information such as latitude, longitude, Speed Over Ground (SOG), Course Over Ground (COG) as derived by the on board GNSS receiver. Such information is broadcasted at a variable transmission rate which depends on the vessel motion: as an example, the rate is increased up to a message every 2 s when the vessel is sailing at high speed or manoeuvring. In addition, every six minutes, vessels transmit a set of data, including their identification numbers (IMO and MMSI, i.e. ship name and call sign), static (size, type of vessel, type of cargo, etc.) and voyage-related information (e.g. Estimated Time of Arrival and destination).

The cooperative nature of AIS and the lack of intrinsic security make it vulnerable to false or missing declarations. Actually, AIS data may be incorrect or intentionally jammed or falsified. As an example, false GNSS tracking information can be produced to simulate specific trajectories [4] or false AIS messages can be generated and transmitted through VHF as recently demonstrated in [5]. Such events may be serious hazards to the safety of navigation, especially in reduced visibility conditions, and can also represent security threats by covering unauthorised activities such as illegal movements of goods and people.

The verification of the trustworthiness of AIS data is becoming a key problem to exploit its full potential not only for safety but also for maritime surveillance and traffic monitoring, enabling far-reaching applications such as traffic knowledge discovery, route prediction and anomaly detection. The latter can target particular low-likelihood motion trajectories [6,7], alerts such as sailing in restricted areas, abrupt changes of direction (an extensive overview of this rules is presented in [8]).

The detection of position reporting anomalies either linked to AIS transponder failures or due to deliberately falsified broadcast dynamic information, can be approached in different ways. One technique is the correlation of AIS data with information provided by additional sensors, which can be used to detect AIS data spoofing; such sensors could be coastal radars (see e.g. [9]), HF Surface Wave Radars [10] or space-based Synthetic Aperture Radar (SAR) [11]. Regrettably, the operational usability of such approaches depends on the availability and persistency of the data provided by such additional sensors, which occurs only in limited areas. Other approaches aim at increasing the trustworthiness of the transponder through the use of additional on-board instrumentation as investigated by [12].

In this work, therefore, we provide the analysis of the AIS as an Intelligent Transportation System (ITS) beyond its conventional applications. As mentioned, the AIS is a smart, cooperative and efficient way to monitor vessels, exploiting long-range radio communication with messages that can be decoded with very-low computational complexity and occupies a tiny fraction of the VHF spectrum (25 kHz). As a consequence, it is a cost-effective way to provide an increased security of all the transports at sea, monitoring and tracking the routes of the vessels. AIS is a vehicle- and infrastructure-based ITS, where the infrastructure can be comprised of ground and at occurrence space stations [13], granting a global coverage. The increasing number of uses of the AIS data makes it more and more important to verify their correctness. To contribute to fulfil this need, we propose a novel mix and use of well-known techniques that is based on a radiolocalisation method exploiting Time Difference of Arrival (TDoA) of the AIS messages at different ground stations and an Extended Kalman Filter (EKF) in order to track vessels in geodetic coordinates. This methodology can be easily applied to an existing AIS network using the messages normally provided by each base station, provided that the receiving base stations are synchronised with one another and significantly separated in space. As a consequence, we hereby provide a demonstration of the potential of analysing in more depth the information gathered from an ITS at the signal level for situational awareness. The final aim is to show that a remarkable amount of information can be gathered processing ITS data not only in a conventional or statistical ways.

Similar approaches, based on multilateration techniques, have been widely used for air traffic surveillance using secondary surveillance radar (SSR) Mode S replies (e.g. [14]) leading to location errors in the order of a few metres (see e.g. [15]). In this work we did not expect to achieve such a high accuracy, as the AIS message has only a 25-kHz bandwidth, which is a fraction of that used in air traffic control, and a much lower repetition frequency. A distinctive feature of the work presented here is that the proposed method has been extensively tested using data provided in real time by a subset of the AIS network managed by the Italian Coast Guard and described in Section 12.2.

When three or more TDoAs are available, a classic multilateration procedure [16] can be applied to estimate the vessel position at each time instant with a certain uncertainty [17]. A discussion on the processing methods based on ToAs (Times of Arrival) and TDoAs, as well as a brief statistical description of their errors, are discussed in Section 12.3, while in Section 12.4 the theoretical foundations of multilateration through TDoAs are presented.

Since the performance achieved in this first step may not be deemed fully suitable for practical applications, Section 12.5 shows that the joint application of a tracker (EKF) makes it possible to recover most of the accuracy in locating a target and hence achieve usable operational results. Actually, the radiolocation accuracy is improved through time integration by means of the EKF in order to fulfil operational requirements. The results show a reduction of the error from dozens of kilometres to just a few hundreds of metres, therefore demonstrating the operational potential of this methodology. In the light of the theoretical analysis and the preliminary results described in Sections 12.4 and 12.5, the algorithms are being applied to all the data collected by the network described in Section 12.2.

The implemented operational design has been optimised to make the system and the processing run in real time and additional features of the signals have been analysed in order to extract as much information as possible and exploit it at the best. For instance, an estimate of the accuracy of the results, based on a variety of the statistics of the received signals, was included as a measurement of the reliability of the results. The description of the signal processing steps and the results of a real-time demonstrator realised within the JRC facilities in Ispra, Italy, is presented in Section 12.6.

The detection and tracking results shown in Section 12.6 are then passed to an additional processing step aiming at detecting anomalies, which highlights inconsistencies between declared and estimated positions. The latter was made possible by the fact that AIS messages transmit the declared vessel's position derived from the GNSS and hence any significant discrepancy between the two can be detected. A description of the ways to identify anomalies and the achieved results are reported in Section 12.7.

Finally, Section 12.8 is finally devoted to the conclusions.

12.2 Operational infrastructure and data collection

The Italian AIS network has been implemented in 2005 to fulfil the requirements of the Directive 2002/59/EC of the European Parliament and of the Council of 27 June 2002, and it has undergone a major upgrade between 2012 and 2013, with the aim to comply with the most recent relevant guidelines and recommendations, such as the IALA Recommendation A 124 on 'The AIS Service' (December 2012) and the ITU Recommendation ITU-R M.1371-4 issued on April 2010. The network currently consists of 60 base stations mainly located to get the best VHF coverage (up to 100 nautical miles, even without the duct effect). The base stations were placed in such a way to get an overlap of the radio coverage (Figure 12.1) in order to:

(i) increase the overall availability of the services provided by the national AIS network and
(ii) enable future adoption of anti-spoofing techniques to improve quality of AIS information received.

The overall architecture of the Italian AIS network is exemplified in Figure 12.2. One of the key elements of the network is the so-called AIS Embedded Server, a fully solid-state device featuring two separate servers thus supporting redundancy or the simultaneous interface to two TCP networks. The AIS Embedded Server, acting as a Physical Shore Station (PSS) Controlling Unit (PCU) according to the IALA Recommendation A-124, allows the integration of one or more base stations and their management from shore systems. The Embedded Server can acquire AIS data from the serial ports and from TCP connections according to the specified configuration.

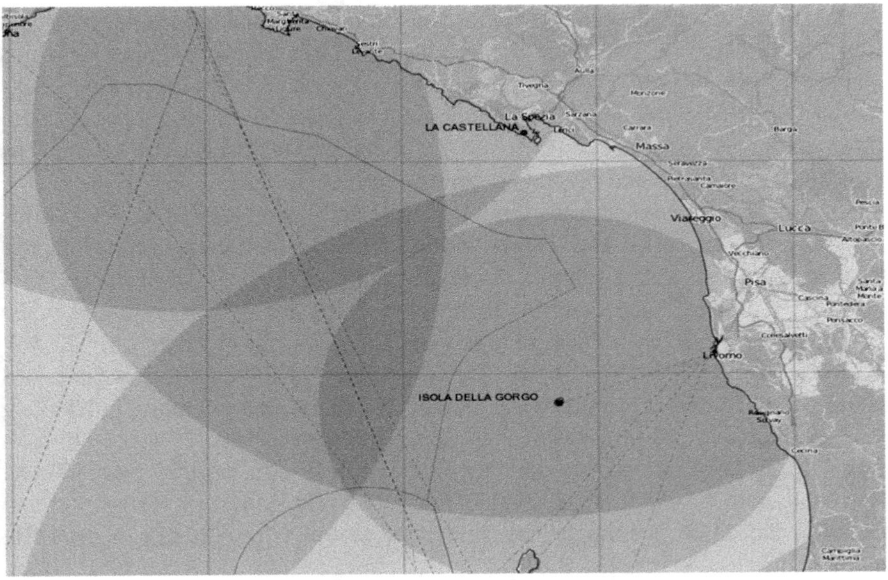

Figure 12.1 AIS coverage overlapping

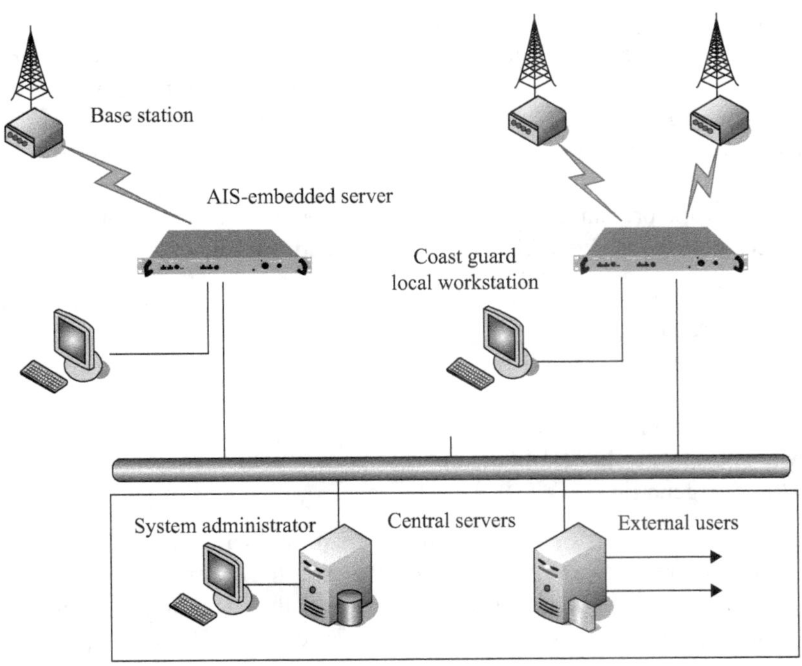

Figure 12.2 The Italian AIS Network architecture

In the same way, the collected AIS data are made available to serial ports and TCP connections. It also features an embedded interface for each web server which allows configuring and monitoring for both the servers and the connected AIS Base Station (AIS-BS), without using any additional software. Authorised operators can log-in on the web interface to configure and monitor every aspect of the server, its ports and connections and the linked devices and users.

The AIS system dynamically configures into cells using a Self-Organised Time Division Multiple Access (SO-TDMA) scheme. The cell size is adjusted to adapt as a function of the traffic density: in highly congested areas for instance, it is necessary to reduce the size of the cell to diminish the number of transmitter in the cell; this is achieved by reducing the power of the AIS transponders from 12.5 W ('high setting') to 1 W ('low setting') thus avoiding message collisions [18]. AIS data are transmitted at a rate of 9.6 kbps using Gaussian minimum shift keying (GMSK) modulation over two channels 161.975 MHz and 162.025 MHz. For each channel, 2,250 slots are allocated within a timeframe of 60 s, starting every minute. The bandwidth of the channels is nominally 25 kHz, although the AIS signal spectrum has to be within an emission mask defined by −25 dBc at +10 kHz and −70 dBc +25 kHz [18].

Moreover, transmissions of AIS devices are synchronised by using a common time reference, Coordinated Universal Time (UTC), provided by the internal GNSS receiver. Transmission timing error including jitter and systematic offsets should be within ±104 μs of the synchronisation source for mobile stations and ±52 μs for base stations, setting the limits of the accuracy of time of flight (ToF) estimation. If the internal GNSS receiver is faulty, the AIS devices are capable of synchronising to secondary timing sources as the received AIS messages; in this case, the timing error may increase up to 312 μs. Anyway, AIS stations are not allowed to discard received messages basing upon timing error, even when it is much bigger than the above specified values.

In the work presented hereby and for the purpose of developing a demonstrator, the data from seven AIS-BSs were selected, as follows:

1. Monte Bignone
2. Forte Castelaccio
3. Monte Castellana
4. Livorno
5. Isola di Gorgona
6. Monte Capanne (Isola d'Elba) and
7. Monte Argentario.

These are shown as orange stars in the map of Figure 12.3. From their geographical positions, it is straightforward that the expected coverage of such a network lies in the jointly covered area, which extends over the Ligurian and northern Tyrrhenian seas, and is mostly delimited by Liguria, Tuscany and Corsica. In particular, in this figure the highest the expected performance, the darker the colour. On the contrary, the lighter the colour, the less accurate the performance. We refer to Section 12.4 for a more complete explanation on how this coverage map has been generated.

Figure 12.3 Displacement of AIS base stations and expected coverage

12.3 ToA and TDoA measurements

In this section we discuss the models used to describe the ToA and TDoA measurements available from the Italian AIS Network. Specific attention is given to the analysis of the noise distributions in order to verify stationary and Gaussian assumptions and therefore obtain reasonable estimates for bias compensation. The standard for AIS-BS s provides for a sentence containing information associated to the time of arrival of received messages, expanding the set of messages defined in the NMEA (National Marine Electronics Association) standard [19]. The specific field in the sentence allows representing a precision up to 1 ns; however, a precision of 1 μs is currently obtained for the Italian Coast Guard National AIS network.

The time of arrival is estimated by detecting the position of a specific marker in the AIS message, the start flag, relative to the starting slot. Low SNR may affect the accuracy of symbol clock recovery performed by the GMSK demodulator, which is currently limited by a resolution of 10.4 μs set by the internal sampling rate. The *ToF* of the electromagnetic wave carrying the AIS message can be calculated as the difference between the ToA and the beginning of the time slot (T_{slot}) during which the message has been transmitted, that is:

$$ToF = ToA - T_{\text{slot}} - \eta_{tx} - \eta_{rx} \tag{12.1}$$

where η_{tx} and η_{rx} are the transmitter and receiver timing errors, respectively. When it is possible to fully characterise the errors η_{tx} and η_{rx}, an unbiased estimate of the *ToF* could be obtained and used to derive distance R from the considered vessel and the receiving AIS-BS, that is, $R = c\ ToF$, where c is the speed of light. Multiple distances from different AIS-BSs could then be used to find an unbiased estimate of the instantaneous vessel position using ToA-based triangulation.

Unfortunately the transmitter timing error η_{tx} is generally unknown and, as mentioned earlier, can be in the order of tens of μs leading to large ranging uncertainties. Moreover, η_{tx} is vessel dependent due to differences in the electronics of different transmission equipment. This means that a full characterisation of η_{tx} is not possible and the *ToA* measurements cannot be robustly used to directly solve the vessel localisation problem. The simplest idea in this case is to compare the received *ToAs* pairwise and use the *TDoAs*. Hence, assuming that the vessel under surveillance is within the coverage area of n AIS-BSs, it is possible to collect $K = \binom{n}{2}$ *TDoAs*, that is,

$$\begin{aligned} T\hat{o}A_i &= ToA_i + \eta_{tx} + \eta_{rx_i} \\ T\hat{o}A_j &= ToA_j + \eta_{tx} + \eta_{rx_j} \\ TDoA_{i,j} &= ToA_i - ToA_j + \eta_{rx_i} - \eta_{rx_j}, \qquad 1 \leq j < i \leq n \end{aligned} \tag{12.2}$$

where $\eta_{rx_i} \sim N(\mu_i, \sigma_i^2)$ and $\eta_{rx_j} \sim N(\mu_j, \sigma_j^2)$ are the receiver timing errors for the i-th and j-th AIS-BS, respectively. We can then rewrite the *TDoAs* vector as follows

$$TDoA = \begin{bmatrix} TDoA_{2,1} \\ \vdots \\ TDoA_{n,1} \\ TDoA_{3,2} \\ \vdots \\ TDoA_{n,n-1} \end{bmatrix} + v, \qquad v \sim N(\mu_{TDoA}, R_{TDoA}) \tag{12.3}$$

$$\mu_{TDoA} = [\mu_2 - \mu_1 \quad \ldots \quad \mu_3 - \mu_2 \quad \ldots \quad \mu_n - \mu_{n-1}] \tag{12.4}$$

$$R_{TDoA} = \text{blockdiag}(R_1, R_2, \cdots, R_K) \tag{12.5}$$

$$R_1 = \begin{bmatrix} \sigma_1^2 + \sigma_2^2 & \cdots & \sigma_1^2 \\ \vdots & \ddots & \vdots \\ \sigma_1^2 & \cdots & \sigma_1^2 + \sigma_n^2 \end{bmatrix} \tag{12.6}$$

$$R_2 = \begin{bmatrix} {\sigma_2}^2 + {\sigma_3}^2 & \cdots & {\sigma_2}^2 \\ \vdots & \ddots & \vdots \\ {\sigma_2}^2 & \cdots & {\sigma_2}^2 + {\sigma_n}^2 \end{bmatrix} \tag{12.7}$$

$$R_K = \sigma_n^2 + \sigma_{n-1}^2 \tag{12.8}$$

As for the parameters (μ_i, σ_i) some *a priori* knowledge might be available from the receivers' characteristics. Since this is not our case, an estimate of (μ_{TDoA}, R_{TDoA}) can be obtained using batch data or over a moving time-window. In addition, if *ToA* measurements are optimally obtained using correlation processing, the standard deviations σ_i can be modelled using the Cramer-Rao Lower Bound [20].

$$\sigma_i \geq \frac{c}{B \bullet \sqrt{SNR_i}} \tag{12.9}$$

where B is the signal bandwidth and SNR_i is the Signal-to-Noise ratio of the i-th channel. SNR_i could be modelled by means of Friis transmission equation as $S\hat{N}R_i = SNR_i(\hat{x}_k)$, where $\hat{x}_k$ is the estimated vessel position at time k.

A noise analysis was performed using real AIS data collected for approximately 9 h over 3 days and is reported extensively in [21]. Both AIS service messages (i.e. base station reports periodically transmitted [5]) and AIS messages from verified vessel trajectories were used to estimate the noise distribution. The results showed that an upper bound of ± 13.5 μs can be selected so that about 95% of noise probability density function (pdf) is included.

12.4 TDoA-based vessel localisation

The problem of estimating an emitter position from *TDoA* measurements occurs in a wide range of applications. In line of principle, a correlation analysis of the received signal from two receivers generates a hyperbolic function. However, due to *TDoA* measurement uncertainty, it is not possible to use multiple hyperbolas to robustly determine a unique intersection [17]. In practical terms, we face a nonlinear estimation problem defined as follows. Let one assume that a vessel transmitting AIS messages is located at position $P_v = (x_v, y_v)$. Let $P = (x, y)$ be a generic point in geodetic coordinates and let $P_i^{RX} = \left(x_i^{RX}, y_i^{RX}\right)$ be the position vector for the i-th AIS-BS. Then a nonlinear estimate of P_v can be found as

$$\hat{P}_v = \arg\min_P \sum_i \sum_{i>j} \left| d\left(TDoA_{i,j}, h\left(P, P_i^{RX}, P_j^{RX}\right)\right)\right|^2 \tag{12.10}$$

where

$$h\left(P, P_i^{RX}, P_j^{RX}\right) = \sqrt{(x - x_i^{RX})^2 + (y - y_i^{RX})^2} - \sqrt{\left(x - x_j^{RX}\right)^2 + \left(y - y_j^{RX}\right)^2} \tag{12.11}$$

and $d(\cdot)$ is a suitable distance. This could be for instance the square of the Euclidean distance (i.e. least squares estimate) or the Mahalanobis distance (i.e. maximum likelihood estimate).

Since we are instead interested in finding the associated estimation error, we can use the following approach. Consider for instance a fixed grid of points P in geodetic coordinates, and set an upper bound on the differential time uncertainty as $K = 13.5$ μs, which corresponds to about $L = 4$ km location uncertainty. This bound was chosen so that *95%* of the noise pdf is covered (see Figure 12.3). Then we can easily identify the set of points P that satisfy the following set of inequalities, that is [21]

$$\left| h\left(P, P_i^{RX}, P_j^{RX}\right) - c \cdot TDoA_{i,j} \right| \leq L, \quad 1 \leq j < i \leq n \tag{12.12}$$

The locus of points where (12.12) is verified is a hyperbola with uncertainty in the flat-earth approximation case [17]. The area contained by all intersections can be thought of as a measure of the estimation error. Examples of this are shown in Figures 12.4 and 12.5, where the AIS message emission is located within the intersection of the hyperbolic functions as derived from the TDoA measurements between three and four stations, respectively.

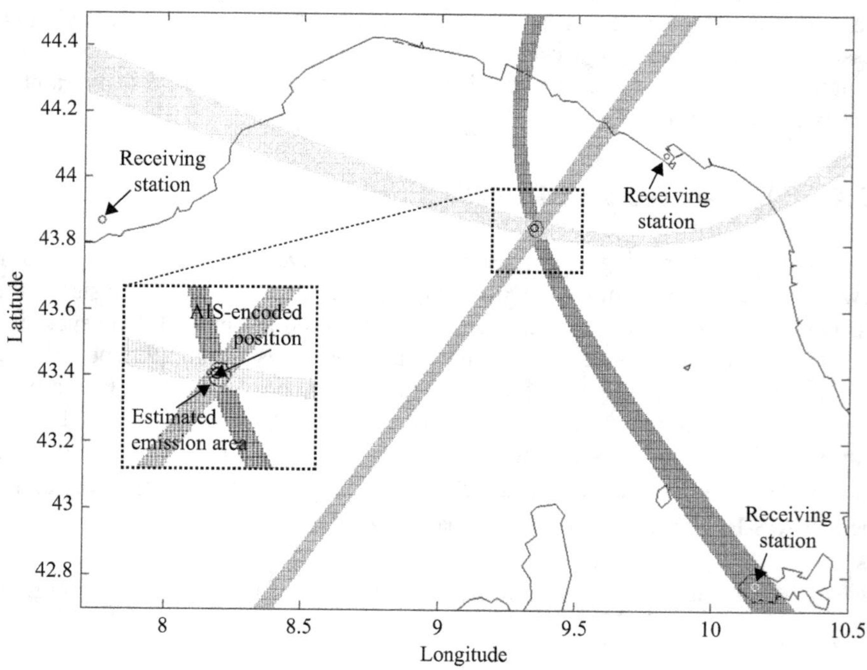

Figure 12.4 Radiolocation of the AIS emission in the Ligurian Sea as the intersection of the TDoAs hyperbolic functions between three receiving stations (circles on land). The AIS-encoded position (circle) falls within the estimated emission area

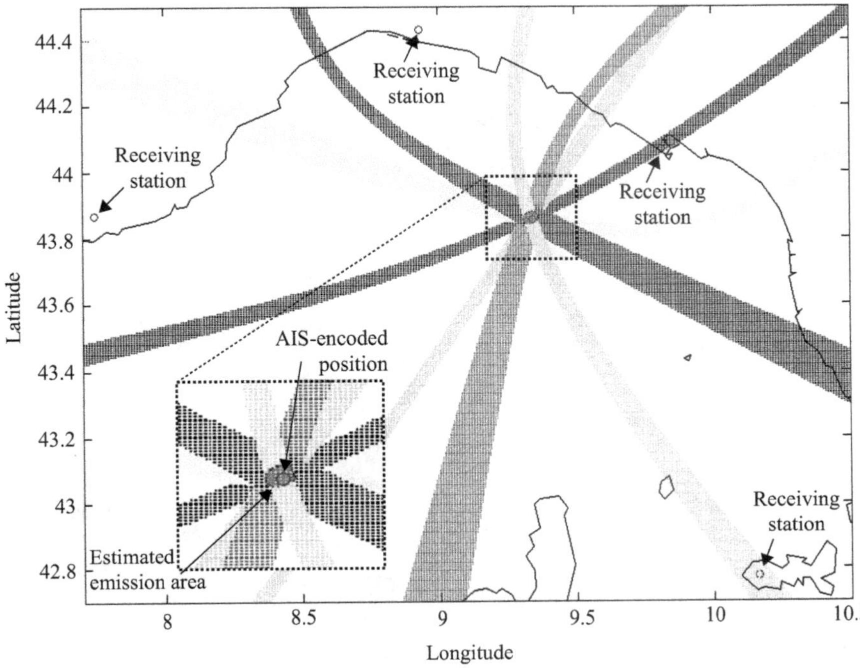

Figure 12.5 *The AIS emission is received by four stations (circles). The additional AIS base station yields a reduction of the intersection area between the available TDoAs, thus reducing the localisation uncertainty*

Then, assuming a Gaussian distribution for the noise terms (as verified in the previous section), the associated estimation covariance is given by the minimum volume ellipse enclosing all points that verified (12.12).

A more rigorous approach uses the Mahalanobis distance:

$$\begin{aligned} d_{i,j}^M &= h\left(P, P_i^{RX}, P_j^{RX}\right) - c \cdot TDoA_{i,j} \\ \Delta_M &= \begin{bmatrix} d_{2,1}^M & \cdots & d_{n,n-1}^M \end{bmatrix} \\ \hat{P}_v &= \arg\min_P \ \Delta_M \hat{R}_{TDoA}^{-1} \Delta_M^T \end{aligned} \tag{12.13}$$

where $\hat{P}_v$ is the maximum likelihood estimate and $\hat{R}_{TDoA}$ is an estimate of the *TDoA* covariance given in (12.6).

The output of the *TDoA*-based localisation procedure is a pair ($\mathbf{z}_k$, R_k). Specifically, $\mathbf{z}_k$ containing the vessel location estimate in geodetic coordinates and R_k is a covariance matrix describing the minimum volume ellipse. Time integration is then performed by using ($\mathbf{z}_k$, R_k) as inputs of a suitable EKF as described in the following section.

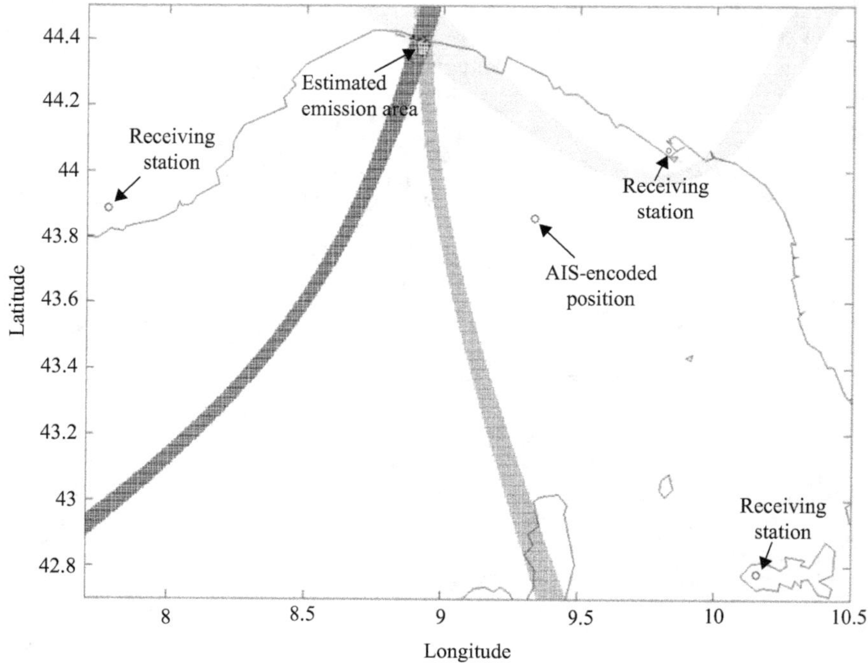

Figure 12.6 The radiolocation results highlight that the message originally transmitted in open seas (AIS-encoded position) is repeated by the base station in Genoa

Another localisation example is reported in Figures 12.4 and 12.5. Here the AIS-based radiolocation and the declared vessel position are sensibly different: this happens when the AIS message from the vessel in open seas (AIS-encoded position) is repeated by a base station,[1] easily located in Genoa in this case (Figure 12.6). The spatial distribution of base stations is central to the performance of the radiolocation process. This is given by the extent of the uncertainty at a specific location in space and varies depending on the transmitter position with respect to the receiving stations. For instance, in Figure 12.7, the message is received by three stations and the emission location can be verified. However, the error area is quite large. This happens because the vessel is far away from all the sensors baselines and the angles of arrivals are similar. To better clarify this point we performed a geometric analysis of the localisation precision with reference to the positions of the Italian Base Stations (Figure 12.8). Results are given in Figure 12.9, where we show how the estimated precision varies in space for different AIS network configurations.

[1]In accordance with AIS standard, a base station can be required to store and forward a position message sent by a vessel.

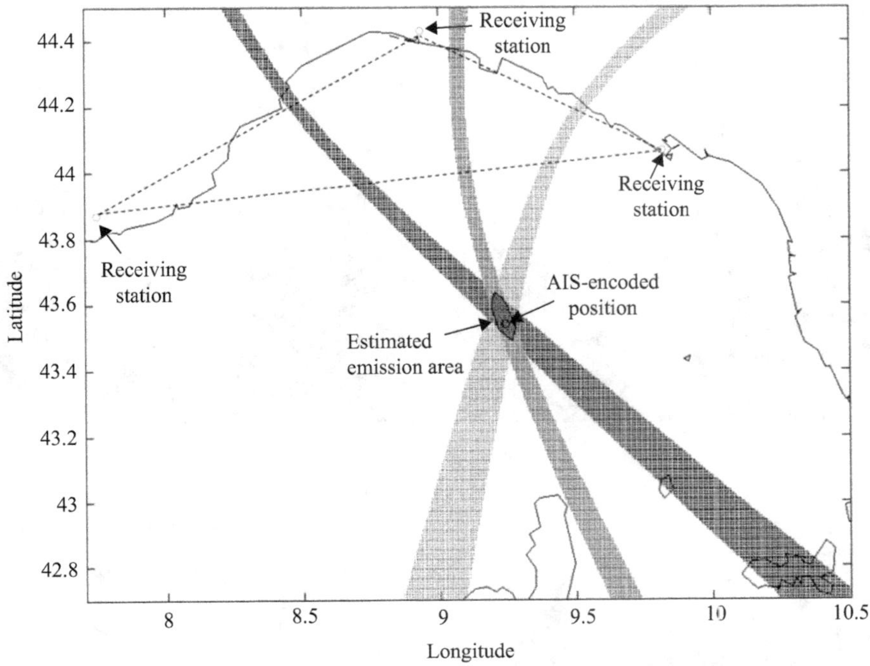

Figure 12.7 The radiolocation uncertainty changes depending on the position of the emission and the receivers' location

12.5 Vessel tracking using an Extended Kalman Filter

As mentioned earlier, at time step k the multilateration procedure gives as output a pair $(\mathbf{z}_k, R_k)$ for each vessel, where $\mathbf{z}_k$ is the estimated vessel position and R_k its associated error covariance. We can then perform model-based time integration by recursively solving the Chapman-Kolmogorov integral and Bayes equation [22]. In fact, Bayesian methods provide a rigorous framework for dynamic state estimation problems. The idea is to construct the pdf of the system state based on all the available information, and then find an approximation of such a posteriori pdf. Classical inference methods for *nonlinear filtering* are the EKF [23], based on linearisation of the system about the current state estimate, and the Unscented Kalman Filter (UKF) [24], based on a deterministic sampling of the a posteriori pdf. Improved tracking accuracy can be generally achieved by means of Sequential Monte Carlo (SMC) methods like the Particle Filter (PF) [25], even if they may require additional computational costs.

We now report the tracking results obtained from processing real AIS data collected by the Italian Terrestrial Network using the methodology duly described in Reference 21. Results for a single vessel are depicted in Figures 12.9–12.11.

Specifically, in Figure 12.9 we report the reference vessel trajectory (squares) from GNSS data, the (lat, lon) estimates from the localisation procedure (crosses) and the EKF position estimates (circles). Notice that we report the reference vessel location only when the AIS message is received by at least three AIS-BS s. This is

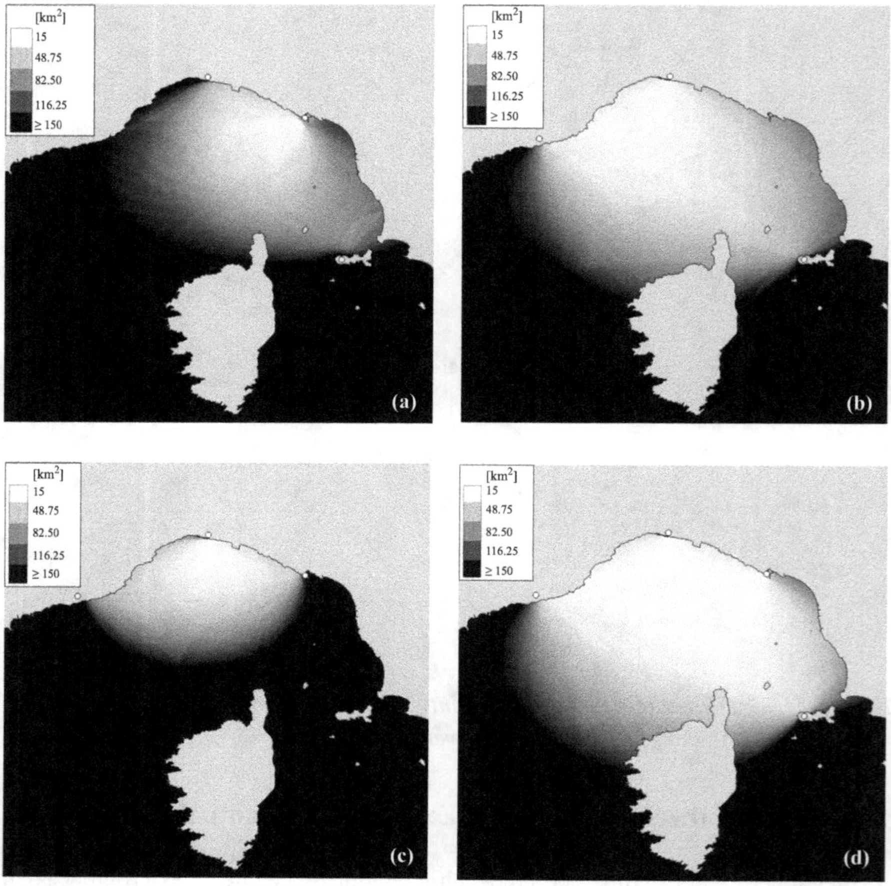

Figure 12.8 Radiolocation uncertainty limits obtained when a single AIS message is received from different base stations using three TDoA measurements (a–c) and considering $K = 13.5$ μs for each TDoA. Resulting performance when the message is received by four base stations, leading to six TDoAs (d)

done in order to highlight the gaps of *TDoA* measurements due to non-perfect coverage. The estimation errors over time for the *TDoA* localisation and after EKF processing are depicted in Figure 12.10. Finally, the EKF results in estimating the vessel kinematics, that is, SOG and COG, are shown in Figure 12.11.

It is worth noting that the number of scans needed by the EKF to converge to the true COG and SOG values sensibly varies from track to track. This is due to a number of factors, chief amongst them the positions of the transmitter and the receivers, the motion and velocity of the vessel and the variable time between scans. The latter depends on (i) the AIS transmission rate, which changes according to the vessel manoeuvre and velocity, and (ii) the rate of reception of the message by three or more stations.

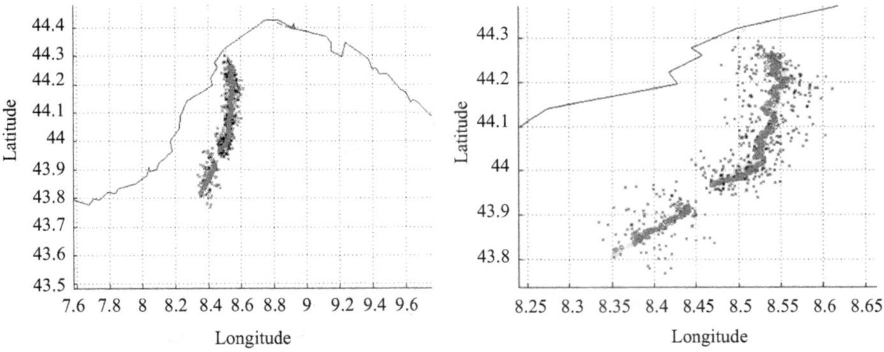

Figure 12.9 Tracking results for a single vessel leaving the port of Savona (left) and zoomed trajectory. True trajectory (squares), estimates from the multilateration procedure (crosses) and EKF position estimates (circles)

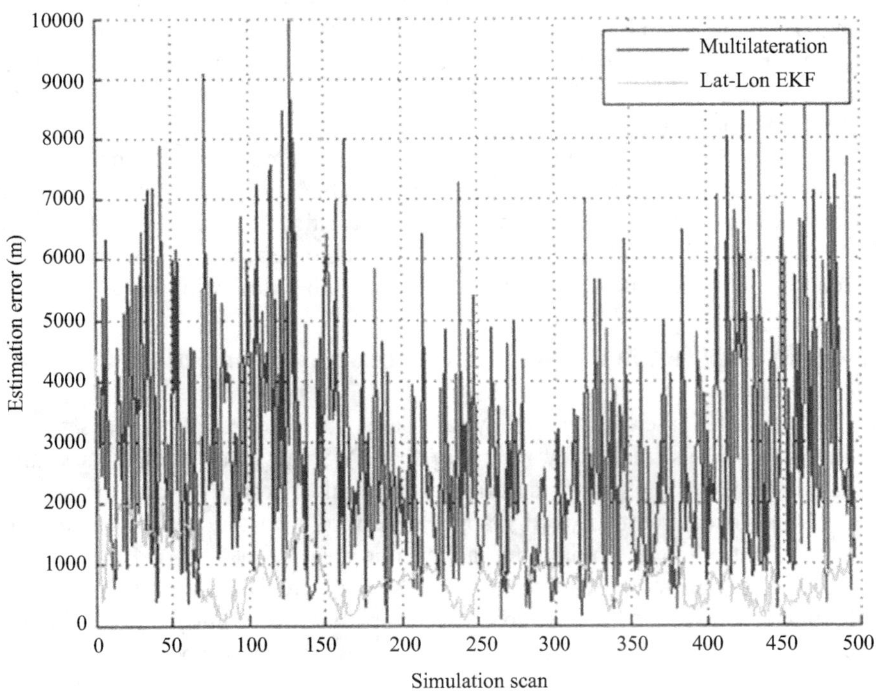

Figure 12.10 Localisation error over time for a single vessel trajectory: multilateration procedure and EKF in geodetic coordinates

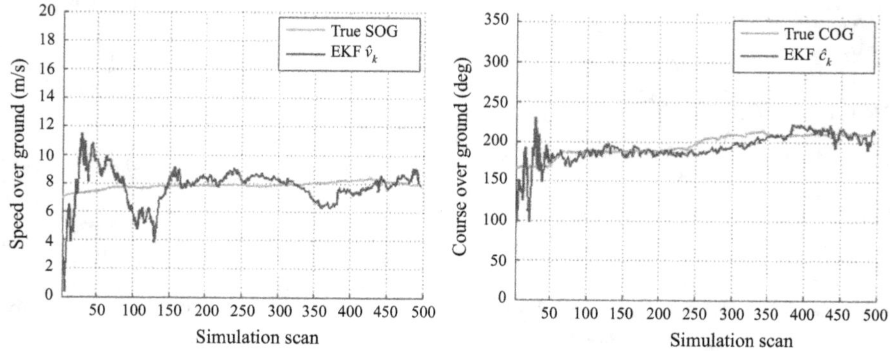

Figure 12.11 EKF results in estimating the vessel kinematics. True and estimated Speed Over Ground (left); true and estimated Course Over Ground (right)

Additional tuning of the filter parameters could lead to improved performance in terms of the estimated COG and SOG. In general terms, reducing the filter process noise leads to more precise estimates for the vessel kinematics. This, however, might reduce the filter robustness and precision in terms of location estimates. Some of the initial/final estimates from the multilateration procedure are far away from the true vessel location: this happens when there is a non-optimal spatial distribution of the AIS-BS-s, leading to multiple solutions (*ghost estimates*) to the *TDoA* localisation problem.

The demonstrated average performance of the proposed radiolocation and tracking technique based on EKF confirms that the approach could be used to perform automatic validation of declared GNSS positions encoded in the broadcast AIS messages. This could be performed using a single step gate validation [26] or a more effective binary hypothesis testing over a moving time window [27].

12.6 Pre-operational results: initial processing and multilateration

So far, we have described the analytical bases for reconstructing the series of positions of vessels tracked by means of their AIS messages. From this section onwards, on the contrary, we describe a practical approach to data processing which aims to deliver trustworthy applicative results to the human operator in charge of controlling and monitoring a maritime area.

First of all, all messages from the base stations are transmitted to a common gathering database (in this case to those reserved at the Joint Research Centre, where the demonstrator/prototype of AIS awareness system has been designed and implemented) as strings. Here, messages are stored and at occurrence read and decoded to feed the multilateration algorithm. The demonstrator is not intended to be a full operational system, but its development aims to verify and validate at a pre-operational stage the feasibility of the exploitation of the AIS signals for awareness and surveillance.

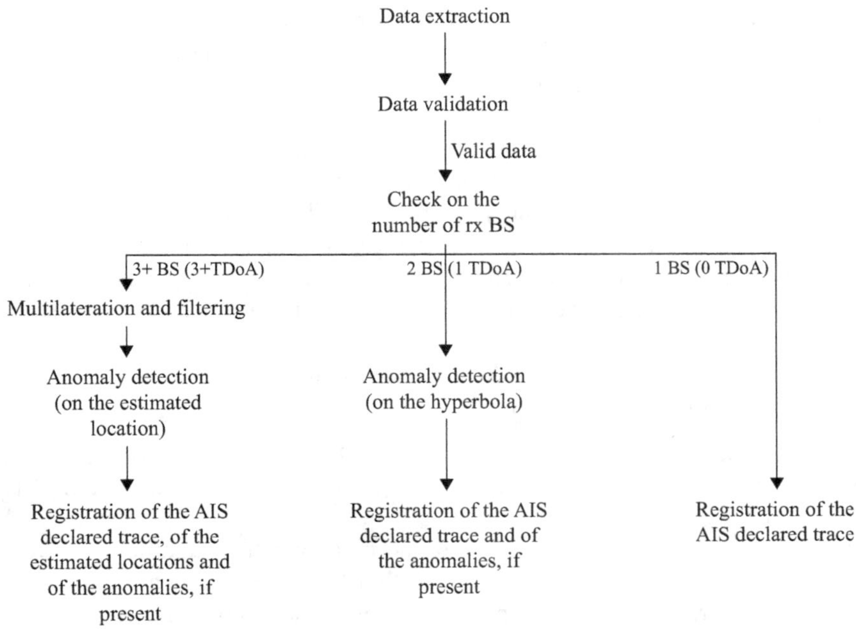

Figure 12.12 General approach to data processing and expected results

The first stage of the processing consists in running a check on the received and decoded data before applying the algorithms. The pre-filter removes all the corrupted and invalid data, in order to ensure the highest degree of confidence with the extracted data and thus to enhance the reliability of the results in the processing chain which follows (Figure 12.12).

It is also clear that it is a remote event that each AIS message is received by all the base stations of the network. Either for geometrical reasons, propagation losses, fading, multipath, temporary faults or any other possible event, base stations do not receive all the AIS messages in the surveilled area. As well, errors in the communication channels may also reduce the number of valid received messages. As a consequence, the processing takes into account three scenarios of interest, as follows:

- Three or more different base stations receive the same AIS message. It is then possible to perform multilateration and therefore estimate the real position of the vessel. It is here important to highlight that single estimates may be noisy, but the intrinsic error can be mitigated by considering more than one measure and performing a second stage of post processing on them. The series in time of the estimated positions, appropriately filtered at occurrence, gives an estimate of the track of the vessel. This has described in the previous parts of this work. At this point, it is possible to compare whether the single measurements or the whole track lie significantly far away from the declared one. Similarly, it is possible to discriminate whether the extracted parameters from the estimated track (e.g. speed or ability of manoeuvring) are consistent with the characteristics of the class of the vessel. This is part of what has been termed as 'anomaly detection' where the system automatically evaluates whether the GNSS declarations contained in the AIS messages are consistent with the estimated tracks at the output of the Kalman filter.

- Only two different base stations receive the same AIS message. It is then not possible to perform multilateration, as explained in Section 12.4. However, by means of the analysis of the TDoA, it is still possible to reconstruct the hyperbola within which the vessel is expected to lie and, through a comparison with the declared position, perform a check on the possibility that an anomaly is occurring (quite obviously the thresholds used to declare an anomaly should be relaxed, in this case).
- The signal is received by one AIS-BS only. In this case, there is no additional processing that can be performed.

An outlook of the processing approach described earlier is reported in Figure 12.13 and the cases of three TDoAs and two TDoAs are reported in detail in the following text.

With respect to the theoretical case, the multilateration algorithm had to be made efficient for working in real time within the pre-operational demonstrator described earlier. As a consequence, the graphical intersection of the hyperbolas shown in Figures 12.4–12.7 was replaced by an iterative minimisation method. In general terms, this method works as follows:

- Given a starting point of the position of the vessel, the multilateration algorithm finds the lat-lon coordinates minimising recursively a cost function in an area

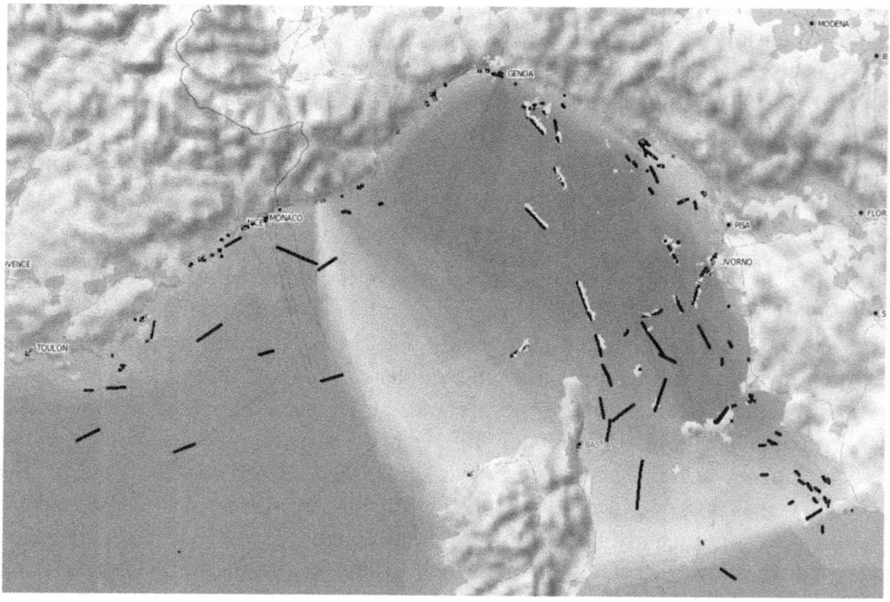

Figure 12.13 Global overview of the output of the tracking algorithm

around the starting point. As mentioned in (12.10) the cost function is based on the comparison between the measured TDoAs. However, in implementing the real time system, it has been observed that taking into account the SNR resulted in more accurate results and therefore (12.10) has evolved to (12.14).

- This is repeated recursively until a condition on the cost function or the maximum number of iterations is satisfied. A maximum number of iterations have been introduced in order to avoid infinite loops. If all iterations are performed, the covered area in which minima are searched is approximately ten times bigger than the expected coverage shown in Figure 12.3. It is important to remark that in all the observed test cases the algorithm has converged quite rapidly and convergence to local minima has not been observed. Should any errors occur during the execution of the minimisation algorithm, an invalid solution is provided in output and removed from the following steps of processing. This is to ensure the maximum degree of confidence in the output of the multilateration. Whilst during the test cases the algorithm returned invalid solutions in extremely rare occasions, the demonstrator is robust to the loss of a few multilaterated positions. This is because the tracking algorithm processes a number of multilaterated positions and as a consequence it is more performing to remove the invalid outputs of the multilateration, rather than having the system processing incorrect data.
- As in all iterative methods, the output of each iteration is the input of the following. A one-off and non-recursive minimisation is run around the final point on a smaller but denser grid. This allows to increase the precision of the estimated lat-lon coordinates of the vessel.
- All the estimated lat-lon coordinates of the vessel are then filtered (hereby with the EKF) in order to minimise the variance of the output estimation error and provide a track to the anomaly detection processing block.

The cost function to minimise has been modified slightly from that in (12.1) and is as follows

$$\left[\hat{lat}, \hat{lon}\right] = \underset{\text{grid}}{\arg\min} \left\{ \frac{\sum_k \sum_{h>k} \alpha_k \alpha_h \left(\Delta d_{h,k}(lat, lon) - c \cdot TDoA_{h,k}\right)^2}{\sum_k \sum_{h>k} \alpha_k \alpha_h} \right\} \tag{12.14}$$

where α_k is the square root of the estimate of the SNR of the received signal at the k-th base station, $\Delta d_{h,k}(lat, lon)$ is the difference of path between the point in $[lat, lon]$ and the two base stations and c is the speed of light. The estimate of the SNR is evaluated at each base station on every received AIS message. The weighting by the square root of the SNR is a way to consider stronger signals more reliable than those weaker. The initial starting point is set in the middle of the polygon comprised by all the AIS-BS-s that received the message which is being processed. Finally, the exit condition of the recursive part of the minimisation algorithm has been set to

$$\sqrt{\frac{\sum_k \sum_{h>k} \alpha_k \alpha_h \left(\Delta d_{h,k}\left(\hat{lat}, \hat{lon}\right) - c \cdot TDoA_{h,k}\right)^2}{\sum_k \sum_{h>k} \alpha_k \alpha_h}} \leq 4 \text{ km} \tag{12.15}$$

where 4 km is a threshold chosen in agreement with four times the expected standard deviation in time of the TDoA, as in Section 12.4 and in more detail in Reference 21.

A reliability flag is also evaluated and provided. We discriminate between three levels of reliability, as follows:

- *Reliability 0*: This is the lowest level of reliability. It is assigned when the output of the multilateration is estimated to have a significant contribution of noise, which may prevent the following stage of processing to perform a correct tracking.
- *Reliability 1*: This is an intermediate level of reliability. It is assigned when the output of the multilateration is estimated to have a moderate contribution of noise, which may affect in part the following stage of processing.
- *Reliability 2*: This is the highest level of reliability. It is assigned when the output of the multilateration is estimated to have a small contribution of noise, and as a consequence all the measurements are considered extremely consistent with one another.

In more detail, the reliability is evaluated on two main parameters:

- the statistical distribution of all the lat-lon positions returned by the multilateration algorithm in a specific time interval (in the all the results presented in this work this has been set to 10 min) and
- the characteristics of the statistical distribution of the corresponding SNR in the same time interval.

The level of the reliability of the measurements is (i) one of the inputs to the filtering stage of processing and (ii) one of the information reported to the human operator using this tool. As anticipated, extended Kalman filtering is then applied to a set of data in order to provide smoother and more reliable results. Tracking results are therefore stored in a database. This allows users to login and connect and hence access the outputs of the demonstrators. Data can also be plotted through a webGIS interface, which allows to display the data over a geographical map. The results provided in the rest of this work take advantage of this kind of representation.

Figure 12.13 shows a generic output of the demonstrator. It has been chosen that, as a convention, dark tracks represent the declared positions of the vessels and bright tracks represent the positions of the vessels estimated by the algorithm. Anomalies, described in the following section, are shown in grey dots. As it can be seen, most of the vessels in the covered area are tracked and their tracks are consistent with the AIS declarations. It is important to highlight that, at the edges of the covered area, some geometric distortions may occur and hence the estimated position is slightly detached from the declared. As well, it is of the utmost importance to remark that at the edges of the covered area or next to the coasts three or more signals are in general not available (due to low SNR and/or propagation issues) and hence only the declared positions can be plotted. In order to limit the effects of geometric distortions and the lack of available signals to process, it is a key factor to gather AIS data from as many base stations as possible. This will also lead immediately to an increase of the coverage and an improvement of the reliability of this kind of passive automatic tool for AIS data validation in support of maritime traffic control.

In more detail, Figures 12.14 and 12.15 show the GPS-declared (dark) and the estimated (bright) tracks of a couple of vessels which are respectively moving on a

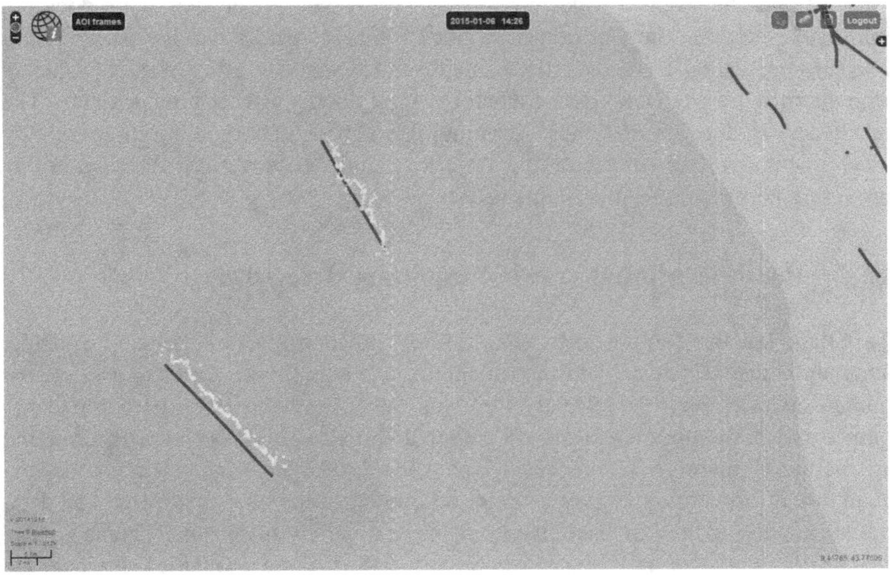

Figure 12.14 Output of the tracking algorithm, linear trajectory

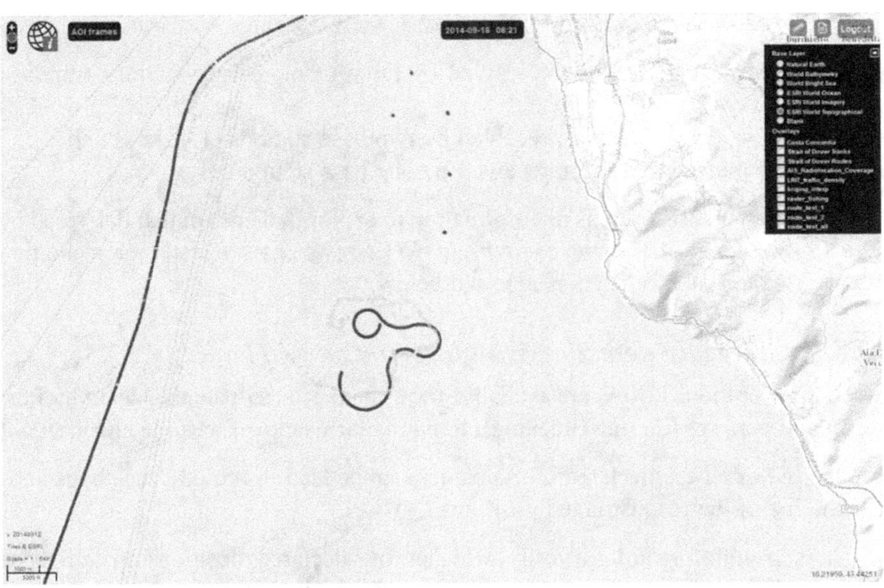

Figure 12.15 General approach to data processing and expected results

straight line (Figure 12.14) and manoeuvring (Figure 12.15). In both cases it is evident that the combination of multilateration and filtering made it possible to achieve a final average accuracy below 500 m in locating the vessel, starting from signals with an accuracy of less than 4 km in the radial direction only. In some cases a minor bias of a few hundred metres has been observed and is mainly, due to non-compensated second order effects, which are being analysed. This does not impact the fact that the algorithm can track robustly vessels moving on a straight line (Figure 12.14). The most remarkable results are shown in Figure 12.15 where a ship performs a very complex trajectory, most likely in a fishing activity. The results shown here are extremely meaningful, whereas most tracking filters rely on local linear approximations of the trajectory and hence would severely fail in providing acceptable results in this scenario.

12.7 Pre-operational results: anomaly detection

This stage aims to demonstrate that additional processing can be applied to detect anomalous inconsistencies of the available data. The ultimate aim is to provide the human operator with a tool that highlights false, erroneous or misleading behaviours and make him/her aware of potential threats. The process of discovering anomalies is therefore a process which should take into consideration all the available information and may also not be limited to the analysis of the AIS data, possibly merging the information coming from a variety of sensors. This is not the case of the work presented here, where only AIS data are available. However we hereby show that, based only on the information gathered by the AIS, it is still possible to add an additional step of processing to have a remarkable increase of the awareness of the area under monitoring.

As it has been anticipated in Figure 12.3, there are three scenarios:

- the AIS message has been received by three or more base stations (three or more TDoAs)
- the AIS message has been received by two base stations (1 TDoA) and
- the AIS message has been received by one base station only.

In the latter case, there is no information to exploit, other than that delivered by the AIS message itself. On the contrary, in the first two cases a first implementation of the detection of anomalies is reported below.

12.7.1 Anomaly detection with three or more TDoAs

When three or more TDoAs are available, the system can estimate the location of the vessel and perform filtering (tracking). In particular we now focus the attention on:

- the list of all the declared GPS positions embedded in the AIS messages and
- the list of the reconstructed positions.

It is straightforward to verify whether the declared positions match those reconstructed. As it has been described before and in Reference 21, an error is expected to lie within the two measures and a consistent threshold has to be set in order to limit the false alarms. In more detail, anomalies are declared when the distance between the declared and estimated positions is greater than 4 km. Once again, a threshold equal to 4 km has been chosen taking into account the statistics of the TDoAs of the AIS signals in Sections 12.3 and 12.4 and in Reference 21. Actually, in a first approximation, assuming to receive a truthful AIS message, the GPS accuracy is such that its position can be considered as the real one of the vessel. The uncertainty in the reconstructed position lies therefore in the errors in the TDoAs, which is diminished by the multilateration and tracking processes. As a consequence, 4 km is a threshold that allows to have a very high certainty that the two positions differ significantly. With regard to this subject, it is important to remark that in most of the pre-operational results the measured distance between the declared and estimated positions is in the order of 500 m or less.

Figures 12.16 and 12.17 show a case where the declared and estimated positions differ substantially and an anomaly is raised. In particular, Figure 12.17 is a zoom of Figure 12.16, with additional explanations. The anomaly is here highlighted through a webGIS interface, which also reports the associated timestamp. As it can be seen, they are approximately 16 km far apart from one another. In order to provide the final user with as much information as possible, when the demonstrator detects an anomaly, there is the possibility to show an area of confidence where the vessel is expected to lie. This is plotted in both the figures below and is marked with label in Figure 12.17. The shown area is evaluated graphically, that is, taking into account the graphical hyperbola as in Section 12.4, and is made available to the final user as an additional

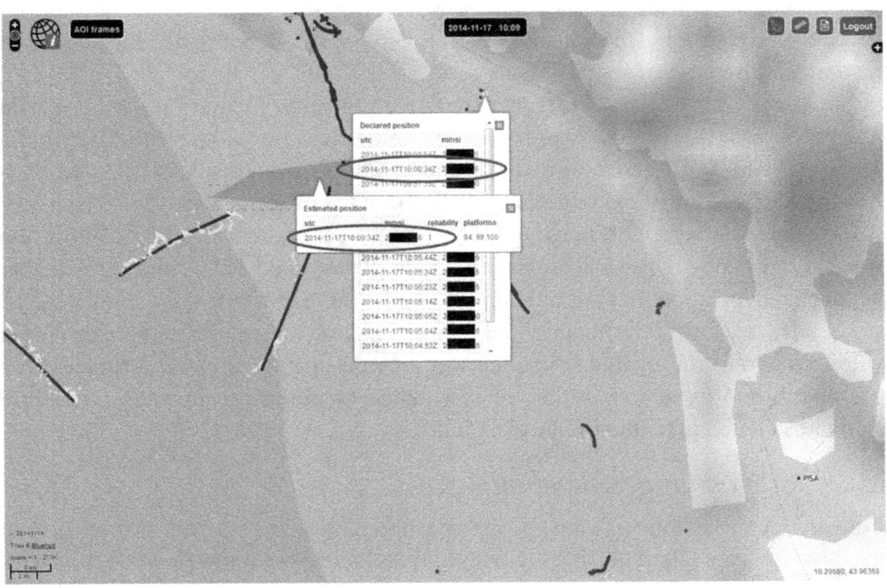

Figure 12.16 Anomaly in the declared position of a vessel (full MMSIs in the area have been partially hidden)

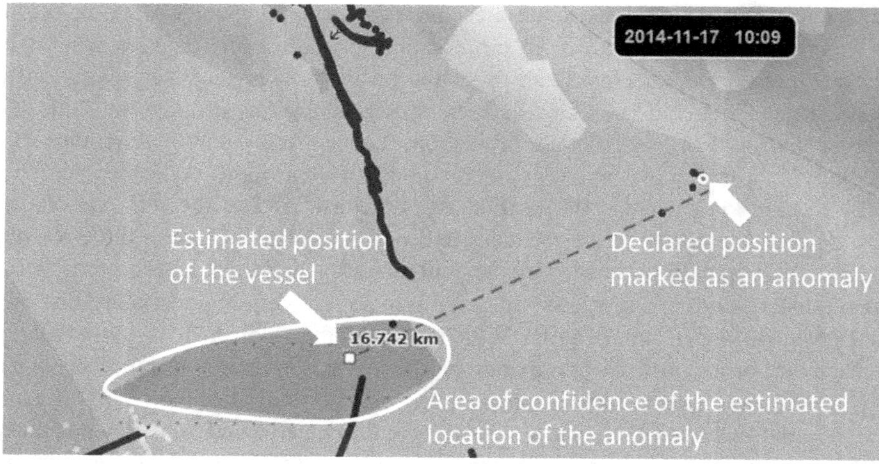

Figure 12.17 Anomaly detection: distance and area of confidence

independent tool to double-check the validity of the provided results. Whilst this exemplifies the detection of anomalies based on a single measure, it is possible to take into account more measures and implement more sophisticated approaches to anomaly detection, as reported in the subsection 'Binomial thresholding'.

12.7.2 Anomaly detection with one TDoA

When only one TDoA is available in processing, it is not possible to compare the estimated position with that declared. However, it is still possible to verify whether the GNSS-declared position lies within the hyperbola of confidence introduced in Section 12.4.

In general terms, we expect that the GNSS position lies inside the hyperbola defined by the TDoA. It is worth recalling here that the width of the hyperbola corresponds to a difference in path of, again, 4 km. An anomaly is declared each time the declared position is significantly far from the expected hyperbola. Similarly to the previous case, additional graphical tools can be developed to allow the operator to plot the hyperbola corresponding to the received TDoA and validate that the declared position lies significantly outside the expected area. Figure 12.18 shows such a case. Unfortunately, having one TDoA to process may determine a wide area of uncertainty, but nevertheless it provides the maximum information that it is possible to extract from the received data. As in the previous case, multiple anomalies can be processed jointly, as follows.

12.7.3 Binomial thresholding

From an operative point of view, noisy measurements and multipath may affect the results of the anomaly detection algorithm. It has actually been observed repeatedly over time that the majority of the anomalies occur in proximity of the coast and that some areas (which are subject to multipath and/or fading for geometrical reasons) present a higher number of anomalies than others. An example is shown in

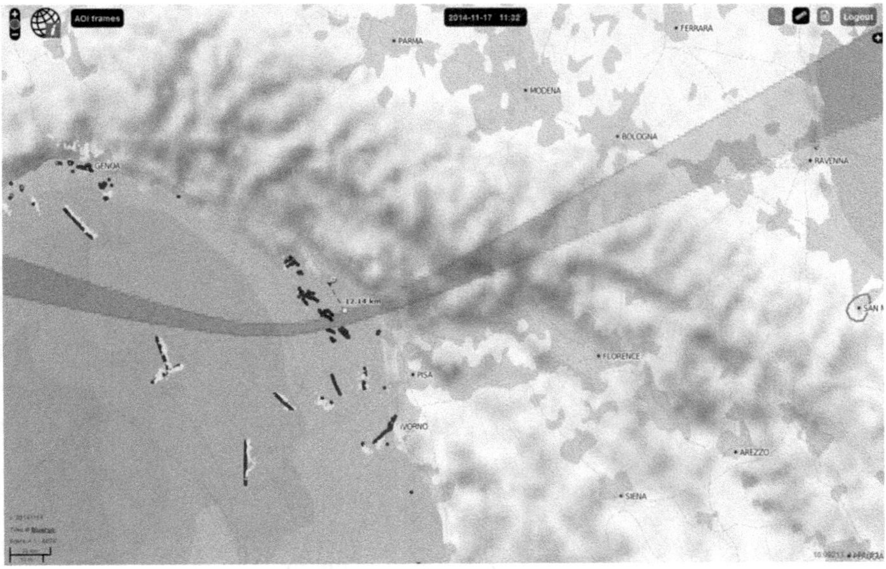

Figure 12.18 Anomaly detection, hyperbola of confidence

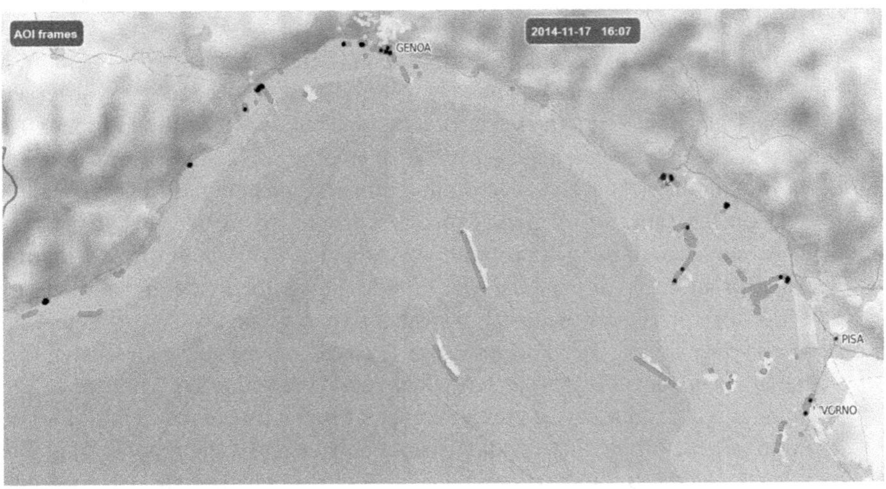

Figure 12.19 Detected anomalies

Figure 12.19. Here it can be seen that vessels at large are multilaterated successfully, whilst there are areas where the rate of anomalies is suspiciously greater.

This has also been confirmed by the example in Figure 12.20. Here the estimated positions of a vessel approaching the coast are consistent with the declared GNSS positions, whilst all of a sudden its declarations are considered as anomalies.

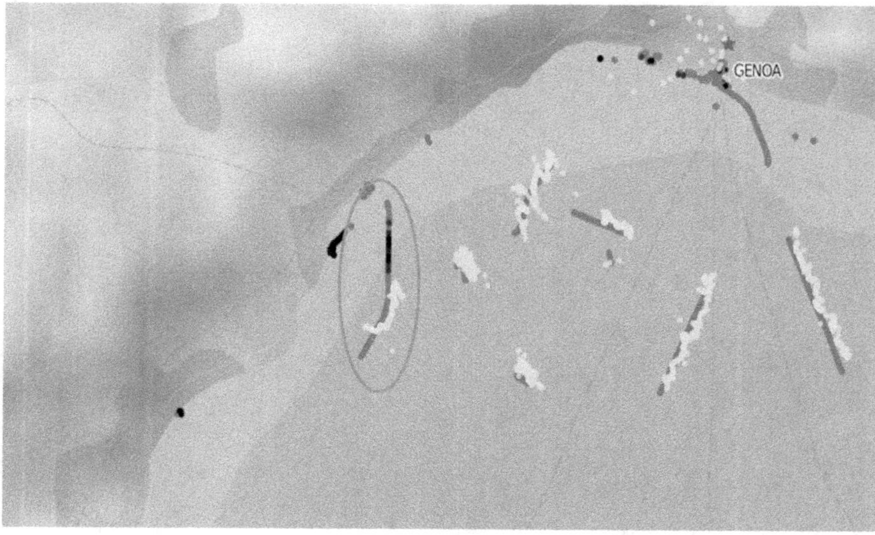

Figure 12.20 From multilateration to anomaly

Because of this and other similar cases which are being observed, two main actions were planned:

- First of all, in order to provide a reliable and efficient service to the operator, a thresholding on the anomalies has been introduced. It was actually observed that not all the points of a track were anomalies, but only a fraction of them. As a consequence, false anomalies can be overcome, or at least mitigated, by declaring an anomaly when at least a significant percentage of anomalies are present on the same set of data. From a statistical point of view, if the percentage is evaluated in an appropriate way, this has the benefit of reducing the declaration of false anomalies, yet introducing a moderate impact on the detection of correct anomalies. For instance, it may be decided to detect an anomaly only when 30% or more of a track is suspect. Obviously, this threshold is a tuneable parameter which may be under the control of the operator to filter out noisy or unreliable data. From a theoretical point of view, assuming P_a the probability that a false anomaly arises, N the number of points of a track, M the minimum number of required anomalies in a dataset, the compound probability P_a^{tot} that a false anomaly is declared can be expressed as

$$P_a^{tot} = \sum_{k=M}^{N} \binom{N}{k} P_a^k (1 - P_a)^{N-k} \tag{12.16}$$

  Figure 12.21 shows the effectiveness of this kind of processing, as a reduction of the number of anomalies is observed.
- Second, as it has been deemed that it is a statistically remote event that most of the anomalies occur in proximity of the coast, the working hypothesis that signals

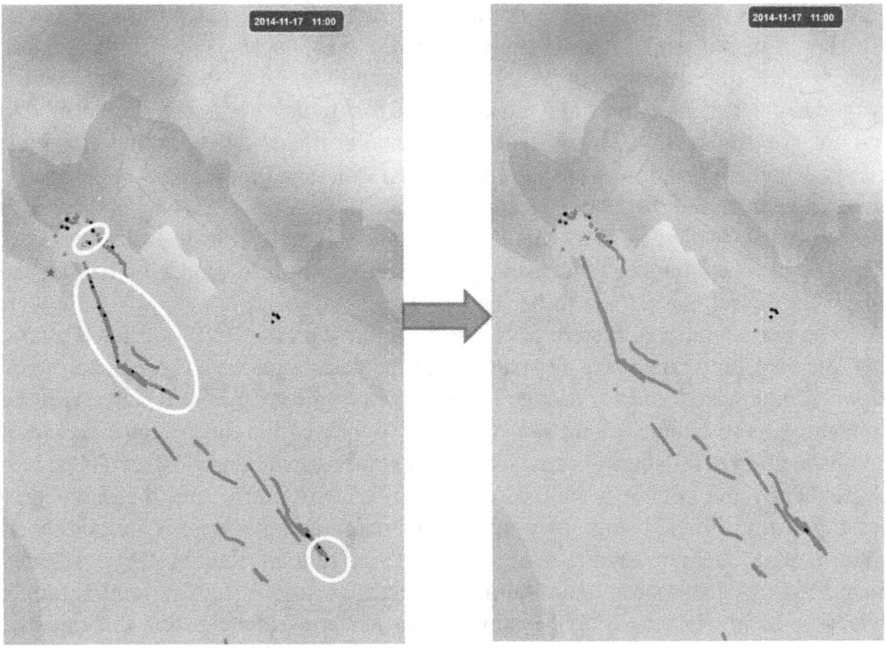

Figure 12.21 Anomaly detection: before and after applying the 30% threshold

may suffer shadowing of direct signal and/or multipath has been implemented. As a consequence, a vast data acquisition of anomalies is taking place, with the ultimate aim to understand the reasons and the extent of this issue.

12.8 Conclusions

In this work the possibility of validating the data provided by a cooperative communication system as the AIS through the analysis of the TDoA of the AIS messages has been shown. Its theoretical background and operational feasibility have also been demonstrated. The analysis of AIS data received by a few base stations makes it possible to perform signal multilateration and vessel tracking enabling AIS data verification.

An introduction on the AIS system and the concept on which this research work is based has been presented. As well, the algorithms used for multilateration and tracking have been reported and exemplified. In addition, the implementation of a real-time pre-operational platform comprised of seven base stations has been introduced and described. This included an analysis of the area under surveillance in order to understand the expected coverage, given the geometry of the pre-operational demonstrator. In a further step, the output of the test system has been presented and analysed. The system is able to provide the estimated tracks of the vessels, with an uncertainty in the position of the targets of in the order of a few hundred metres. At this stage, it is important to recall that the intrinsic TDoA of AIS has a resolution of 4 km, so all the processing stages yield a radiolocation performance improvement of an order of magnitude. In a second instance, the system can detect anomalies and it can be used as a tool to verify the shipborne AIS stations are trustworthy. This is the basis to counter illegal AIS spoofing activities, that is, when the positions transmitted in the AIS message are altered intentionally. It was demonstrated that anomaly detection can be performed as long as there is at least one TDoA, that is, the same AIS message is received at least by two base stations, by checking that the message is located within the area of confidence where the vessel is expected to lie.

The pre-operational system gave results in line with the GNSS-declared tracks. Yet, the very high rate of the anomalies near the coast requires a deeper analysis in order to understand and tackle this issue. A working hypothesis, which has to be confirmed, is that the combination of the shadowing of the direct signal and multipath or ground reflections causes the base stations to register offset ToAs, and consequently the system to calculate erroneous TDoAs, affecting the entire processing chain. This is being analysed in more detail and therefore a thorough analysis is part of future work. On average, as a rule of thumb, the developed demonstrator can process (data validation, multilateration, tracking and anomaly detection as in Figure 12.12) 15 min of data in approximately 100 s. Typically, 500–700 different vessels are present in the monitored area. As it is evident, therefore, the required computational resources make the demonstrator suitable for real-time applications. At the same time, it is far from reaching its limit and, and hence is expected to be robust against an increasing number of monitored vessels.

From an operational point of view, a number of activities (traffic monitoring & management, Search & Rescue, etc.) at sea are based upon information provided by vessels through AIS. The AIS communication system is vulnerable to tampering and spoofing, thus the validation of the data provided through the AIS is an important issue faced by authorities involved in the different areas of maritime surveillance. It is not known the level of actual alteration of AIS cooperative data and therefore, it is important to improve the assessment of the reliability of the position data transmitted by ships in an automated way to increase the safety of the seas.

References

[1] Angrisano, A., Gaglione, S., Gioia, C. 'Performance assessment of aided global navigation satellite system for land navigation'. *IET Radar, Sonar & Navigation.* 2013;**7**(6):671–80

[2] Safety of Life at Sea (SOLAS) convention Chapter V, Regulation 19

[3] Directive 2002/59/EC of the European Parliament and of the Council establishing a Community vessel traffic monitoring and information system, as amended by Directive 2009/17/EC and Commission Directive 2011/15/EU

[4] Kroener, U., Dimc, F. 'Hardening of civilian GNSS trackers'. *Proc. 3rd GNSS Vulnerabilities and Solutions Conference*, 2010

[5] 'Ship Tracking Hack Makes Tankers Vanish from View', www.technologyreview.com/news/520421/ship-tracking-hack-makes-tankers-vanish-from-view [Accessed Jul 2006]

[6] Pallotta, G., Vespe, M., Bryan, K. 'Vessel pattern knowledge discovery from AIS data: a framework for anomaly detection and route prediction'. *Entropy*. 2013;**15**(6):2218–45

[7] Ristic, B., La Scala, B., Morelande, M., Gordon, N. 'Statistical analysis of motion patterns in AIS data: anomaly detection and motion prediction', *Proc. 11th IEEE Int. Conf. on Information Fusion*, 2008

[8] Roy, J. 'Anomaly detection in the maritime domain'. *SPIE Defense and Security Symposium*, International Society for Optics and Photonics, 2008

[9] Guerriero, M., Willett, P., Coraluppi, S., Carthel, C. 'Radar/AIS data fusion and SAR tasking for maritime surveillance'. *Proc. 11th IEEE Int. Conf. on Information Fusion*, 2008

[10] Katsilieris, F., Braca, P., Coraluppi, S. 'Detection of malicious AIS position spoofing by exploiting radar information'. *Proc. 16th IEEE Int. Conf. on Information Fusion*, 2013. pp. 1196–203

[11] Vespe, M., Sciotti, M., Burro, F., Battistello, G., Sorge, S. 'Maritime multi-sensor data association based on geographic and navigational knowledge'. *Proc. IEEE Radar Conference*, 2008

[12] TRITON (TRusted vessel Information from Trusted On-board iNstrumentation) project, funded by the EU within the Seventh Framework Programme (FP7)

[13] Høye, G.K., Eriksen, T., Meland, B.J., Narheim, B.T. 'Space-based AIS for global maritime traffic monitoring'. *Acta Astronautica*. 2008;**62**(2):240–5

[14] Pourvoyeur, K., Mathias, A., Heidger, R. 'Investigation of measurement characteristics of MLAT/WAM and ADS-B'. *Proc. IEEE Tyrrhenian Int. Workshop on Digital Communications-Enhanced Surveillance of Aircraft and Vehicles*, 2011. pp. 203–6

[15] Galati, G., Gasbarra, M., Magaro, P., De Marco, P., Mene, L., Pici, M. 'New approaches to multilateration processing: analysis and field evaluation'. *Proc. 3rd IEEE European Radar Conference*, 2006. pp. 116–119

[16] Chan, Y.T., Ho, K.C. 'A simple and efficient estimator for hyperbolic location'. *IEEE Transactions on Signal Processing*. 1994;**42**(8):1905–15

[17] Gustafsson, F., Gunnarsson, F. 'Positioning using time-difference of arrival measurements'. *Proc. IEEE Int. Conf. on Acoustics, Speech, and Signal Processing*. 2003, vol. 6. pp. VI–553

[18] Recommendation ITU-R M 1371-4, 'Technical characteristics for a universal shipborne automatic identification system using time division multiple access in the VHF maritime mobile band'

[19] IEC 62320-1, Maritime navigation and radiocommunication equipment and systems – Automatic Identification System (AIS) – Part 1: AIS Base Stations – Minimum operational and performance requirements, methods of testing and required test results

[20] Kay, S.M. *Fundamentals of statistical signal processing: estimation theory*. Prentice Hall Inc., Upper Saddle River, NJ, USA, 1993, ISBN:0-13-345711-7

[21] Papi, F., Tarchi, D., Vespe, M., Oliveri, F., Borghese, F., Aulicino, G., Vollero, A., 'Radiolocation and tracking of automatic identification system signals for maritime situational awareness'. *IET Radar, Sonar & Navigation*, 2014, DOI: 10.1049/iet-rsn.2014.0292

[22] Anderson, B.D., Moore, J.B. *Optimal filtering*. Courier Dover Publications, Mineola, NY, USA, 2012

[23] Bar-Shalom, Y., Fortmann, T. '*Tracking and data association', mathematics in science and engineering*, vol. 179. Academic Press; 1988

[24] Julier, S.J., Uhlmann, J.K.: 'Unscented filtering and nonlinear estimation'. *Proc. IEEE*. 2004;**92**(3):401–22

[25] Gordon, N.J., Salmond, D.J., Smith, A.F. 'Novel approach to nonlinear/non-Gaussian Bayesian state estimation' in *IEE Proc. F (Radar and Signal Processing)* (Vol. 140, No. 2, pp. 107–113); 1993, April, IET Digital Library, Savoy Place, London, UK

[26] Bar-Shalom, Y., Willett, P.K., Tian, X. *Tracking and data fusion: a handbook of algorithms*. YBS Publishing, Storrs, CT, 2011

[27] Van Trees H.L. '*Detection, estimation, and modulation theory: radar-sonar signal processing and Gaussian signals in noise*'. Krieger Publishing Co., Malabar, Florida, U.S.A., 1992

Chapter 13

The impact of Satellite AIS to the environmental challenges of modern shipping

Nicole Schill and Margaret Browning**

Automatic Identification System (AIS) is a mandatory navigation safety communications system under the provisions of the Safety of Life at Sea (SOLAS) Conventions. The convention requires ships of 300 gross tons and upwards engaged on international voyages, cargo ships of 500 gross tons and upwards not engaged on international voyages and all passenger ships irrespective of size to be fitted with AIS.[1] The ability to decode AIS messages from these vessels using a constellation of satellites has been continuously demonstrated by exactEarth since 2010 exploiting big data management expertise to deliver clear, global near real-time maritime vessel tracking information to government authorities worldwide.

The superior detection technology of these satellites supports the rapid build-up of verifiable maritime domain awareness (MDA), which is now used to support a number of maritime applications including environmental protection, detecting and curbing illegal fishing and monitoring marine pollution. In an ever-changing environment where there is a vital need for the persistent monitoring of human activity to protect the world's waters and marine life from ship disasters and illicit activity, Satellite AIS (S-AIS) has become a very powerful tool in the delivery of MDA to a variety of maritime and geospatial users around the world. This chapter aims to explore the current applications of S-AIS data in environmental protection as well as potential future uses of S-AIS in this arena.

13.1 Background into AIS

Since 2004, the International Maritime Organization (IMO) has required AIS transponders to be aboard most vessels. Over 120,000 ships worldwide have installed these transponders at a combined cost of several hundred million dollars, making AIS one of the most successful maritime technology deployments of all time. Additionally, equipment employing AIS technology is increasingly being deployed

*exactEarth Ltd, Canada

[1]SOLAS Ch. 5 Regulation 19 Para 2.3 www.navcen.uscg.gov/pdf/AIS/AIS_Regs_SOLAS_MTSA_FR.pdf accessed 14 September 12.

in smaller vessels as well as radio/AIS enabled Aids-to-Navigation (AtoN) and it is also being installed on Search and Rescue (SAR) vessels and aircraft.

The prevalence of AIS transponders has resulted in the increased value of AIS data that can be used not just by ships, but by naval forces, port authorities, coast guards and other competent maritime authorities to enhance the SOLAS convention and improve maritime situational awareness. While AIS has been deployed successfully, it suffers from a major limitation in that, due to the curvature of the Earth, its range is limited to approximately 40–50 nautical miles under most atmospheric conditions. However, this can vary greatly depending on the site topology, density of ships in the area and atmospheric conditions. Maritime agencies wish to gain greater visibility into vessel traffic over a much broader area in order to enhance their operational effectiveness.

13.2 AIS and how it works

AIS transponders automatically broadcast information, such as their position, course, speed, rate of turn and navigational status, at regular intervals via a VHF transponder. The information originates from the ship's navigational sensors, typically its global navigation satellite system (GNSS) receiver and gyrocompass. Other information, such as the Maritime Mobile Service Identifier (MMSI), vessel name and VHF call sign, are programmed when installing the equipment and are also transmitted regularly. The signals are received by AIS transponders fitted on other ships and on land based systems, such as Vessel Traffic Systems (VTS). The received information can be displayed on the radar or chart plotter, showing the other vessels' positions and call sign information, in order to augment these systems and improve safety of navigation world-wide. The additional data provided by AIS can be added to radar and chart plotting systems, making these critical systems much more effective.

AIS is Radio Frequency (RF) based communications system designed originally as a collision avoidance system for large SOLAS class vessels. AIS communications are based on time division multiple access (TDMA); similar to the first digital communications protocol used by cell phones, employing two VHF marine radio frequencies dedicated worldwide to the service. TDMA requires each device to communicate within a given slice of time, or slot. However unlike cell phones, the AIS does not have continuous communication with a controlling device for slot assignment, that is, tower. Therefore, the AIS architecture had to develop a strategy for defining how to keep ships within close proximity from using the same time slot to transmit a position report.

Self-Organizing TDMA (SOTDMA) fits this requirement as every AIS device will preannounce the time slots it intends to use for position reports and will not use the slots reserved by other AIS devices, effectively organizing around each other and creating a communication cell. These communication cells cover nine "one minute time frames," each frame consists of 2,250 26.6 ms time slots per radio frequency channel (Figure 13.1).

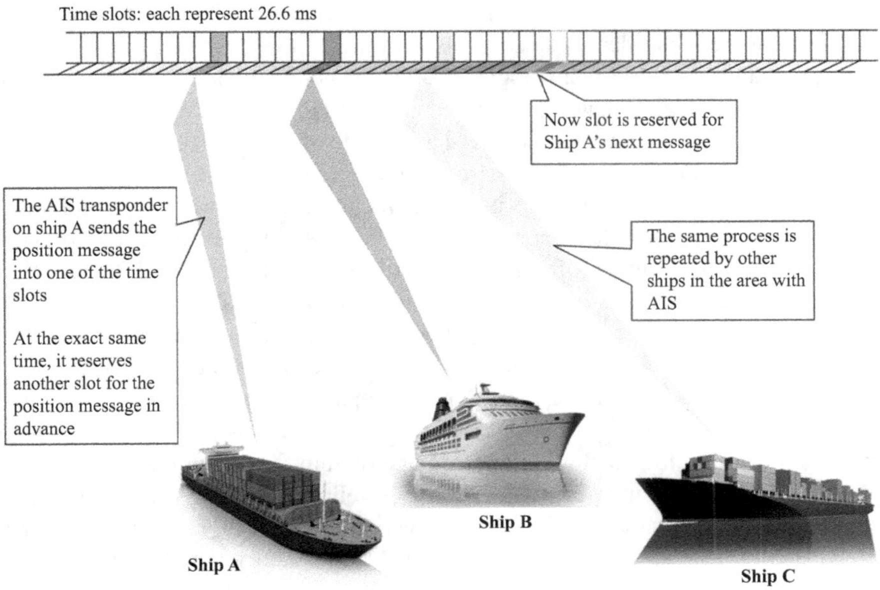

Figure 13.1 Representation of slot selection, by multiple ships within reception distance, over a single one minute frame

Slot selection by an AIS unit is randomized within a communication cell and tagged with a random timeout across the nine frames of the communication cell. When an AIS unit changes its slot assignment, it announces both the new location and the timeout for that location. Each AIS unit chooses slots within the communication cell depending on its reporting rate, which varies between 2 s and 3 min as determined by the speed and heading of the vessel. As vessels pass in and out of communication cells there is a high probability of slot collision, but with two channels providing redundancy and the fact that the closer vessel will always be heard, these collisions do not affect the primary purpose of AIS which is collision avoidance. However, these collisions do impact receiver stations that may hear multiple communication cells simultaneously, such as terrestrial AIS base stations and satellites.

13.3 The importance of first pass detection

With the advent of new space-based sensors, namely S-AIS, the possibility of detecting and identifying all vessels regardless of location has become a reality however it poses additional challenges. First, with the abundance of maritime traffic information readily available, the search for behaviors of interest in the vast amounts of data must be examined in order to observe behavioral anomalies.

Second, combining information from multiple space-based sensors (S-AIS and Space-based Radar satellites) provides an opportunity to create a fused data set which can be used to identify vessels of interest (both the behavior of interest such as an oil slick and the probable identity of the vessel through AIS); however, this capability has yet to mature as S-AIS technologies have had generally poor detection capability in dense shipping areas resulting in limited ability to identify the vessels.

Terrestrial AIS is used for ship to ship, ship to shore, and shore to ship communications ranging in applications from collision avoidance to security. Terrestrial AIS surveillance or monitoring systems have the benefit of continuous coverage and detection rates that approach 100% close-in to shore. These systems also have the disadvantage of very limited range and high cost per square mile covered. S-AIS addresses both of these disadvantages as it is able to provide complete global coverage as well as a low cost per square mile covered. However, to achieve effective operational use of S-AIS data, and since AIS was never originally designed for satellite reception, the capability must effectively mitigate the challenges presented by time slot collisions inherent to the AIS system, which impedes the ability to detect a high percentage of ships via satellite.

An effective measurement of a S-AIS constellation's ability to detect a high percentage of ships is called First Pass Detection (FPD). FPD is the ability for a single satellite to maximize the reception of AIS position report broadcasts from ships in order to satisfy operational maritime requirements. Given that ship populations are not static it is of paramount importance to maximize the number of ships detected during each and every satellite pass. A pass is defined as a subset of a complete satellite orbit over which a given ship is within the satellite's field of view (nominally a maximum of 10 min per pass), as it is likely that when another satellite passes, any given ship may have moved in to, or out of, the satellite field of view.

The deployment of a highly effective S-AIS constellation capable of excellent FPD faces many challenges due to the fact that AIS was primarily intended for local sea-level transmission and reception. The saturation of the satellite receiver due to the high amount of AIS message broadcasts, particularly in highly dense shipping areas such as the Mediterranean and Baltic Seas, and the Straits of Malacca can create a significant technical barrier to developing an accurate picture for MDA using S-AIS only.

Recent advancements in satellite technology have led to the creation of higher detection performance AIS satellites that are able to detect a significantly higher number of ships during a single satellite pass than any other system, even in dense shipping regions of the world. The completeness of the maritime traffic domain for any given Area of Interest (AOI) can be defined as the ability to detect ships and the latency for any given ship position report. Latency is defined as the time interval between the reception of the ship's position report on the satellite to delivery of that report to the end user. Latency is impacted not only by the mechanics of satellite constellation (i.e., the time it takes for the satellite to find and downlink data to an Earth Station) but also by the ability to detect as many ships as possible within a single pass over the AOI. If multiple satellite passes are required to detect all ships, the true latency is significantly higher than any S-AIS service with a high FPD rate as it will take a substantial longer period of time to obtain a complete picture of the AOI. Additionally, ships are moving targets therefore each satellite may have only a single opportunity to detect the ship before it travels out of the AOI.

The likelihood of FPD can be impacted by a number of factors including ship density, AIS message broadcast frequency and in-band RF interference. Understanding that there are many factors that can impact detection, AIS time slot collisions and satellite receiver saturation are, by far, the main factors impacting an AIS satellite's FPD performance and the detection methodology has the largest impact on mitigating both message collision and receiver saturation.

There are two detection methodologies supported by S-AIS constellations currently in orbit.

13.3.1 On-board processing (OBP)

This processing mode essentially decodes AIS messages directly on the satellite using narrow band filters and stores the messages on the spacecraft for later downlink to the nearest Earth Station. It does not require any special processing and is effective in very low density areas, such as the middle of the Pacific Ocean. However, the detection probability is low in areas where the satellite footprint (~5,000 km in diameter) contains a ship density exceeding about 500 ships as it become likely that all of the AIS message slots are being used by more than one ship at a time. This effect results in a time slot collision and it becomes much more difficult to successfully decode.

Statistical analysis has shown that the FPD performance for OBP in high dense areas is much lower than spectrum de-collision processing over the same area. OBP suffers this poor detection performance partially because of its inability to deal with these time slot collisions inherent to the AIS system when received from space. With a low per pass probability of detection across high density ship populations, effective operational capability is seldom achieved with S-AIS OBP only.

13.3.2 Spectrum de-collision processing (SDP)

SDP requires the capture of the AIS RF spectrum and processing of that spectrum using highly specialized algorithms to successfully decode AIS reports. With SDP the FPD ability is high even within high ship density areas, thus quickly achieving effective operational maritime capability. Statistical analysis has shown that the improvement in FPD for this detection methodology in highly dense shipping areas can result in significantly higher ship detection, number of position reports received (for behavioral modeling and predictive algorithms) and reception of multipart (multi-slot) messages containing static and voyage related information. If the AOI contains higher ship densities, SDP is required to achieve FPD performance that will enable operational use of S-AIS to develop an accurate picture of maritime traffic.

13.4 Comparison of varying methods of Satellite AIS reception techniques: OBP and SDP

The figures below effectively demonstrate a measured and absolute detection performance improvement based on analysis conducted using onboard AIS signal decoding with OBP and using AIS RF spectrum capture with SDP. This analysis was conducted on October 23, 2012, using exactEarth satellites operating in both modes (Figure 13.2).

Example area: North-East United States

Processing mode	AIS messages processed	Vessels detected
SDP	9,321	1,969
OBP	272	227

Figure 13.2 The area shown is the North-East United States highlighting the differences in ship detection levels when utilizing Spectrum Decollision Processing vs. On-Board Processing

S-AIS is an emerging technology that provides advantages across a wide variety of applications.

However, without high rates of FPD provided by SDP, those applications are limited to a smaller sampling of the maritime traffic rather than providing a more complete picture. After an in-depth comparison analysis, OBP is adequate in low density shipping areas of the world but as the number of ships utilizing AIS continues to increase, those areas will become increasingly rare. Operational FPD must maximize the number of ships detected during each and every pass, and only through SDP can the constellation effectively increase the reception levels for all AIS messages.

FPD using SDP provides a complete and comprehensive view into shipping activity across the globe regardless of ship density. SDP proves to be a critical enabler for achieving true MDA which is so imperative to be effective in ongoing efforts to protect our ocean environment from the perils of shipping.

13.5 The challenges to environmental protection

With more than 90% of global trade being carried by sea,[2] the amount of industry – petrochemical, timber, cargo, fishing – traversing the world's waters on any given day

[2]IMO and the Environment 2011 www.imo.org/OurWork/Environment/Documents/IMO%20and%20the%20Environment%202011.pdf.

is staggering. Although shipping is the anchor of the global economy, this intense level of vessel traffic can leave detrimental marks on the marine environment. Petrochemical spills, ships run aground, debris and illicit activities such as bilge dumping and overfishing have all contributed to an ailing ocean world and the flora, fauna and aquatic species that call it home.

Recent environmental disasters such as the Rena sinking off New Zealand in 2011 and the oil spill catastrophe in the Gulf of Mexico in 2006 shed new light on the critical need for protection of the ocean environment as the ability for ships to transport more commodities, and potentially hazardous materials, form risks that are too great to ignore.

13.6 Satellite AIS for environmental protection

Oceans are regularly victim to pollution caused by oil slicks and the dumping of rubbish resulting from accidents and illegal activities. Nearly half of the pollution at sea is caused by ships' accidental or deliberate discharge of crude oil and other refined products. The inevitable trend is now such that marine protection officers have begun to change their fundamental protection strategy from one of enforcement to prevention.

S-AIS data can play a critical role in not only assisting to identify polluters post-incident but also preventing dumping by providing authorities with the ability to spot deviations in predefined routes indicating suspicious activity. S-AIS provides tracks and speed of ships to assist governments to determine whether or not a ship is in violation of existing regulations surrounding prohibited discharges.

The FPD performance realized through SDP processing of S-AIS data has proven itself to be highly valuable for leveraging imaging gathered from radar satellites. SDP processing guarantees the maximum first pass yield for ship detection, therefore is much more likely to properly correlate ship positions with images contained within the radar capture. Once positional correlation is achieved, SDP provides an additional advantage because it is able to not only capture many position reports that are key to effective predictive algorithms, but also much more likely to receive the static and voyage information transmitted by ships. This information can then be used to correlate to other ship registry information allowing the authorities to confirm the ship is in fact the ship it claims to be. The voyage data can be used to cross check the position and course over ground to confirm that the ship is reporting accurately.

High FPD performance is fundamentally important to fusion of radar/optical imagery with S-AIS and therefore the identification of ships. The radar image provides virtually 100% detection of large ships (by design) as an anonymous target, but identifying those ships and using that information to identify a ship of interest can only be achieved quickly if the S-AIS data contains a report from all (or most) ships captured in the image.

The number of AIS position reports acquired during the satellite pass are used in predictive algorithms and to fully identify any given ship. The static data that identifies the IMO hull number and ship name and callsign is contained in a multi-slot message, as discussed earlier, the most likely detection of multi-slot messages is achieved using SDP. The accuracy of predictive algorithms is dependent upon the amount and quality of the data received, the more AIS position reports the more likely the predicted position will be.

Fusing S-AIS with radar and optical imagery can assist enormously in creating an accurate picture of the maritime domain to provide near real-time data eliminating the need to wait for interpolation or forecasting. It also helps to improve ship validation by implementing measurement tools with radar images as well as improve accuracy for ship identifications.

To protect offshore installations and reduce the risk of a maritime casualty and resulting marine pollution, traffic exclusion zones and precautionary areas have been established. Precautionary areas might envelop wind farms or oil and gas fields, and apply to vessels of a certain size. Vessels may be prohibited from entering certain sea areas due to environmental concerns. Other areas may be prohibited or tightly controlled due to ecological disasters, or to allow endangered marine species and organisms to thrive or recover.

S-AIS provides accurate information on ship behavior and position such that determining, or definitively concluding offenses such as pollution violations, route violations, or irresponsible navigation is made that much more probable for the purposes of prosecution or establishing innocence. While S-AIS can only report the position information during a given pass, the quantity of those position reports can establish a pattern of ship behavior where activity during the time when the ship is not being monitored can be fairly conclusive.

The high FPD provided by SDP maximizes the probability that a ship of interest will be identified when authorities are investigating incidents. In the event of a pollution violation it is important to quickly determine all ships that may be responsible for the violation. In the event of incursion violations, it is again important to maximize the probability of vessel detection either during the actual incursion or with enough data to prove incursion. The number of position reports is absolutely critical when evaluating behavior of a ship, whether on a particular route or over time in an AOI. Proving route violations and irresponsible navigation requires accurate modeling which can only be achieved through S-AIS and SDP, as there is no other practical wide area sensor over the ocean areas. The richness of multiple position reports can provide the data necessary to accurately predict and confirm behavior during coverage gaps.

S-AIS provides information on vessels that have deviated from their predefined route which might indicate they are trying to avoid detection of prohibited discharge. It also serves as a checks-and-balances tool because it can validate the actual position of a vessel against the declared position or time entered into a particular area to identify discrepancies in a ship's oil, cargo and garbage record book. Authorities can use S-AIS data to provide historic tracks to backtrack vessel positions to identify probable pollution culprits, enabling more violator prosecutions for previously undetected oil spill coating on the ocean surface.

13.7 Illegal ballast water exchange

Since the introduction of steel hulled vessels around 120 years ago, water has been used as ballast to stabilize ships at sea. This ballast water is pumped in to the hull to reduce stress on the hull, provide stability, improve propulsion and maneuverability, and compensate for weight lost due to fuel and water consumption. However, this practice poses serious ecological, economic and health risks due to the multitude of marine species carried in ships' ballast water. The transferred species may survive to establish a reproductive population in the host environment, becoming invasive, out-competing native species and multiplying into pest proportions.

As a result National Governments will impose strict limits from their shorelines in which ballast water may not be discharged and exchanged. For the Canadian/USA Great Lakes area for example, Federal regulations call for vessels to conduct mid-ocean ballast water exchange during ballast-laden voyages outside the 200 nautical miles (nm) limit from any shore prior to arrival in port (Figure 13.3).[3]

13.8 Oil spill identification

Radar imagery (both from airborne and satellite sources) is used extensively in maritime surveillance and is especially useful not just for identifying surface vessels but also for locating other ocean surface features such as oil slicks as these radically change the ocean surface reflection characteristics and therefore are easily identifiable. However, unfortunately the radar imagery is only an unattributed picture and does not allow for the identification of the surface objects detected and so requires augmenting from other sources to give a complete picture.

Combining Radar imagery with S-AIS allows for both the identification of surface pollution as well as the ships in the vicinity and therefore for an immediate short-listing of potential offenders (Figure 13.4).

13.9 Tracking illegal fishing

Several decades of overfishing in most of the world's major fisheries has created large declines in many commercially important fish populations across the world. Illegal, unreported and unregulated (IUU) fishing today is recognized as a major threat to achieving sustainable fisheries. IUU vessels may or may not be using AIS, but they will usually be found in areas populated with legitimate fishing vessels, and often around a mother ship for transshipment. Traditionally vessel monitoring systems (VMS) are used for monitoring fishing activities by National and regional Fishery Management agencies to prevent IUU fishing, but since its inception crews have tried to defeat VMS by spoofing the information sent.

S-AIS can be used to validate vessel positions received by VMS to eliminate false reporting while also extending the surveillance range beyond the limits of

[3]www.dfo-mpo.gc.ca/science/publications/article/2011/06-13-11-eng.html.

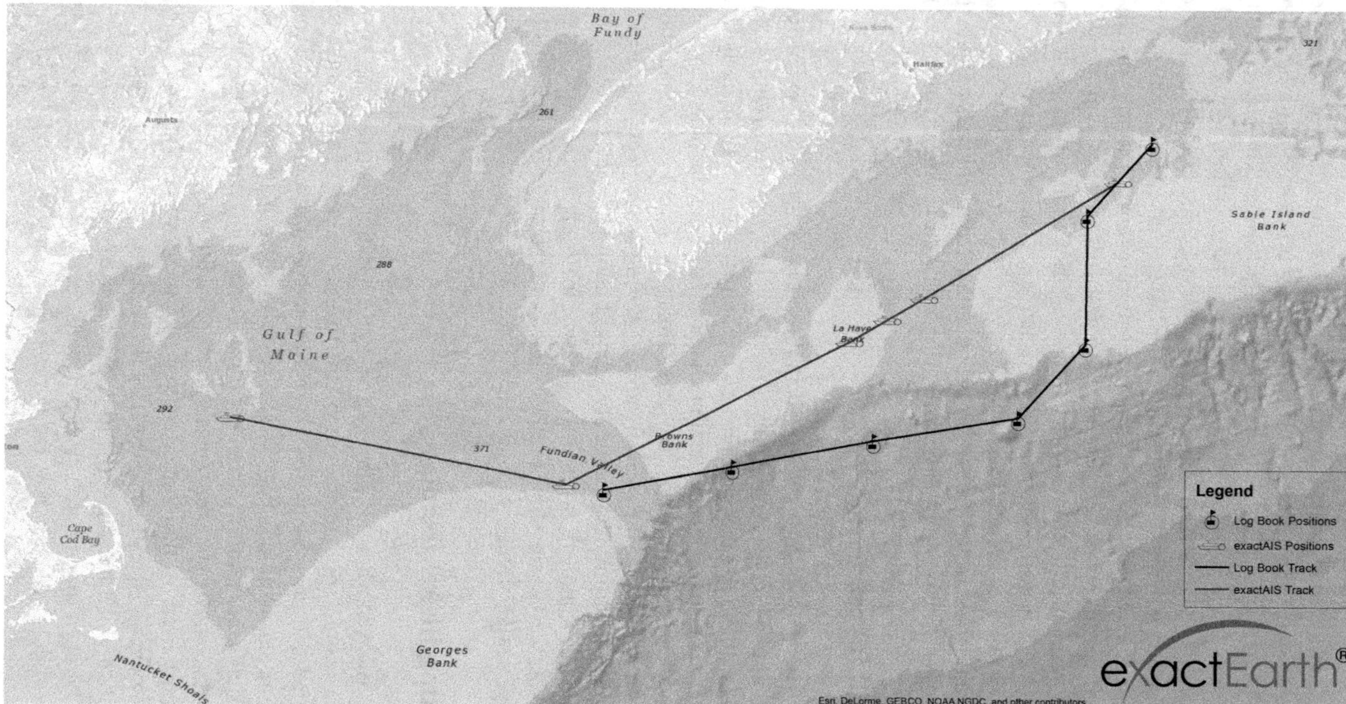

Figure 13.3 *In this example from 2011, S-AIS data was used to prove that a ship's log had been falsified in an attempt to avoid voyaging out beyond the set limit for ballast exchange. The S-AIS provides unequivocal proof of the real track of the ship allowing for monetary penalties to be administered*

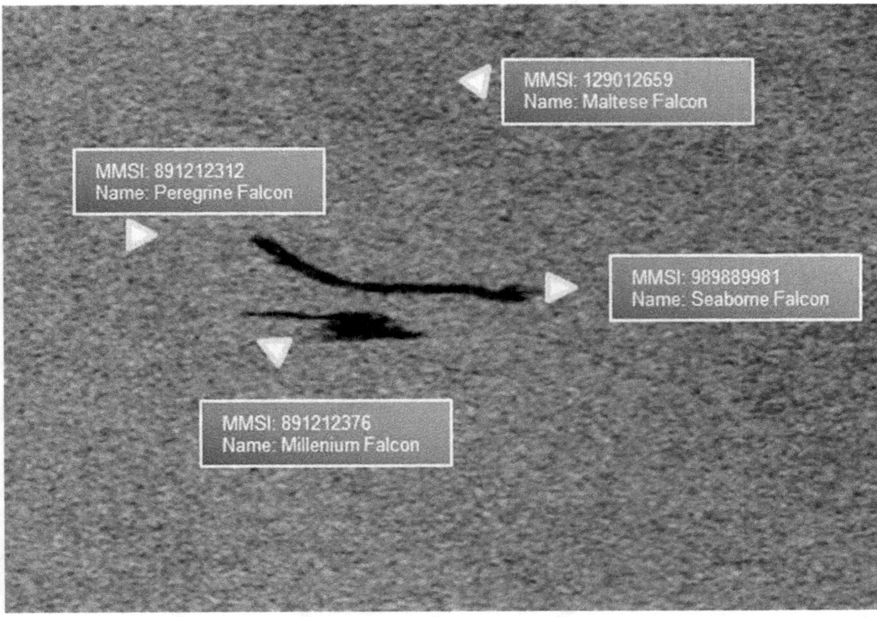

Figure 13.4 *Oil spills are clearly visible on the radar image of the ocean surface. By overlaying the ship position and track information collected by S-AIS operators are able to immediately identify vessels in the vicinity for further investigation*

VMS to include areas where vessels are not intended to be fishing. It has become increasingly common for ships involved in transshipment to be found in Polar waters where only S-AIS can effectively monitor. By extending the surveillance area, S-AIS is able to detect non-cooperative vessels allowing authorities to shift their attention to these ships and prosecute more violators.

With all the advances made in the monitoring of the world's shipping, there still remains a gap in the effective tracking of every boat on our busy waterways. A particular need has arisen around the thousands of boats currently traversing our oceans without the necessary equipment installed to allow for accurate monitoring and surveillance, as AIS is only mandatory on ships of a certain gross tonnage. Developments in S-AIS technology have allowed for this critical piece of the ship monitoring puzzle to come together as of late with a revolutionary new system that essentially tracks every boat on the oceans, regardless of size. This is crucial in the monitoring of illegal fishing as the majority of IUU vessels are under the size limitations of AIS.

13.10 Monitoring ship emissions

S-AIS data was selected for the Satellite Monitoring of Ship Emissions in the Baltic Sea (SAMBA) project, which assessed the technical feasibility as well as the economic viability of an integrated service that can be used by national maritime and port authorities to monitor air pollution from maritime emissions and take relevant actions.[4] The project was carried out in the framework of the "Integrated Applications Promotion (IAP)" program of the European Space Agency, which called for a combination of earth observation (EO) instruments, air quality (AQ) and emission modeling. The study also aimed to illustrate the potential for integration of emissions and S-AIS.

With the overall goal to develop a system to track ship emissions on a large scale, the first phase of SAMBA aimed to show the viability to use S-AIS, EO instrument data and modeling to achieve such a system. The service was built on GPS signals received by on-board AIS equipment on ships, the transmission of the AIS information from the ships through S-AIS or ground VHF and the conversion of AIS-based navigational information from all ships in a sea area into an emission grid. Early results were positive in proving that the emission compliance shipping lanes can be monitored from space.

Combining AIS with GPS-based maritime emissions and pollution monitoring using atmospheric remote-sensing instruments allows for deeper analysis into maritime emission compliance in the heavily traversed Baltic Sea. Through collaboration with the Finnish Meteorological Institute (FMI), S-AIS data was selected to strengthen the EU Strategy for the Baltic Sea Region and its project on clean shipping as the data can provide vessel specific emission statistics (Figure 13.5).

13.11 Satellite AIS for environmental planning

As critical and beneficial as the near-real time (NRT) feature of S-AIS is, so too is the ability for maritime authorities to go back in time to acquire and analyze the historical movements and patterns of transiting vessels. Hindsight can not only provide key evidence for building present-day cases of illicit activity, but it also can prove instrumental in helping to develop better planning, monitoring and navigation strategies for the future.

With access to historical S-AIS data, authorities can gain improved historical insight and knowledge to better protect the marine environment. In particular, authorities can definitively identify ships in breach of MPAs, enabling them to retroactively enforce the violation. They can establish patterns of vessel behavior and flag suspicious ships for continued tracking to identify illegal acts such as oil dumping (with the added benefit of being able to trace a detected spill back to the likely offender), unjust fishing or encroaching into protected areas.

S-AIS data lends itself to the creation of shipping density maps, an effective tool used in the study of trends within the shipping industry and for performing in-depth global vessel pattern analysis. Density maps are particularly useful in assessing ship strike risk, helping to protect the marine environment and the ocean life within it while also developing mitigation measures as well as analysis into the efficiency of existing shipping routes (Figure 13.6).

[4] http://artes-apps.esa.int/projects/samba

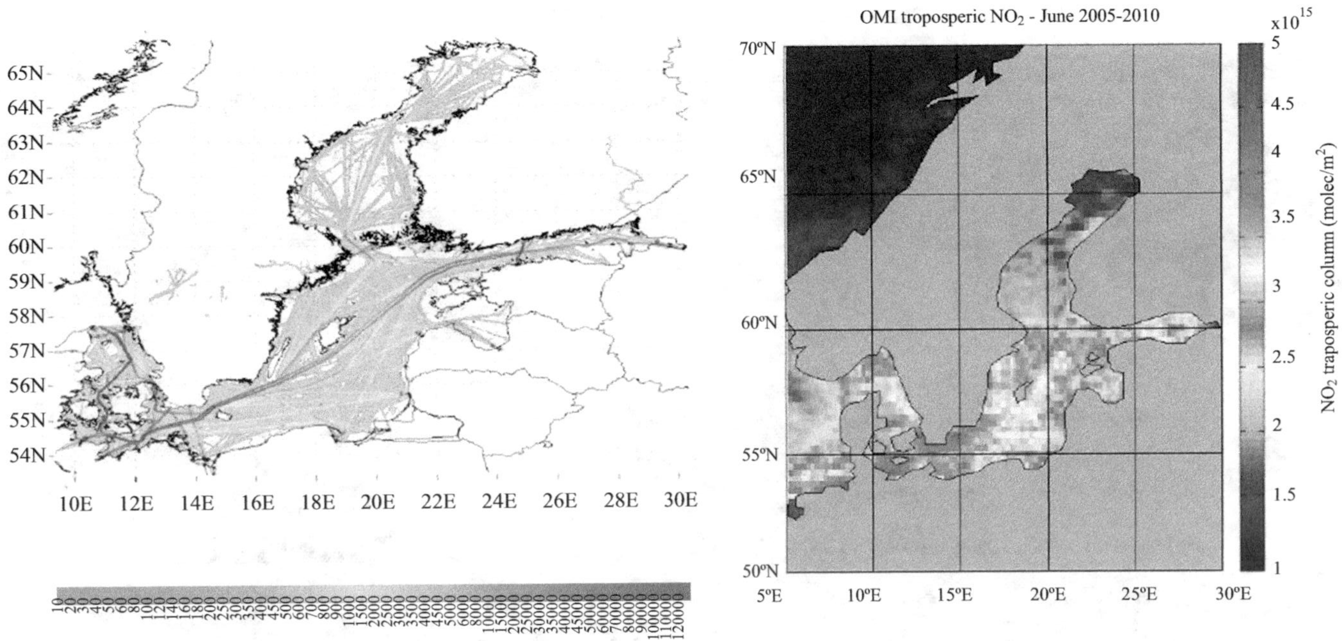

Figure 13.5 *S-AIS was able to help improve Air Quality forecasts through the SAMBA project with the Finnish Meteorological Institute*

Figure 13.6 S-AIS density map showing global shipping traffic from August 2013

13.12 Conclusion

For maritime authorities, protecting and preserving the delicate marine ecosystem is a complicated endeavor as each day they tackle the elusive crimes of illegal fishing, breaching MPA zones and unlawfully dumping oil into the water. Officials can analyze maritime traffic patterns in geo-fenced areas and develop effective refinements and routing changes to ensure vessels avoid the protected zones in the future.

The more accurate MDA offered by S-AIS provides maritime authorities with a better understanding of the who, what, when and where of shipping, which promotes a more comprehensive situational awareness of ship activities, and enables authorities to develop more watertight, preventive strategies to reduce environmental risk.

Chapter 14
How 'green' is e-Navigation?

*Jan-Hendrik Oltmann**

14.1 The challenge – 'What do you mean by "protection of the environment", specifically?'

The International Maritime Organisation (IMO) launched their e-Navigation strategy in 2008 by a decision of their Maritime Safety Committee (MSC) [1]. There, e-Navigation is technically defined as the 'harmonised collection, integration, exchange, presentation and analysis of marine information on board and ashore by electronic means to enhance berth to berth navigation and related services for safety and security at sea *and protection of the marine environment*' ([1], para 1.1; emphasis added). When I was giving an introduction to e-Navigation before a chamber of commerce audience recently using the above definition, I was challenged by a representative of the World Wildlife Fund (WWF) who was asking what 'protection of the environment' is supposed to mean in the context of e-Navigation *in specific terms.* In return, being a professional in the domain of Vessel Traffic Services (VTS) which is also 'designed to improve the safety and efficiency of vessel traffic and the protection of the environment' ([2], Annex 1, para 1.1.1), I presented the *'classic theory'* of that profession: namely, that the protection of the environment is achieved by the prevention of accidents at vessel traffic, which in turn would prevent the unintended and undesired release of any substance harmful to the marine environment, and that the improvement of efficiency of vessel traffic 'certainly' has a 'green' side to it. The point was not further pursued in the ensuing discussion of that event. However, when reflecting on my reply afterwards, I recognised that the 'classic theory' may not be fully convincing any more: the prevention of vessel traffic accidents seems to be taken as self-evident today by many. Also, the 'improvement of efficiency of vessel traffic' may be too unspecific regarding the optimisation goals driving it – efficiency for the benefit of whom or what? – and regarding the actual benefit for the protection of the environment even if it were meant to benefit it. Further, 'greenness by heart' seems to look differently. Therefore, within the context of a book dealing with the maritime side of 'intelligent' transportation systems it might be appropriate to look into the question how the maritime transportation system today and/or envisaged for the foreseeable future *specifically* targets current 'green' topics.

*Senior System Engineer at German Federal Waterways and Shipping Administration; compare Disclaimer.

As it would be both imprudent and impractical to not consider what is there already in terms of 'potentially open doors', this chapter starts to find these in existing or proposed relevant regulations and/or strategies. I shall do this in this chapter by the following steps: first, I shall look into how the present maritime transportation system is envisaged to be tied into the 'Beyond Rio+20' process which seems to reflect the present 'cutting edge' of environmental protection. Second, I shall investigate how the environmental concerns of people living ashore express themselves to shipping by Marine Spatial Planning (MSP). Third, I shall scrutinise an internationally recognised candidate strategy at hand to introduce 'green intelligence' into the envisaged 'green' maritime transportation system, namely, the IMO e-Navigation strategy introduced earlier. Finally, I shall conclude with some postulates derived from the discussion until then.

14.2 Sustainability in maritime transportation – the larger context

In short, being 'green' today in the global context of the United Nation's (UN) flow of related conferences starting with the UN's Conference on Human Environment (1972, Stockholm) over UN Conference on Environment and Development (1992, Rio de Janeiro) over the World Summit on Sustainable Development (2002, Johannesburg) to the UN Conference on Sustainable Development (2012, Rio de Janeiro) means to stress the *environmental* dimension of the three fundamental dimensions underpinning the concept of 'Sustainable Development', namely the economic, social and environmental dimensions.[1] Turning towards the maritime domain in this context requires to turn towards the UN Specialized Agency in this domain, namely the IMO, because it 'regulates all technical aspects of international shipping, delivering 53 treaty instruments, supported by hundreds of codes and guidelines, covering the entire life span of commercial ships from cradle to grave' ([3], p. 8).

And indeed, IMO, in the past, has undertaken major steps towards the prevention of *immediate* damage to the environment from the *vessel itself*, namely by implementing measures against the release of marine pollutants and against oil spills, measures for cleaner exhaust gases, measures against introducing alien species into marine ecosystems by ballast water release, to name some. Also, regarding the *navigation of vessels*, a certain degree of awareness for environmental sensitivity has been introduced by IMO into shipping, namely, for example, the implementation for certain no-go-areas for environmental reasons. IMO also committed itself to continuous work for the protection of the marine environment by running a dedicated committee with that name, that is, the Marine Environment Protection Committee (MEPC). Further, there must be acknowledged the legitimate interests and efforts of ship owners and operators to improve the *operation of their respective fleets for cost savings*, resulting in operational modes like 'slow steaming', which have benefits for the environment as well. Finally, it has been

[1]For a more elaborate introduction refer to Reference 3, pp. 4–5.

remarked that maritime transportation, by its very nature, is the *most efficient of all modes of transportation*,[2] resulting in a relatively positive environmentally balance, 'obviously' and 'by default', in comparison with other modes of transport and projected on the amount of goods transported per unit of energy. This is certainly a true statement, but it is a comparison assessment, that is, being better than another mode of transport may not mean to qualify *to be 'good enough'* in absolute terms.

Apparently, despite these efforts for the protection of the environment mentioned earlier and recognising the fundamental efficiency of maritime transportation compared to other modes of transport at large, it was deemed necessary to specifically address the contribution of global shipping to 'Sustainable Development' as envisaged by the UN's 'Beyond Rio+20' process. Hence, on the occasion of the World Maritime Day 2013 (23 September 2013) the Secretary General (SG) of IMO, Mr Koji Sekimizu, presented and published his concept of how to transform the present Maritime Transportation System into a *Sustainable* Maritime Transportation System (SMTS) [3], thereby implicitly and explicitly recognising the need for 'sustainability' to move closer to the focus of attention of the stakeholders involved, in particular those assembled at IMO. He also expressively identifies the *challenge* incurred by this move: 'The movement of goods by the Maritime Transportation System is subject to economic, social and environmental responsibilities and requirements on many levels. The challenge lies in how these can be translated equitably and fairly across the chain of actors in order to make the whole System sustainable. This is particularly difficult because coordination between shore-side maritime actors and the international shipping sector is not well-established. This is due to a prevalent tendency towards profit-maximizing by each of the actors, who may succeed in shunting costs to other actors, and this may in turn affect the sustainability of those other actors' operations, and so affect the logistics chain as a whole. In other words, a loss of resilience, and of sustainability, in one link risks degrading the chain over time' ([3], p. 7).

The concept of the SMTS then goes on to explore how this challenge should be met, namely first by establishing ten *goal domains* and defining one or several *goal(s)* for each goal domain ([3], pp. 12–21) and second by breaking those goals down into specific actions for both IMO and specifically identified partners of IMO ([3], pp. 22–32). There are several goal domains which contain goals and actions relevant to the present topic, namely 'Safety Culture and Environmental Stewardship', 'Education and Training in Maritime Professions', 'Energy Efficiency and Ship Port Interface', 'Energy Supply for Ships', 'Maritime Traffic Support and Advisory Systems', 'New Technology and Innovation' and 'Ocean Governance'.

In particular, SMTS' goal domain '*Maritime Traffic Support and Advisory Systems*' is more relevant to the present topic ([3], p. 16) where it is stated as an

[2]Compare, for example, 'International shipping transports about 90 per cent of global trade to people and communities all over the world. Shipping is the most efficient and cost-effective method of international transportation of goods, providing a dependable, low-cost means of transporting goods globally, facilitating commerce and helping to create prosperity among nations and peoples' ([3], p. 1).

explicit goal that these systems may help solve some of the challenges of MSP, an aspect to be further considered in Section 14.3, and that e-Navigation, an aspect to be further considered in Section 14.4, is expected to support the stated goal of 'the development of optimal systems for navigation' and 'the use of intelligent routeing systems and aids for weather routing' with the goal to 'optimize safety and fuel efficiency' (all [3], p. 16).

While the IMO SG's concept of a SMTS thus has a specific 'green' side to it, thus truthfully trying to create a balance between the three named dimensions of sustainable development, the *one major weakness* of the document cannot be overlooked, namely, that it is not (yet) an official publication of the IMO. In order to achieve the 'green' goals of the envisaged SMTS in reality eventually, they need to be 'translated' into effective instruments of relevant international and regional bodies, with IMO leading the way.[3]

14.3 The 'message' of Marine Spatial Planning – a reflection on the values involved

In the previous section it was mentioned that Maritime Traffic Support and Advisory Systems could potentially solve some of the challenges of MSP. What these challenges might be specifically is further investigated in this section.

On one hand, 'green' initiatives, such as switching to renewable energy collected in power plants off-shore as well as temporary or permanent marine nature reserves, have gained priority allocations in certain quarters of the sea by the process of MSP.[4] The 'message' of this is *that people living ashore* make it clear, by means of their laws and regulations, that they are *not prepared to accept just any environmental risk or damage or even 'only' certain environmental side-effects of the current way of living any longer,* as in the case of energy supply.[5] They are also not prepared to accept replacement of nuclear energy with fossil energy ressources to any degree because of its negative side effects on the world climate. Hence, the acceptance of the *usage of off-shore sea areas for the construction and operation of off-shore renewable energy sources*. Also, sensitivity to the value of wildlife species in their respective (marine) habitat has grown as has the understanding for the necessity of a balance of priorities when it comes to development of projects that have a potential tremendeous impact on wildlife. To provide appropriate nature reserves for marine wildlife also uses sea area.[6]

[3]Compare discussion in Reference 3, p. 10.

[4]Sometimes also called *Maritime* Spatial Planning and also abbreviated MSP. UNESCO has endorsed the concept internationally on the level of the United Nations [4]. A reader-friendly and practical introduction into the fundamental ideas of MSP at the example of the Baltic Sea in Northern Europe is given in Reference 5.

[5]Compare, for example, the 'Energiewende' in Germany as an example in the wake of the severe Fukushima nuclear power plant accidents in March 2011.

[6]Compare, for example, the EU Directive 92/43/EEC on fauna, flora and habitat.

On the other hand, people living ashore want to reap the benefits of the current way of living: this results in the following *value derivation chain:* the current way of living necessitates and incurs *(global) trade* which in turn creates *(overseas) maritime transport* which again prompts *maritime traffic.* Maritime traffic can again be construed as consisting of the 'berth-to-berth navigation' of many vessels which immediately prompts the most relevant *attributes of vessel navigation and vessel traffic,* namely, *safety, efficiency, security* and *environmental friendliness or 'greenness'*, amongst others.

Only recently shipping recognised the challenge of MSP in full, namely that it is no longer 'alone at home' on the – now no longer so – 'open' sea: thus, vessel traffic has been confined to *shipping 'lanes' – literally* – in those quarters. This became evident in particular in sea areas where there is already a dense vessel traffic today, for example, in the North Sea Region in Northern Europe.[7] The challenge may be even larger if forecasts will come true in the foreseeable future (up to mid-2020s) that both vessel traffic volume and vessel sizes increase. The shipping industry has recognised this, and guidance has been issued to mariners when sailing in areas with shipping lanes created by renewable off-shore energy installations, as a first and immediate measure.[8]

While the *need* to resolve the above challenges is obvious, the task now lies in *how* to reconcile those apparently opposing requirements within the maritime domain with *means* available to an envisaged 'Sustainable Maritime Transportation System': obviously more 'green intelligence' is needed in the domain of maritime transportation and vessel traffic. e-Navigation is considered by some as a prime means to achieve this.

14.4 Investigating the IMO e-Navigation strategy for its 'greenness'

In Section 14.1 the notion of 'potentially open doors' to introduce 'green intelligence' into an envisaged 'green' maritime transportation system was presented, and I set forth the intent to identify these 'potentially open doors' in existing or proposed relevant regulations and/or strategies, one of which has been identified as the IMO e-Navigation strategy. This I shall now scrutinise as *to what degree it intends to introduce 'green intelligence'* and thereby determine whether e-Navigation in itself has got a 'green heart'. Remember, this would require in particular *to go beyond the 'classic theory' for the protection of the environment.*

There are now two major IMO endorsed documents available on e-Navigation, namely the 'Strategy for the Development and Implementation of e-Navigation' adopted by IMO in 2008 [1] and the 'e-Navigation Strategy Implementation Plan', abbreviated SIP, which was adopted by IMO in November 2014 [8]. I shall investigate both in this order of appearance.

[7]Compare analysis of this situation in Reference 6.

[8]Compare, for example, Reference 7 which also provides an introduction to MSP from the shipping industry's perspective and which references a wealth of relevant international resources on the topic.

The 'Strategy for the Development and Implementation of e-Navigation' [1] provides a definition of e-Navigation (compare Section 14.1), justifies the need for e-Navigation, projects a vision of looking at the shipboard side, the shore side and the necessary communications in-between in a holistic way, specifies the 'core objectives' of e-Navigation, lists perceived benefits of it, collects the high level user needs supposed to determine the direction of the implementation of e-Navigation, names key strategy elements and then discusses implementation issues in general terms.

Regarding the topic of this chapter, it first needs to be noted that there is *not a single 'core objective'* ([1], p. 5) specified which would take up the notion of the protection of the environment expressed in the definition of e-Navigation. The protection of the environment is expected to benefit expressively, *however*, by 'improving navigation safety (...), thereby reducing the risk of collisions and groundings and the associated spillages and pollution' ([1], para 6.1.2.1) as well as by the 'enhancement of ability and capacity in responding and handling of emergencies such as oil spills' ([1], para 6.1.2.3), that is, *objective improvements in the sense of the 'classic theory'.*

e-Navigation also indirectly recognises as a 'user' people living ashore who expressed themselves by MSP, as discussed earlier. They are represented in the e-Navigation strategy by their '*marine environment managers*' ([1], Annex 2, Table 'Shore-based users'). This may be significant as e-Navigation claims that it should be implemented based on 'user needs' ([1], para 7.1.1).

When looking for content of e-Navigation regarding the protection of the environment *going beyond the 'classic theory'* the following statements can be identified:

- The environment is expected to benefit from e-Navigation by '*reducing emissions by using optimum routes and speed*' ([1], para 6.1.2.2; emphasis added).
- 'Automated and standardized reporting functions for optimal communication of ship and voyage information' ([1], para 8.2.2) are identified as *a high level user need*, including also '*information pertaining to (...) environmental protection to be communicated amongst all users*' ([1], para 8.2.2; emphasis added). In addition, this information should be communicated by 'effective and robust communications means' ([1], para 8.2.3), which are recognized as a high level need of shore-based users 'to facilitate safety, security and environmental protection' ([1], para 8.2.3). Thus, this may become instrumental when considering future applications to protect the environment which go beyond the 'classic theory' and require such communication connectivity 'amongst all users'.
- A whole new field of expertise and application which clearly goes beyond the 'classic theory' (while not excluding it) is recognised as a high level user need of the shore side in the field of situational data analysis and decision support to be performed by shore-based systems: 'E-navigation systems should support good decision making, improve performance and prevent single person error. To do so, (...) *shore-based systems should support environmental impact analysis*, forward planning of vessel movements, hazard/risk assessment,

reporting indicators and incident prevention. Consideration should also be given to the use of *analysis for (...) environment protection measures*' ([1], para 8.2.7; emphasis added).
- Finally, IMO asserts the right to govern the implementation of the e-Navigation strategy recognising also the appropriate IMO instruments: '*Being responsible for establishing mandatory standards* for enhancing the safety of life at sea, maritime security and *protection of the marine environment* as well as having a global remit, IMO is the only organization that is capable of meeting the overall governance' ([1], para 9.2; emphasis added). Here is a clear indication that the above IMO identified means to protect the marine environment beyond the 'classic theory' may eventually also benefit from mandatory instruments issued by IMO.

In summary, the above analysis of the IMO e-Navigation strategy *revealed a promising start in terms of 'greenness' of e-Navigation beyond the 'classic theory' in no unprecise terms.* The question is now, how this has been translated into IMO's specific implementation plan.

Turning towards the second major, younger and much more specific document of the IMO e-Navigation strategy, namely, the 'e-Navigation Strategy Implementation Plan' (SIP) [8], the following specific references to the protection of the environment can be found:

- The first mentioning of the protection of the marine environment is in the introduction section, where the 'expectation' is expressed that 'as shipping moves into the digital world, *e-navigation (...) provides digital information and infrastructure for the benefit of* maritime safety, security and *protection of the marine environment,* reducing the administrative burden and increasing the efficiency of maritime trade and transport' ([8], para 1; emphasis added).
- The only other mentionings appear in the context of the *Maritime Service Portfolios* (MSPs) ([8], p. 17ff and Annex 2) where specific maritime services provided to shipping are described. According to the plan, they will be further developed as a dedicated 'task' No. 17 ([8], p. 17 and Table 7) to provide the so-called prioritized solution No. 9 of the SIP, namely, the 'Improved Communication of VTS Portfolio (Not limited to VTS stations)' ([8], p. 11). There, the protection of the marine environment is mentioned in the classic way as one of the goals of the VTS Information Service (MSP1), the Pilotage Service (MSP6), the Tugs Service (MSP7), Vessel Shore Reporting (MSP8) and the Real Time Hydrographic and Environmental Service (MSP15) each ([8], Annex 2).

In comparison with the above IMO e-Navigation strategy document *the SIP appears to be sobering in regards to the protection of the environment beyond the 'classic theory'* as it does not take up the 'green' momentum of the first document. However, it does not preclude the pursuit of the identified benefits and of the fulfilment of the relevant high level user needs of the IMO e-Navigation strategy document either.

14.5 Concluding postulates

I shall now pull together the findings and intermediate conclusions of the above sections into a number of postulates.

1. The 'classic theory' of the protection of the environment, namely by the prevention of incidents and accidents, does not appear to be sufficient any more. While it must be stressed whole-heartedly that safety of vessel traffic still is and needs to be the prime target – even to the extent of zero accidents – the protection of the environment in the maritime domain needs to move beyond the 'classic theory', i.e. 'green intelligence' needs to be introduced in the Maritime Transportation System.
2. The IMO SG's concept of a SMTS contains a substantial goal base capable of introducing 'green intelligence' in several of its goal domains. Therefore, the SMTS concept should be further considered, refined and translated into appropriate IMO instruments to become effective in due course.
3. The challenge of the MSP to shipping should be positively and proactively taken up by those carrying responsibility in shipping and particularly in vessel traffic management.
4. The IMO e-Navigation strategy as described in the two major IMO adopted documents introduced in this essay is indeed an 'open door' to the introduction of 'green intelligence' into maritime traffic. Therefore, it is rightly recognised as a contribution to the SMTS within its goal domain 'Maritime Traffic Support and Advisory Systems'.
5. The momentum in regards to the protection of the environment beyond the 'classic theory', as contained in the IMO's e-Navigation strategy vision of 2008, should be revived. Solutions developed in that spirit should be specifically introduced in the context of MSPs, where appropriate, during the implementation phase of e-Navigation now triggered by the adoption of the IMO e-Navigation SIP.
6. The extended use of environmental impact analysis of traffic situations by shore-based systems, which was identified by IMO as a high-level user need of the shore side for e-Navigation, should receive appropriate attention by further studies and research, leading eventually to amendments of operational processes accordingly.
7. For full effectiveness of 'green intelligence', it must become a parameter of daily conduct of users and stakeholders, that is, it must be mapped into practical applications which must exhibit a high degree of usability at their human–machine interfaces.
8. The VTS is one of the prime candidates for the incorporation of such operational processes on the shore side. To that end, criteria for 'green tagging' of traffic situations to be resolved by the appropriate VTS service as well as supporting algorithms providing such de-conflicting solutions taking into account 'green tags' should be developed and incorporated eventually in a revised international description for VTS.
9. By default, if so desired by the users and stakeholders of the shipboard side, there is no fundamental principle preventing the introduction of

shipboard 'environmental impact analysis' for the navigation berth-to-berth, for example, by 'green tagging' of navigational options by shipboard decision support tools.

10. As 'Sustainable Development' comprises the economic, social and environmental dimensions in a balance, there must be demonstrated economic and social benefits by those wishing to introduce even the 'apparently best green intelligence' solution available.

Disclaimer

The opinions and propositions expressed in this chapter are solely those of the author in his own scientific and professional judgement and are not to be construed as official statements of any administration or authority of Germany.

Abbreviations used and glossary of terms

e-Navigation	a holistic strategy of the IMO and a new paradigm of thinking in the domain of maritime navigation, that is, spanning the 'three sides of the coin' – shipboard, shore-side and communications in-between – with the main goal to harmonise the usage, evaluation, presentation and exchange of digitally represented data and information and the associated digital systems and their interfaces for data exchange on a global scale. e-Navigation is not a technical system or device; technical systems or devices may be 'fit for e-Navigation' or 'e-Navigation compliant' if designed in accordance with the e-Navigation paradigm, however.
IMO	International Maritime Organisation
MEPC	Marine Environmental Protection Committee of the IMO
MSC	Maritime Safety Committee of the IMO
MSP	Marine Spatial Planning; sometimes also called Maritime Spatial Planning
MSPs	Maritime Service Portfolios
SG	Secretary General
SIP	IMO e-Navigation Strategy Implementation Plan
SMTS	Sustainable Maritime Transportation System
UN	United Nations
UNESCO	United Nations Educational, Scientific and Cultural Organization
VTS	Vessel Traffic Services; a set of three operational services, namely Information Service (IS), Navigational Assistance Service (NAS), and Traffic Organisation Service (TOS), which are 'implemented by a competent authority, designed to improve the safety and efficiency of vessel traffic and the protection of the environment' (IMO Resolution A.857(20), Annex 1, 1.1.1).
WWF	World Wildlife Fund

References

[1] International Maritime Organisation, Maritime Safety Committee. 'Strategy for the Development and Implementation of e-navigation'. In: IMO/MSC. *Report of the Maritime Safety Committee on its Eighty-Fifths Session* (26 November to 05 December 2008). MSC85/26. Add. 1, Annex 20, published 6 January 2009. London: IMO. *Note 1:* IMO documents of past sessions are available on IMO's website (www.imo.org) using the public user access and the document reference given. *Note 2:* Statements in IMO documents are referenced here by their paragraph number

[2] IMO, Assembly. *IMO Resolution A.857(20) on Guidelines for Vessel Traffic Services*. Adopted during its 20th session on 27 November 1997. London: IMO

[3] Sekimizu, Koji (Secretary General of IMO). *World Maritime Day. A Concept for a Sustainable Maritime Transportation System.* (World Maritime Day, 23 September 2013). London: IMO. Can be retrieved via the IMO website; compare Reference 1

[4] Ehlers, Charles, and Fanny Douvere. 'Marine Spatial Planning. A Step-by-Step Approach toward Ecosystem-based Management'. UNESCO, Intergovernmental Oceanographic Commission. *Manual and Guides* No. 53, ICAM Dossier No. 6. 2009 (Ed. Rahel Dahl). http://www.unesco-ioc-marinesp.be [Accessed 28 Nov 2014]

[5] Lamp, Jochen, Michèle Stoltz and Karl Wagner. Illustrations by Erik Liebermann. *Maritime Spatial Planning in the Baltic Sea. Become a Maritime Spatialist within 10 minutes*. Frankfurt am Main: WWF Germany. 2010

[6] Oltmann, Jan-Hendrik, Thomas Porathe, Mads Bentzen et al. *ACCSEAS Baseline and Priorities Report.* Edition 3. 2015. http://www.accseas.eu/content/download/2743/25720/ACCSEAS+Baseline+and+Priorities+Report+v3.0.pdf; accessed 25 June 2015

[7] Patraiko, David, and Paul Holthus. *The Marine Industry and Marine Spatial Planning. A professional Approach. Version 1.* London: The Nautical Institute; World Ocean Council; in association with IALA, St-Germain-en-Laye, France. November 2013

[8] IMO, Sub-Committee on Navigation, Communication, Search and Rescue (NCSR). 'e-Navigation Strategy Implementation Plan'. In: IMO, NCSR. *Report to the Marine Safety Committee*. NCSR1/28, Annex 7. 16 July 2014. As adopted by MSC95 (November 2014)

Chapter 15

Optimal ship operation: monitoring technology of ship overall heat balance

*Yung-Ho Yu**

15.1 Introduction

15.1.1 IMO MEPC Circ. 684

International Maritime Organization (IMO) Marine Environment Protection Committee (MEPC) Circular 684 (Circ. 684) is a detailed explanation of the Energy Efficiency Operational Indicator (EEOI). So MEPC Circ. 684 is cited as it is in this section, including the annex, because IMO official documents must be transmitted unchanged. MEPC Circ. 684 is an invaluable reference for developing a real-time EEOI on board [1].

The MEPC of the IMO, at its 59th session (13–17 July 2009), agreed to circulate the guidelines for voluntary use of the ship EEOI as set out in the annex. As a result member governments are invited to bring the Guidelines (MEPC.1/Circ. 684) to the attention of all parties concerned and to recommend that they use the Guidelines on a voluntary basis.

IMO Assembly resolution A.963 (23) is related to the reduction of greenhouse gas (GHG) emissions from ships and it urges the MEPC to identify and develop such a mechanism or mechanisms as are needed to achieve first, the limitation or reduction of GHG emissions from international shipping, giving priority, in doing so, to the establishment of a GHG baseline; and secondly, the development of a methodology to describe the GHG efficiency of a ship in terms of a GHG emission indicator for that ship.

As urged by the Assembly, MEPC 53 approved Interim Guidelines for Voluntary Ship CO_2 Emission Index for Use in Trials.

These Guidelines can be used to establish a consistent approach for voluntary use of an EEOI, which will assist ship-owners, ship operators, and parties concerned in the evaluation of the performance of their fleet with regard to CO_2 emissions. As the amount of CO_2 emitted from a ship is directly related to the consumption of bunker fuel oil, the EEOI can also provide useful information on a ship's performance with regard to fuel efficiency.

*Division of IT Engineering, Korea Maritime and Ocean University, Korea

These Guidelines may be updated periodically, to take account of:

- Operational experiences from use of the indicator for different ship types, as reported to MEPC by industry organisations and administrations.
- Any other relevant developments.

Industry organisations and interested administrations are invited to promote the use of the attached Guidelines or equivalent approaches and their incorporation in company and ship environmental management plans. Furthermore, they are invited to report their experience in applying the EEOI concept back to MEPC. In addition to these Guidelines, account should be taken of the pertinent clauses within the International Safety Management (ISM) Code on a voluntary basis, along with reference to relevant industry guidance on the management and reduction of CO_2 emissions.

This document constitutes the guidelines for the use of an EEOI for ships. It sets out:

- The objectives of the IMO CO_2 emissions indicator.
- How a ship's CO_2 performance should be measured.
- How the index could be used to promote low-emission shipping, in order to help limit the impact of shipping on global climate change.

15.1.2 Objectives of these guidelines

The objective of these Guidelines is to provide users with assistance in the process of establishing a mechanism to achieve the limitation or reduction of GHG emissions from ships in operation. These Guidelines present the concept of an indicator for the energy efficiency of a ship in operation, as an expression of efficiency, in the form of CO_2 emitted per unit of transport work. The Guidelines are intended to provide an example of a calculation method which could be used in an objective, performance-based approach to monitoring the efficiency of a ship's operation. These Guidelines are recommendatory in nature and present a possible use of an operational indicator. However, ship-owners, ship operators and parties concerned are invited to implement either these Guidelines or an equivalent method in their environmental management systems and to consider adoption of the principles herein when developing plans for performance monitoring.

15.1.3 Definition

15.1.3.1 Indication definition

In its most simple form the EEOI is defined as the ratio of mass of CO_2 (M) emitted per unit of transport work:

$$\text{Indicator} = M_{CO_2}/(\text{transport work})$$

15.1.3.2 Fuel consumption

Fuel consumption, FC, is defined as 'all fuel consumed at sea and in port or for a voyage or period in question, for example a day, by main and auxiliary engines, including boilers and incinerators'.

15.1.3.3 Distance sailed

'Distance sailed' means the actual distance sailed in nautical miles (deck log-book data) for the voyage or period in question.

15.1.3.4 Ship and cargo types

The Guidelines are applicable for all ships performing transport work.

- Ship types:
  - dry cargo carriers;
  - tankers;
  - gas tankers;
  - container ships;
  - ro-ro cargo ships;
  - general cargo ships;
  - passenger ships, including ro-ro passenger ships.
- 'Cargo' includes but is not limited to:
  - all gas, liquid and solid bulk cargo;
  - general cargo;
  - containerised cargo (including the return of empty units);
  - break bulk;
  - heavy lifts;
  - frozen and chilled goods;
  - timber and forest products;
  - cargo carried on freight vehicles;
  - cars and freight vehicles on ro-ro ferries and passengers (for passenger and ro-ro passenger ships).

15.1.3.5 Cargo mass carried or work done

In general, cargo mass carried or work done is expressed as follows:

- Dry cargo carriers, liquid tankers, gas tankers, ro-ro cargo ships, and general cargo ships – metric tonnes (t) of the cargo carried should be used.
- Container ships carrying solely containers – number of containers (TEU) or metric tonnes (t) of the total mass of cargo and containers should be used.
- Ships carrying a combination of containers and other cargoes – a TEU mass of 10 t could be applied for loaded TEUs and 2 t for empty TEUs.
- Passenger ships, including ro-ro passenger ships – number of passengers or gross tonnes of the ship should be used.

In some particular cases, work done can be expressed as follows:

- Car ferries and car carriers – number of car units or occupied lane metres.
- Container ships – number of TEUs (empty or full).
- Railway and ro-ro vessels – number of railway cars and freight vehicles, or occupied lane metres.

For vessels such as certain ro-ro vessels, which carry a mixture of passengers in cars, foot passengers, and freight, operators may wish to consider some form of weighted average based on the relative significance of these trades for their particular service or the use of other parameters or indicators as appropriate.

15.1.3.6 Voyage

'Voyage' generally means the period between a departure from a port to the departure from the next port. Alternative definitions of a voyage could also be acceptable.

15.1.4 Establishing an EEOI

The EEOI should be a representative value of the energy efficiency of the ship operation over a consistent period which represents the overall trading pattern of the vessel. Guidance on a basic calculation procedure for a generic EEOI is provided in the Appendix.

In order to establish the EEOI, the following main steps will generally be needed:

- Define the period for which the EEOI is calculated*.
- Define data sources for data collection.
- Collect data.
- Convert data to appropriate format.
- Calculate EEOI.

15.1.5 General data recording and documentation procedures

Ideally, the data recording method used should be uniform, so that information can be easily collated and analysed to facilitate the extraction of the required information. The collection of data from ships should include the distance travelled, the quantity and type of fuel used, and all fuel information that may affect the amount of carbon dioxide emitted. For example, fuel information is provided on the bunker delivery notes that are required under regulation 18 of International Convention for the Prevention of Pollution from Ships (MARPOL) Annex VI. If the example formula given in the Appendix is used, then the units used for distance travelled and quantity of fuel should be nautical miles and metric tonnes. The work done can be expressed using units appropriate for the ship type, as in Section 15.1.3.5.

It is important that sufficient information is collected on the ship with regard to fuel type and quantity, distance travelled, and cargo type, so that a realistic assessment can be generated. The distance travelled should be actual distance travelled, as contained in the ship's log-book. Amount and type of fuel used (bunker

*Ballast voyages, as well as voyages that are not used for transport of cargo, such as voyages for docking service, should also be included. Voyages made for the purpose of securing the safety of a ship or saving life at sea should be excluded.

delivery notes) and distance travelled (according to the ship's log-book) could be documented by the ship based either on the example described in the Appendix or on an equivalent company procedure.

15.1.6 Monitoring and verification

15.1.6.1 General

Documented procedures to monitor and measure, on a regular basis, should be developed and maintained. Elements to be considered when establishing procedures for monitoring could include:

- Identification of operations/activities with impact on performance.
- Identification of data sources and measurements that are necessary, and specification of the format.
- Identification of frequency and personnel performing measurements.
- Maintenance of quality control procedures for verification procedures.

The results of this type of self-assessment could be reviewed and used as indicators of the system's success and reliability, as well as identifying those areas in need of corrective action or improvement.

It is important that the source of figures established is properly recorded, as well as the basis on which figures have been calculated and any decisions on difficult or grey areas of data. This will provide assistance in identifying areas for improvement and be helpful in any later analysis.

In order to avoid unnecessary administrative burdens on ships' staff, it is recommended that monitoring of an EEOI should be carried out by shore staff, utilising data obtained from existing required records, such as the official and engineering log-books and oil record books. The necessary data could be obtained during internal audits under the ISM Code, routine visits by superintendents, etc.

15.1.6.2 Rolling average indicator

As a ship energy-efficiency management tool, the rolling average indicator, when used, should be calculated using a methodology whereby the minimum period of time or a number of voyages that is statistically relevant is used as appropriate. 'Statistically relevant' means, for the period set as standard for each individual ship, that it should remain constant and be have a wide enough range that the accumulated data mass reflects a reasonable mean value for operation of the ship in question over the selected period.

15.1.7 Use of guidelines

The methodology and use of EEOI, as described in these Guidelines, provide an example of a transparent and recognised approach for assessment of the GHG efficiency of a ship with respect to CO_2 emissions. The Guidelines are considered to be suitable for implementation within a company environmental management system. Implementation of the EEOI in an established environmental management system should be performed in line with the implementation of any other chosen indicator and follow the main elements of the recognised standards (planning, implementation and operation, checking and corrective action, management review). When using the EEOI as a performance indicator, the indicator could provide a basis for consideration of both current performance and trends over time. One approach could be to set internal performance criteria and targets based on the EEOI data.

15.1.8 Calculation of EEOI based on operational data

15.1.8.1 General

The object of the Appendix is to provide guidance for calculation of the EEOI based on data from the operation of the ship.

15.1.8.2 Data source

Primary data sources selected could be the ship's log-book (bridge log-book, engine log-book, deck log-book and other official records).

15.1.8.3 Fuel mass to CO_2 mass conversion factors (C_F)

C_F is a non-dimensional conversion factor between fuel consumption measured in g, and CO_2 emission, also measured in g, based on carbon content. The value of C_F is given in Table 15.1.

15.1.8.4 Calculation of EEOI

The basic expression for EEOI for a voyage is defined as

$$\text{EEOI} = \frac{\sum_j \text{FC}_j \times C_{Fj}}{m_{\text{cargo}} \times D} \tag{15.1}$$

where an average of the indicator for a period or for a number of voyages is obtained, the Indicator is calculated as

$$\text{Average EEOI} = \frac{\sum_i \sum_j (\text{FC}_{ij} \times C_{Fj})}{\sum_i (m_{\text{cargo},i} \times D_i)} \tag{15.2}$$

Table 15.1 Conversion factor for fuel mass to CO_2 mass

Type of fuel	Reference	Carbon content	C_F(t-CO_2/t-Fuel)
Diesel/gas oil	ISO 8217 Grades DMX to DMC	0.875	3.206000
Light fuel oil (LFO)	ISO 8217 Grades RMA to RMD	0.26	3.151040
Heavy fuel oil (HFO)	ISO 8217 Grades RME to RMK	0.85	3.114400
Liquefied petroleum gas (LPG)	Propane	0.819	3.000000
	Butane	0.827	3.030000
Liquefied natural gas (LNG)		0.75	2.750000

where

- j is the fuel type
- i is the voyage number
- FC_{ij} is the mass of fuel consumed j at voyage i
- C_{Fj} is the fuel mass to CO_2 mass conversion factor for fuel j
- m_{cargo} is cargo carried (tonnes) or work done (number of TEU or passengers) or gross tonnes for passenger ships
- D is the distance in nautical miles corresponding to the cargo carried or work done.

The unit of EEOI depends on the measurement of cargo carried or work done, for example tonnes CO_2/(tonnes·nautical miles), tonnes CO_2/(TEU·nautical miles), tonnes CO_2/(person·nautical miles).

It should be noted that (15.2) does not give a simple average of EEOI for a given number of voyages i.

15.1.8.5 Rolling average

A rolling average, when used, can be calculated in a suitable time period, for example one year following the end of a voyage, or number of voyages, for example six or ten voyages, agreed as statistically relevant to the initial averaging period. The Rolling Average EEOI is then calculated for this period or number of voyages by (15.2) above.

15.1.8.6 Data

For a voyage or period, for example a day, data on fuel consumption/cargo carried and distance sailed in a continuous sailing pattern could be collected as shown in the reporting sheet (Table 15.2).

15.1.8.7 Conversion from g/tonne-mile to go g/tonne-km

The CO_2 indicator may be converted from g/tonne-mile to g/tonne-km by multiplication by 0.54.

Table 15.2 CO_2 indicator reporting sheet

Name and type of ship:					
Voyage or day (*i*)	**Fuel consumption (FC) at sea and in port in tonnes**			**Voyage or time period-data**	
	Fuel type ()	**Fuel type ()**	**Fuel type ()**	**Cargo (*m*) (tonnes or units)**	**Distance (*D*) (NM)**
1.					
2.					
3.					

NOTE: For voyages with $m_{cargo} = 0$, it is still necessary to include the fuel used during this voyage in the summation above the line.

Table 15.3 Example of CO_2 indicator reporting sheet

Name and type of ship:					
Voyage or day (*i*)	**Fuel consumption (FC) at sea and in port in tonnes**			**Voyage or time period data**	
	Fuel type (HFO)	**Fuel type (LFO)**	**Fuel type ()**	**Cargo (*m*) (tonnes or units)**	**Distance (*D*) (NM)**
1.	20	5		25,000	300
2.	20	5		0	300
3.	50	10		25,000	750
	10	3		15,000	150
Total	100	23			

15.1.8.8 Example

A simple example including one ballast voyage, for illustration only, is provided below. The example illustrates the application of the formula, based on the data reporting sheet (Table 15.3).

$$\text{EEOI} = \frac{100 \times 3.114 + 23 \times 3.151}{(25{,}000 \times 300) + (0 \times 300) + (25{,}000 \times 750) + (15{,}000 \times 150)}$$

$$= 13.47 \times 10^{-6}$$

Unit: tonnes CO_2/(tonnes·nautical miles)

15.2 Present heat balance of marine diesel engine

15.2.1 Wärtsilä waste heat recovery (WHR) [2]

This case study is illustrated by a 12-cylinder Wärtsilä RT-flex96C engine installed in a container ship. The engine can be operated at an average of 85% load for about 6,500 h a year on bunkers coasting US$ 250 per tonne, with an average total electrical load of 5,350 kW. The annual operating costs for main and auxiliary engines, including fuel, maintenance, and lubricating oil, would be US$ 19.54 million without a WHR plant and US$ 17.29 million with a high-efficiency WHR plant. Wärtsilä insist payback time is less than five years. Table 15.4 shows the specification of a High-Efficiency WHR case study with a Wärtsilä 12RT-flex96C.

Figure 15.1 shows a schematic diagram of a WHR plant such as might be employed with a Wärtsilä 12RT-flex96C-D engine for a large tanker with the turbo-generator supplying all of the ship's service electricity and assisting propulsion while the ship is at sea.

Figure 15.2 shows a comparison of Sankey diagrams for Wärtsilä 12RT-flex96C engines without WHR and with WHR. Wärtsilä changed the intake of scavenge air from the engine room to outside the ship, to reduce heat loss due to scavenge-air

Table 15.4 Specification of case study of WHR

Basic ship		
Main engine	Wärtsilä 12RT-flex96C	
CMCR power, kW	80,080	
Annual operating hours	6,500	
Average engine load, %	85	
Total electrical load, kW	5,350	
Auxiliary engines, kWe	4 × 3,000	
Bunker fuel price, US$/tonne	250	
Cylinder lubricating oil price, US$/tonne	1,500	
Annual engine operating costs	Basic ship	With WHR
Fuel costs, US$	18,449,000	16,359,000
Maintenance costs, US$	486,000	333,000
Lubricating oil costs, US$	602,000	559,000
Total costs, US$	19,537,000	17,287,000
Annual savings, US$		
at US$ 250/tonne fuel	–	2,250,000
at US$ 400/tonne fuel	–	3,482,000
Investment cost, US$	–	9,500,000

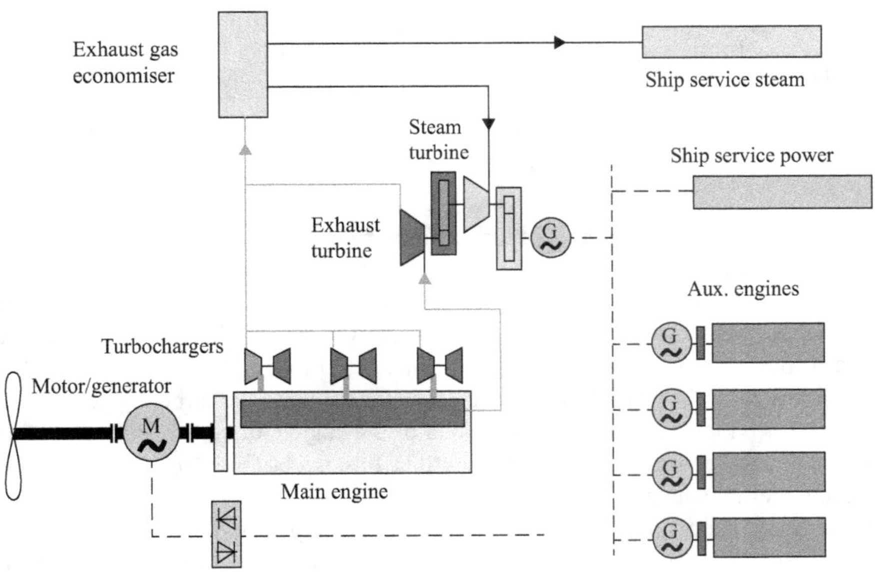

Figure 15.1 Schematic diagram of WHR with turbo-generator

cooling water by decreasing the temperature of air intake, while normally an engine is designed for a temperature of intake air of 45°C, in case of air intake from the engine room. If the intake air is drawn from outside, the maximum possible intake temperature can be assumed to be 35°C.

These endeavours recover 1.2% of the total energy from scavenge-air cooling water. They improve the flow of exhaust gas from the main engine to the exhaust-gas

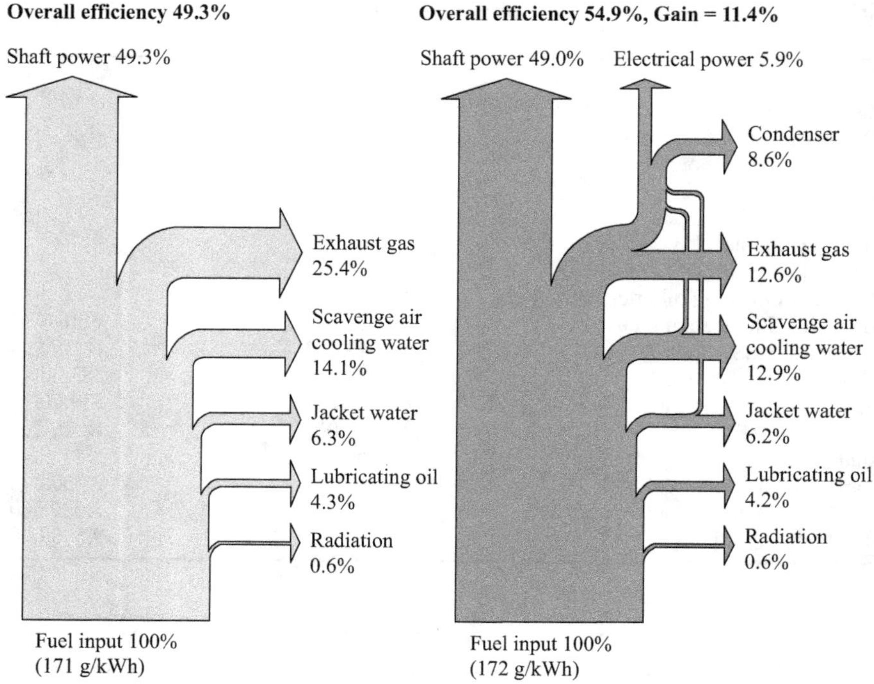

Figure 15.2 Comparison of Sankey diagram of Wärtsilä 12RT-flex96C engine, without WHR (left) and with WHR (right)

economiser to generate steam for a turbine-driven generator via a stable and efficient arrangement of conduits.

Even with a certain quantity of exhaust gas siphoned off to the power turbine and therefore not available for the turbocharger, the thermal loading of the engine remains the same as in a conventional engine.

In this case, the lower temperature of air flowing into the turbocharger, means there is a need to rematch the thermal load of the engine, to benefit fully from this low intake-air temperature. Wärtsilä say that this is possible because the special turbocharger matching, in combination with the exhaust power turbine, allows full utilisation of the available efficiency of the turbocharger, and also because of the ambient suction tuning. Even if this adapted tuning incurs a penalty in the form of slightly increased fuel consumption at ISO reference conditions, the gain in recovered energy more than compensates for this.

Heat recovered from the engine's scavenge air and jacket cooling water is used to heat the boiler-feed water close to evaporation temperature. The overall result of the application of these concepts is that the quantity of energy recoverable in an exhaust-gas economiser and in the exhaust-power turbine is increased without affecting the air flow through the engine.

Also the change in engine air intake has benefit of reducing the electric power required to supply air for the engine. Wärtsilä say that in the case of a container

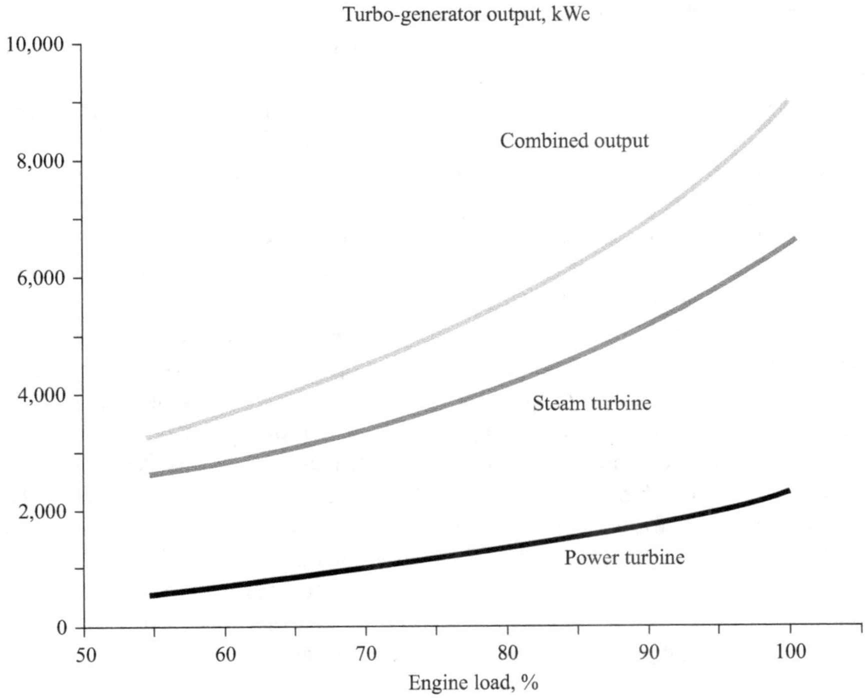

Figure 15.3 Energy recovered by the turbo-generator set in a high-efficiency WHR plant on a 12RT-flex96C engine with Delta tuning (for improved part-load fuel savings) and ambient air intake

with a 12RT-flex96C engine, the total power for ventilation fans may be cut from 400 to 200 kW.

Figure 15.3 shows turbo-generator and steam-generator output according to engine load.

The high-efficiency WHR plant consists of a dual-pressure economiser, a multiple-stage dual-pressure steam turbine, an exhaust-power turbine, an alternator driven by both the steam turbine and the power turbine, a feed water pre-heating system, and a shaft motor/alternator system. In the rematched system, about 10% of the engine's exhaust gas flow can be employed to drive an exhaust-power turbine which is incorporated in the steam turbo-generator.

The exhaust-gas economiser consists of a high-pressure part with HP evaporator and superheating section and a low-pressure part with LP evaporator and superheating section. The pressure of the high-pressure steam drum is at about 9.5 bar and that of the low-pressure drum 3.8 bar. The temperature of the economiser outlet is not less than 160°C, to prevent sulphur corrosion in the economiser outlet. Saturated steam is drawn from the HP steam drum for ship service heating. The feed water is heated by the engine's jacket cooling water to a temperature of 85°C as the first stage. The feed water for the high-pressure section is further heated by the engine's scavenge-air cooler to about 150–170°C.

The rpm of the dual-pressure steam turbine is about 6,750. The high-pressure side works at about 8.0–9.0 bar inlet pressure. This requires three stages in high-pressure side turbine at a 0.065 condenser pressure. The outlet temperature of the economiser is 160°C and the steam pressure in the low-pressure steam turbine inlet is 3.0–3.5 bar. This requires six stages for a low-pressure side turbine with 0.065 bar of condenser. A speed reduction gear between the steam turbine and the generator reduces speed at the turbine to 1,800 rpm in the generator.

An exhaust-gas turbine uses a part of the exhaust-gas stream (about 10%) from the diesel engine to generate shaft power which can be conveyed to the steam turbine driving the generator.

A special matching of the power turbine is necessary for the application in a waste heat recovery system because the turbine operates in a constant speed operating profile as it is coupled to the generator, in contrast to the comparable process in a turbocharger with a free-running rotor. The torque of the power turbine is transmitted to the steam turbine rotor through a reduction gear and an over-running clutch. The power turbine operates only between 55 and 100% of engine load. The flow of exhaust gas from the exhaust-gas manifold is controlled by an orifice at the outlet of the exhaust-gas manifold. If the engine load is less than 55%, the gas flow to the power turbine is shut off, because the power turbine is not sufficiently efficient. As the expansion ratio of the power turbine and that of the turbocharger are nearly equal, the outlet temperature of the exhaust gas of the power turbine is the same as that of the turbocharger.

The engine is tuned to operate up to a permissible maximum pressure in the temperature of intake between −5 and 35°C. If the ship is operated at below −5°C of ambient temperature, the engine should be protected from excessive maximum cylinder pressure because of the high specific density of the cold air. The waste gate acts as a blow-off valve in the flow of scavenge air or exhaust gas. The waste gate valve operates as a simple ON/OFF to divert the air intake.

The shaft motor/alternator is directly mounted in the propeller shaft line and operated at one of a range of frequencies, its selection controlled by a frequency control system to and from the electrical supply. The system operates as either a motor which can be used to boost propulsion or a generator which can be used to supply electricity to the bus on 6,600 V.

Figure 15.4 shows the piping system for a high-efficiency WHR system.

Figure 15.5 shows the steam pressure and temperature of the HP evaporator and LP evaporator in an economiser.

15.2.2 Waste Heat Recovery System (WHRS) of MAN B&W [3]

MAN says that it is possible to generate an electrical output of up to 11% of main engine power by utilising this exhaust gas energy in a waste heat recovery system comprising both steam and power turbines, and combined with utilising scavenge-air energy for exhaust boiler feed water heating. Even if exhaust-gas energy from main diesel engine relatively low after turbocharger, but just high enough for producing electricity by means of WHR system. In this case the result will be an improvement in total efficiency but a slight reduction of the efficiency of the main engine will be seen.

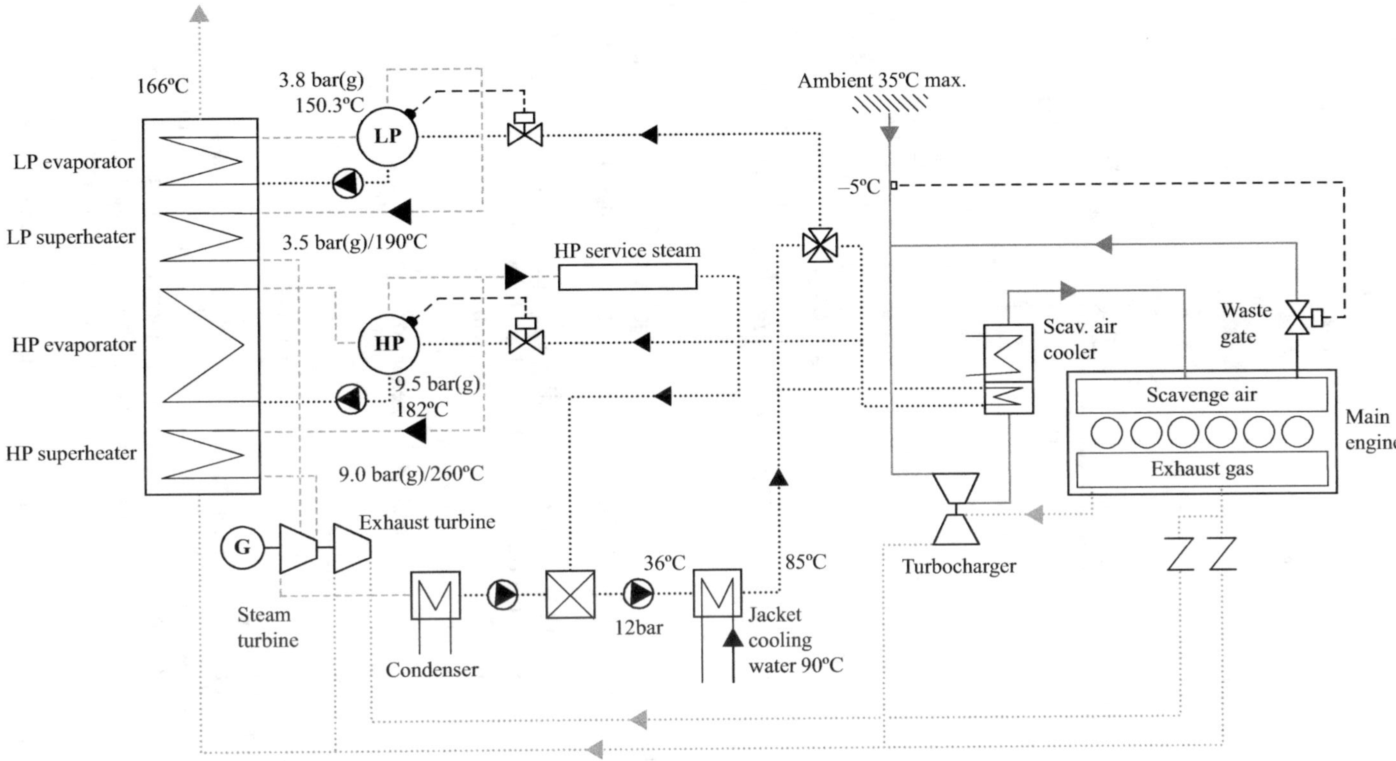

Figure 15.4 Process diagram for the high-efficiency waste heat recovery system

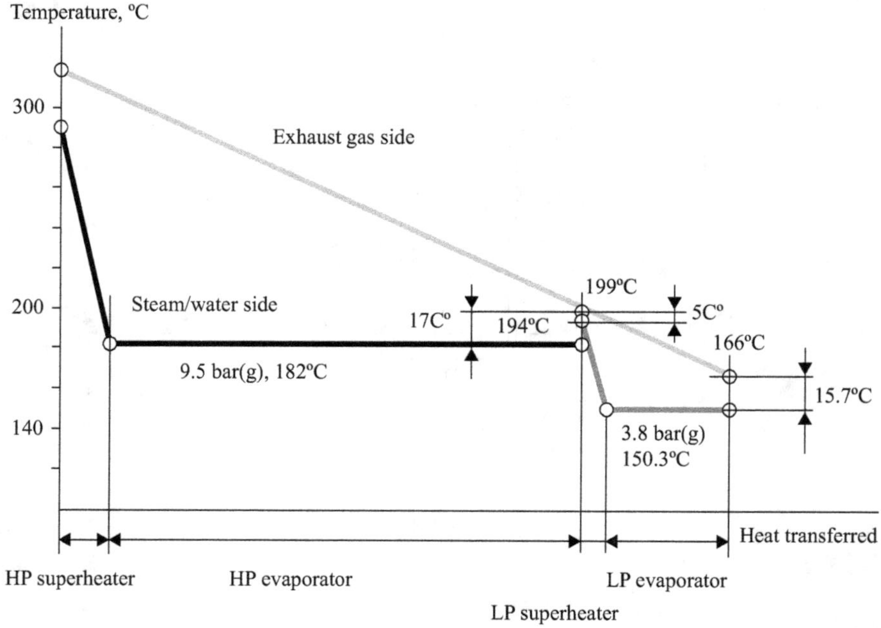

Figure 15.5 Typical economiser diagram for the high-efficiency waste heat recovery system

Figure 15.6 shows a comparison of engine heat balance with and without a WHRS system. The total efficiency of a diesel engine when combined with a WHRS increases by up to about 55%.

There are several different WHRSs available. Depending on the degree of complexity acceptable to the owner and shipyard, and the actual electrical power consumption on board, a WHRS of any of the following can be selected. The steam turbine may be single- or dual-pressure.

- ST–PT: Steam Turbine–Power Turbine generator
- STG: Steam Turbine Generator
- PTG: Power Turbine Generator

Combinations of this kind will cope with Tier III, concerning NO_x, from 2016 regulations and others thereafter.

The principle of the WHRS-matched MAN B&W low-speed diesel engine is that part of the exhaust-gas flow is directed past the main engine turbochargers through an exhaust-gas bypass. As a result the total amount of intake air and exhaust gas is reduced; and so after the bypass of the main engine turbochargers exhaust-gas temperature increases and the power that can be generated with an exhaust-gas-fired boiler and steam turbine is maximised. The increased pressure drop owing to the exhaust-gas bypass, which is part of the WHRS, can be utilised to produce electricity with a power turbine. Figure 15.7 shows the operation of a WHRS.

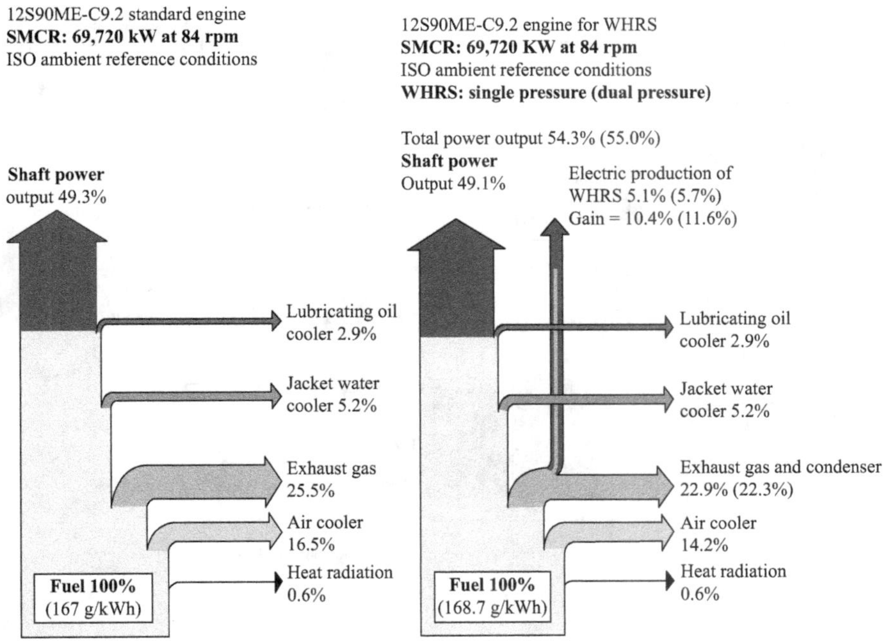

Figure 15.6 Heat balance for large-bore MAN B&W engine type with and without WHRS

As seen in Figure 15.7, a WHRS will be able to supply the total electricity need of the ship as a standalone power source, but it can be also run in parallel with a shaft generator, shaft motor and auxiliary diesel generating sets. In this case the most important thing is an optimal power management system (PMS) in which the main diesel engine and components of the WHR system, such as the power turbine, steam turbine, and shaft PTI/PTO, are designed to communicate and exchange information to establish the best combination.

- Power turbine generator (PTG):
  This, the simplest and cheapest system, consists of an exhaust-gas turbine, called a power turbine, installed in an exhaust-gas bypass. The exhaust-gas bypass valve will be closed at an engine power lower than about 50% of the specified maximum continuous rating (SMCR). Using this arrangement will provide a 3–5% recovery ratio depend on the main engine size.
- Steam turbine generator (STG):
  This uses the whole of the exhaust gas, without using a power turbine, to generate steam from the exhaust-gas boiler and supply steam to a steam turbine, called a turbo-generator.

  When part of the exhaust-gas flow bypasses the turbocharger, the total amount of air and gas will be reduced, and the exhaust-gas temperature will increase. This will increase the obtainable steam production power for the exhaust-gas fired boiler.

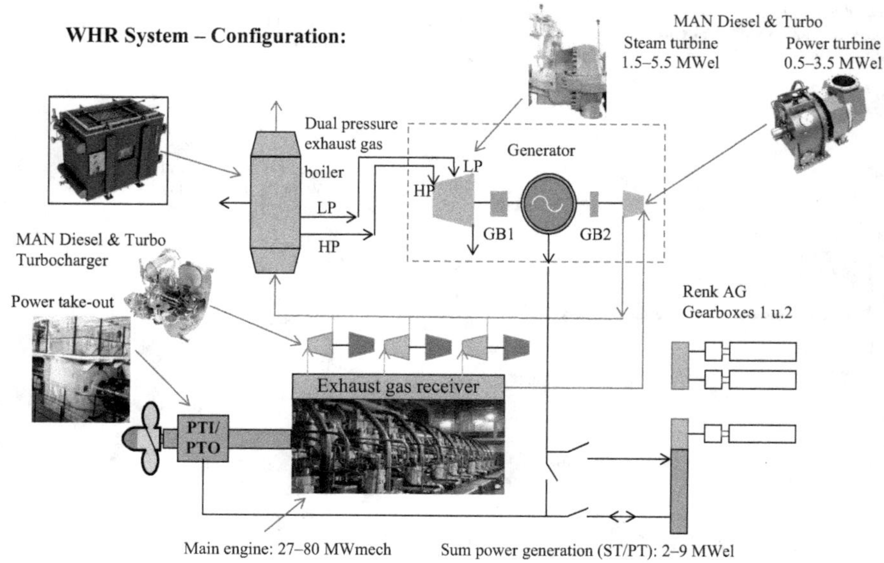

Figure 15.7 Principles of waste heat recovery system

- Steam turbine, power turbine, and generator:
  If the electric power demand on a ship is very high, for example in a container ship, the power turbine and the steam turbine can be applied together to form a combined system. The power output from the power turbine can be added to the generator via reduction gear with a special clutch. However, the steam turbine will start first at 30–35% SMCR of main engine power, followed by the power turbine, which starts power production at 40–50% of SMCR. Figure 15.8 shows a WHRS ST–PT schematic diagram. This arrangement reduces the fuel cost of the ship considerably, by covering the total electric power demand on board. If this arrangement is not enough to cover total electric demand on board, the shaft motor/generator (PTI/PTO) connected to the main engine shaft could be deployed.

Even though conditions will vary from case to case, the following guidelines may be very useful to decide which WHRS is most beneficial.

- As a rule of thumb, MAN recommend the following:
  - Main engine power > 25,000 kW → Combined ST and PT
  - Main engine power < 25,000 kW → PTG or STG (e.g. with super heater)
  - Main engine power < 15,000 kW → PTG or ORG (Organic Rankine Cycle).

The main engine control and the WHRS control must function as an integrated part of the total control system of the ship. The ME and MC types of MAN engine

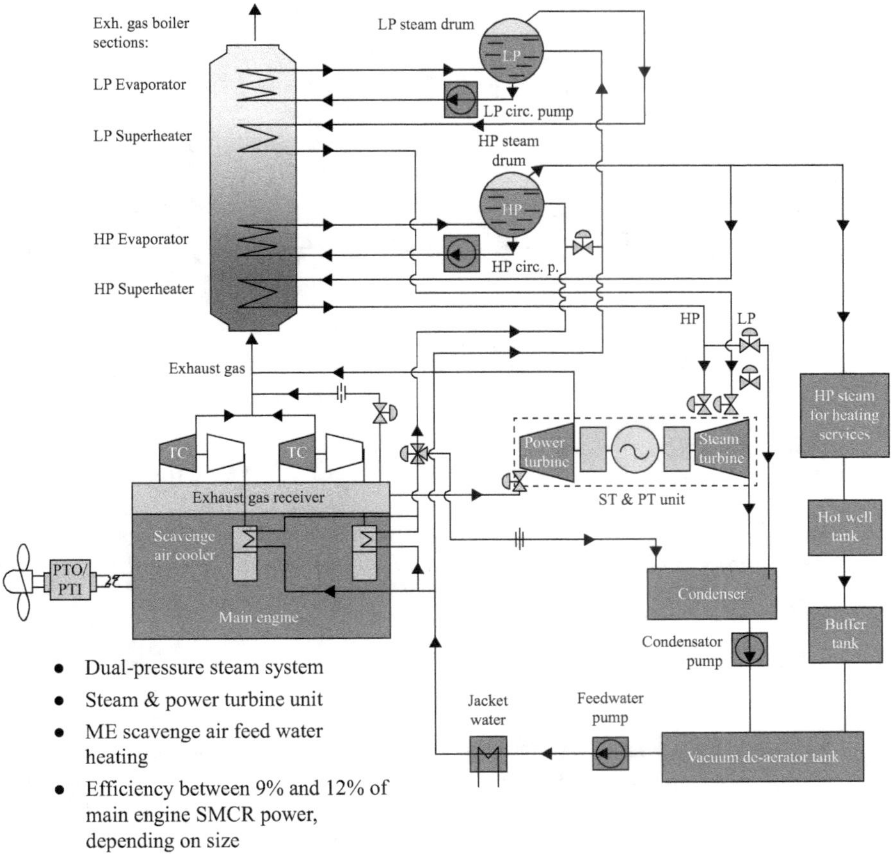

Figure 15.8 Schematic diagram of WHRS ST–PT system

can install the WHRS control and MAN B&W main engine control system. Figure 15.9 shows a generalised ship control system in a large vessel. The owner's insistence that a ship must be economical in all operational ranges, owing to the high price of fuel oil, has driven the development of a combined control system of WHRS and main engine. Because the WHRS steam turbine and power turbine use energy from exhaust gas, the engine control system and WHRS control system are closely linked. If the energy recovery from WHRS is large, then the efficiency of the main engine may decrease and the total efficiency will drop. So a control system combining a WHRS and the main engine should be considered with regard to the possibility of overall optimisation.

Figure 15.10 shows the connection of a WHRS and a main engine, featuring exhaust-gas flow.

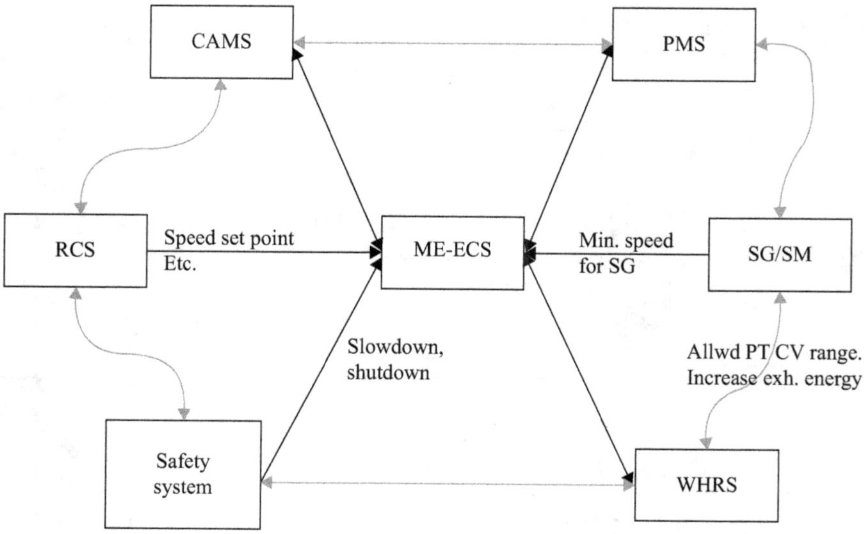

Figure 15.9 General ship control system in a large vessel

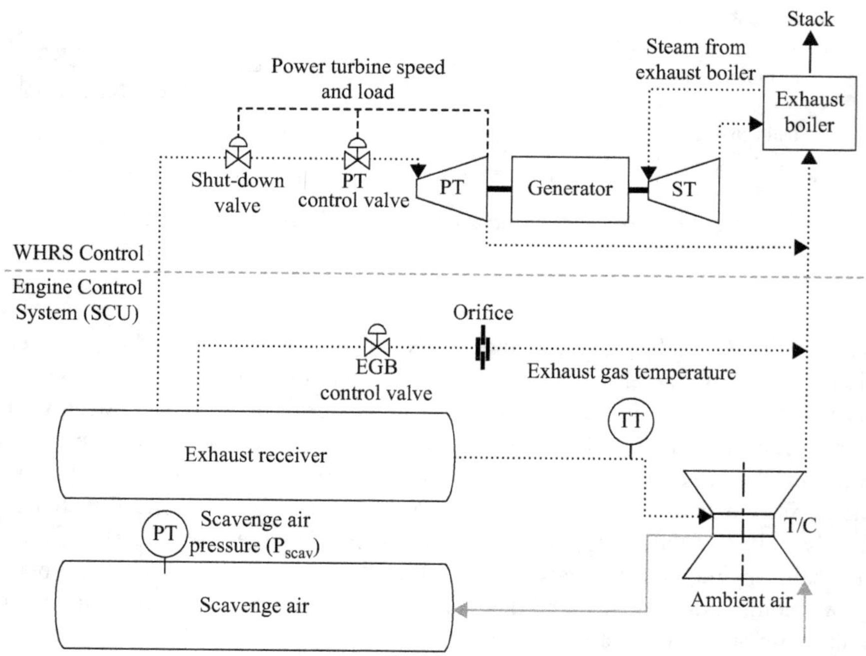

Figure 15.10 Control strategy for the WHRS ST–PT system

Based on this layout, each side of the control system should protect the engine against unacceptable conditions and ensure maximum available power for the WHRS.

The WHRS ST–PT system is controlled by controlling the amount of the exhaust gas bypass (EGB) and keeping it within acceptable limits. The EGB valve is controlled via the engine control system and the power turbine control valve. The sum of the exhaust bypass valve must always be within the maximum allowed bypass amount selected for the particular engine with WHRS tuning. Figure 15.11 shows the control principle of these valves. In Figure 15.11 the black curve slopes downward past 100% main engine load. This is to balance the thermal protection of the main engine while at the same time maximising the energy for the power turbine, keeping both within the maximum allowed turbine pressure P_{scav}. The WHRS control system must give the position of the power turbine control valve, so that the engine control system can compensate by controlling the EGB valve position. The control valve of the power turbine controls the amount of exhaust gas to the power turbine, and the output of the power turbine.

In the same way, the steam turbine governing system controls the amount of steam and output of the steam turbine, and also the frequency of the generator. When the steam turbine generator is operating in parallel with auxiliary diesel generators, the governing system operates in the normal way to effect load sharing.

- If PT control valve closed, setting will follow MinBP curve
- PT control valve can operate freely between 0 and MaxBP
- If PT control valve setting < MinBP, EGB control valve will open, so the bypass area sum will be MinBP
- Increased signal may increase EGB control valve setting to above MinBP

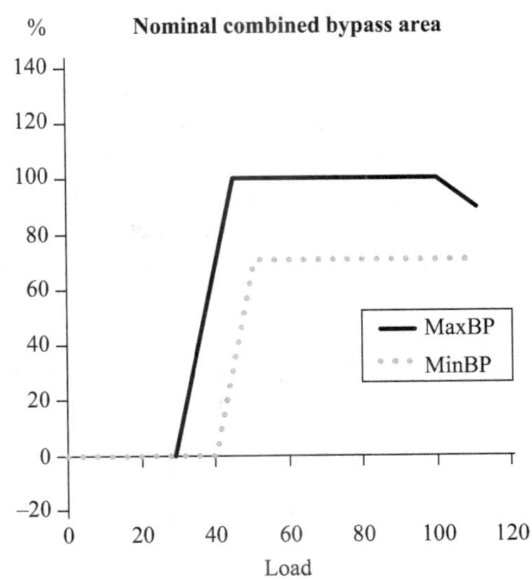

Figure 15.11 Control strategy of WHRS ST–PT system – bypass control

The power management system (PMS) in the ship is configured so that it prioritises the power sources as follows:

- WHRS generator
- PTO generator if installed
- Auxiliary diesel generators
- PTI motor if installed

When using the dual-pressure steam system, it is not possible to install an exhaust gas low-pressure preheater section in the exhaust-gas boiler, because the exhaust-gas boiler outlet temperature would be too low and increase the risk of wet oily soot deposits on the boiler tubes. That may be the cause of corrosion in exhaust piping.

The more complex dual-pressure steam system needs a supplementary waste heat recovery (WHR) source such as jacket-water and scavenge-air heat for preheating feed water. That will increase the steam and electric power production potential of the WHRS. Figure 15.12 shows the processing diagram of a dual-pressure exhaust-gas boiler system of a WHRS and Figure 15.13 shows the corresponding temperature/heat transmission diagram.

The exhaust-gas boiler has to be designed in such a way that the risk of soot deposit and fires is minimised. It is further recommended to install a bypass that allows exhaust gas to bypass the exhaust boiler when the engine load is below 30% of SMCR and in case of other malfunctions of the steam system. Today the dual-steam-pressure system is on the way to becoming standard on large container ships applying WHRSs.

The WHRS utilises not only wasted energy, in exhaust gas, but also the energy in the main engine jacket water and scavenge-air cooling systems. Heat from both of these is used to heat the steam system feed water to a temperature level just below evaporation temperature for the selected steam pressure.

Figure 15.14 shows the connection between the LP and HP economisers, their circulation pump, feed water, vacuum condenser, LP and HP drums, hot well tank, etc. Control valves in front of the steam turbine cover both start-up functions and the possibility of dumping steam if required.

MAN Diesel & Turbo offers engine data for engines tuned for WHRS. They define a steam production power (SPP) guarantee, guaranteeing the energy level available for the exhaust boiler.

$$\begin{aligned}\text{SPP[kW]} = {} & 1.60[\text{kj/kg}^\circ\text{C}] \times (\text{MixedExhaustGasTemp}[^\circ\text{C}] - 160[^\circ\text{C}]) \\ & \times \text{ExhGasAmount[kg/h]} \times \text{h}/3{,}600\ \text{s}\end{aligned}$$

The SPP guarantee is provided with a tolerance of ±7% points.

The above amounts and temperature of exhaust gas are valid under the following conditions:

- ISO ambient conditions, ISO 3026/1 – 1995(E)
- Scavenge-air coolant temperature 25°C
- Ambient air pressure 1,000 mbar

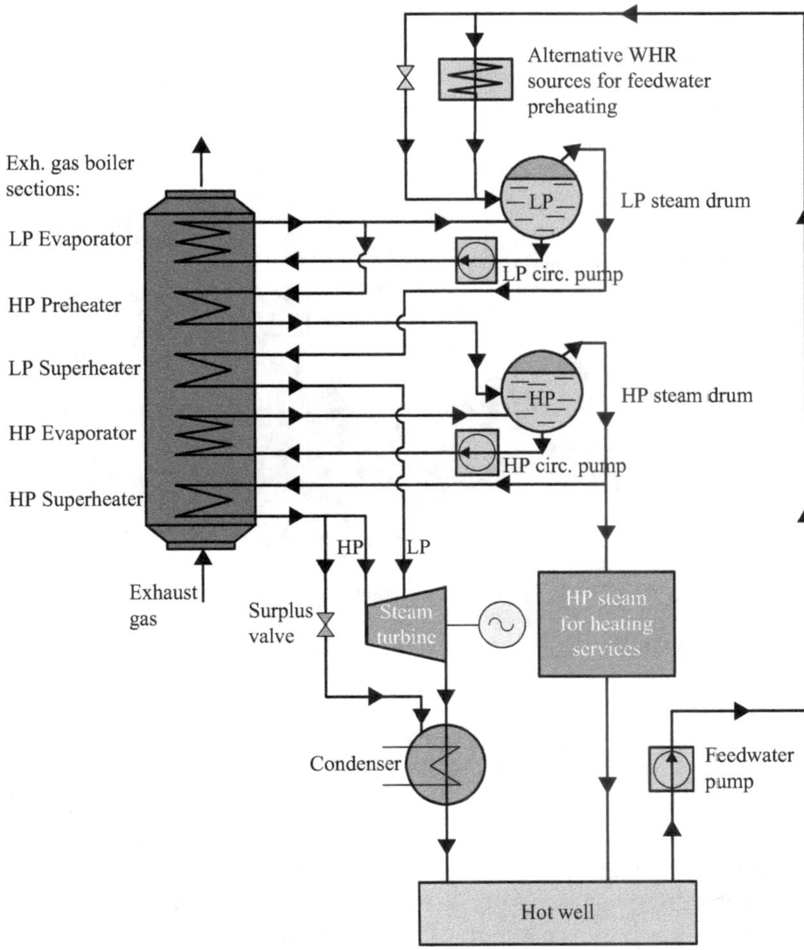

Figure 15.12 Process diagram for the dual-pressure exhaust-gas boiler system

- Ambient air temperature at turbocharger intake 25°C
- Exhaust-gas back pressure 30 mbar

MAN B&W say that WHRS recovery ratios according to the combination of WHRS components are as follows, for ant engine type, size, and rating:

- TCS-PTG: Max. electrical recovery 3–5%
- STG-single-pressure system: Max. electrical recovery 4–7%
- STG-dual-pressure system: Max. electrical recovery 5–8%
- Full WHRS (ST–PT): Max. electrical recovery 8–11%

Table 15.5 shows WHRS recovery output when a large container ship is installed with full WHRS according to the engine load. Figure 15.15 shows the WHRS recovery output data of the same ship according to the seasons.

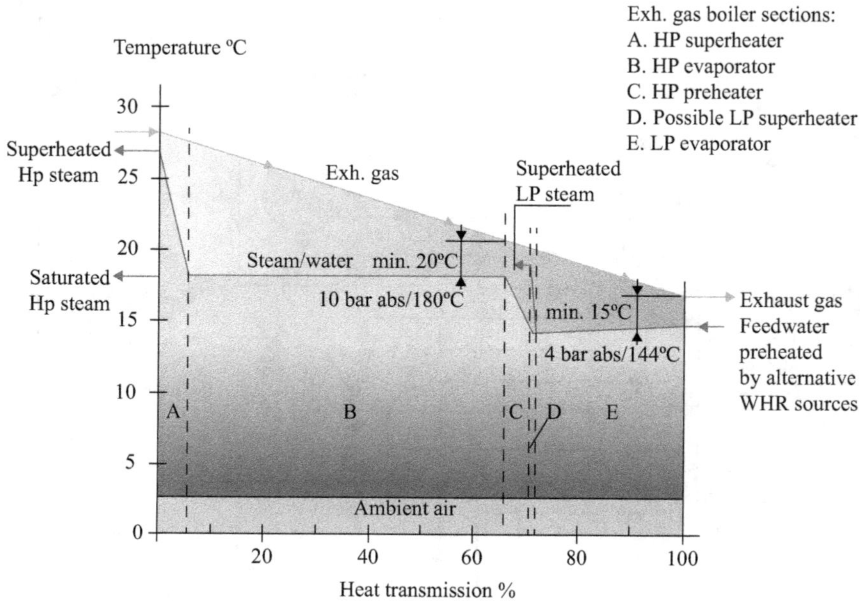

Figure 15.13 Temperature/heat transmission diagram for the dual-pressure steam system

15.3 Monitoring system for ship's heat balance

In Section 15.2 existing WHR systems were surveyed as a case study. As mentioned in the previous section, a WHR system has a strong relationship with the engine, because the WHR system is utilising energy contained in the diesel engine's exhaust gas. Although many components of a WHRS are mechanical systems, it is necessary not only to monitor heat recovery from each component continuously in real time but also to control the WHRS-related valves to maintain optimal operating conditions. In this section the monitoring of heat recovery from each component of a WHRS is described in detail. If the heat recovery of each component is monitored every few minutes, then it is possible to investigate how to operate the overall system, including the main engine, for maximising overall efficiency.

It is assumed that shaft horse power and fuel consumption can be measured. It is very difficult to measure the flow of exhaust gas in a standard exhaust trunk, even a large one, because the measured value of flow is affected by pulse flows from the cylinder outlet in the exhaust trunk. So estimation by calculation from the mass flow of fuel and air can be recommended, with the assumption that fuel consumption and engine output can be measured. Then the Mass flow ($\dot{m}_{\mathrm{f}}$) rate of fuel on the basis of specific fuel consumption (sfc) can be calculated as follows [4].

$$\text{s.f.c} = \frac{\dot{m}_{\mathrm{f}}}{\text{output power}} \tag{15.3}$$

$$\dot{m}_{\mathrm{f}} = \text{s.f.c} \times \text{output power} \tag{15.4}$$

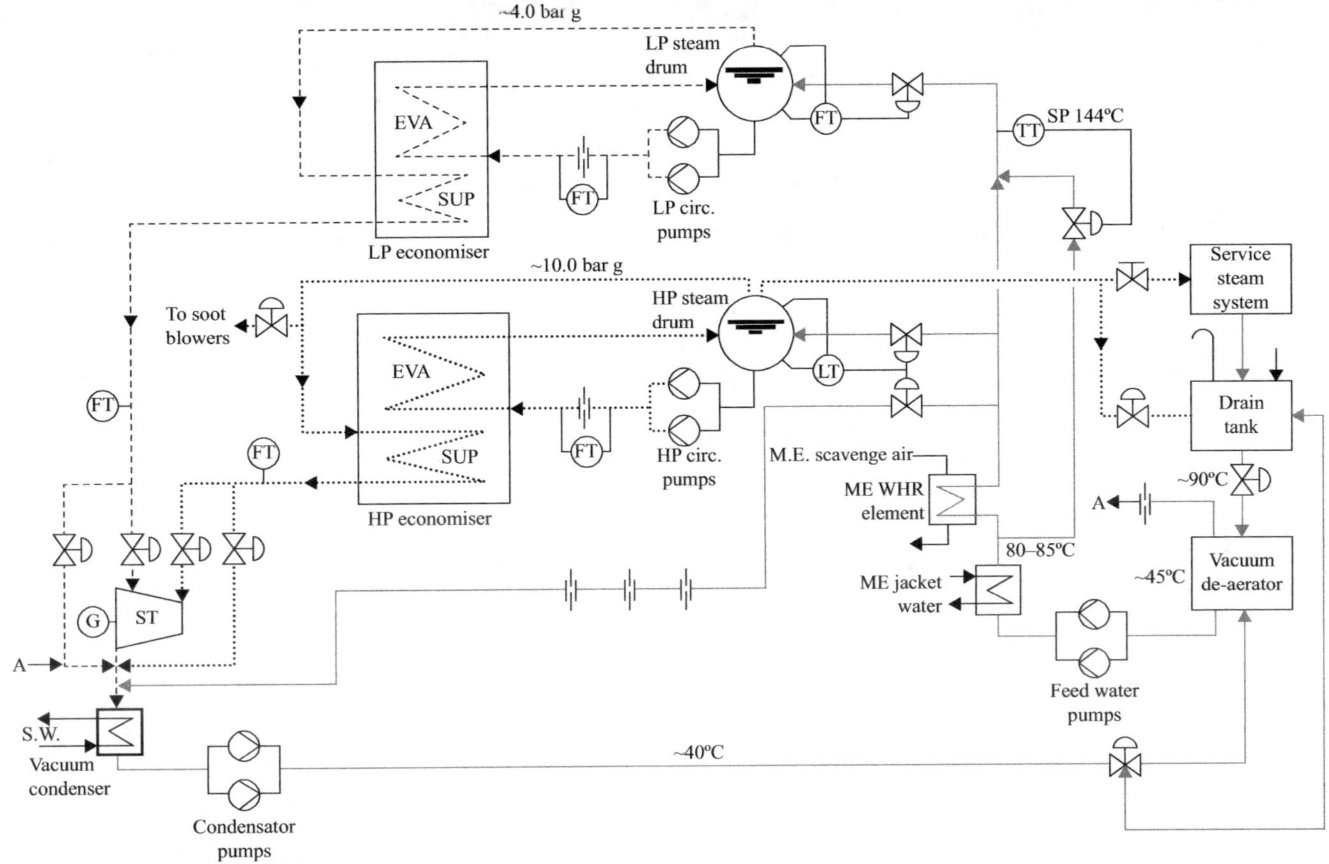

Figure 15.14 Recommended steam and water diagram for a dual-pressure WHRS

Table 15.5 WHRS recovery output data for a large container ship

WHS data based on:

MAN B&W 10S90ME-C9.2

Power: 48,510 kW

Bypass: 11.6%

Recoverly rate: 8.6% at 90% SMCR

Service steam: 1.0 ton/h

At ISO conditions

Load Point		100% SMCR	90% SMCR	85% SMCR	70% SMCR	60% SMCR	50% SMCR
HP Steam at Turbine Inlet							
Pressure	bar(a)	10.0	9.2	8.6	7.4	6.7	6.5
Temperature	°C	259	258	257	254	254	256
Flow	t/h	11.83	10.88	10.31	8.99	8.10	6.97
HP Service Steam							
Flow	t/h	1.00	1.00	1.00	1.00	1.00	1.00
Pressure	bar(a)	10.7	9.7	9.2	7.7	6.9	6.8
LP Steam at Turbine Inlet							
Pressure	bar(a)	4.5	4.5	4.5	4.5	4.5	4.5
Temperature	°C	148	148	148	148	148	148
Flow	t/h	3.92	3.39	3.09	2.28	1.787	1.45
Condensing Steam							
Pressure	bar(a)	0.057	0.052	0.049	0.045	0.045	0.045
Temperature	°C	35.3	33.6	32.5	31.0	31.0	31.0
Flow	t/h	15.75	14.27	13.40	11.27	9.88	8.42
Output							
Steam turbine	kW	2,477	2,248	2,108	1,766	1,539	1,280
Power turbine	kW	1,836	1,515	1,360	1,766	754	526
Total Generator Output	kW	4,313	3,763	3,468	2,747	2,293	1,806

Main engine load – % SMCR

MAN 10S90ME-C9.2

Power: 48,510 kW

Bypass: 11.6%

Recovery rate: 8.6% at 90% SMCR

Service steam: 1.0 tonne/h

At ISO condition, Tropical and Winter

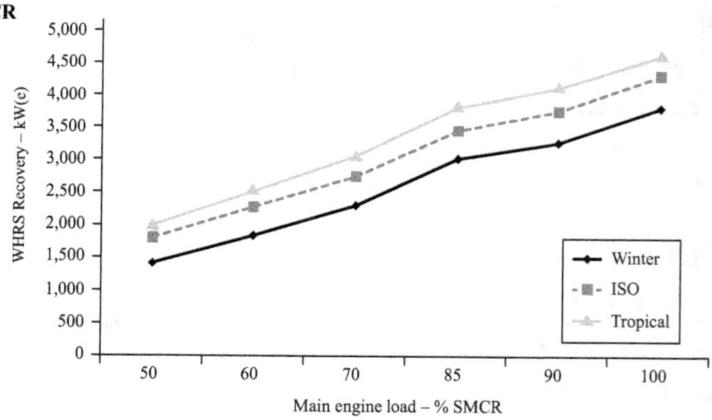

Figure 15.15 WHRS recovery output data for a large container ship according to the seasons

As volume rate ($\dot{v}$) can be calculated by stroke volume (V_s) multiplied by speed,

$$\dot{v} = V_s \times \text{rps} \tag{15.5}$$

As volume efficiency (η_v), which can be referenced in the design factor, is volume of air (V_a) divided by stroke volume (V_s), and volume of air is calculated as mass flow rate of air multiplied by specific density of air (ρ_a),

$$\eta_v = \frac{V_a}{V_s} = \frac{\dot{m}_a/\rho_a}{V_s \times N} \tag{15.6}$$

$$\dot{m}_a = \eta_v \times \rho_a \times N \times V_s \tag{15.7}$$

Mass flow rate of exhaust gas ($\dot{m}_E$),

$$\dot{m}_E = \dot{m}_f + \dot{m}_a \tag{15.8}$$

So we can calculate the energy in exhaust gas,

$$Q_E = \dot{m}_E \times C_p \times \Delta T \tag{15.9}$$

Figure 15.16 is based on the WHRSs of Wärtsilä and MAN B&W as references. This figure includes almost all WHRS components and present methods. If a flow transmitter is installed in front of the power turbine, turbocharger, and EGB line, then the exhaust-gas energy of each exhaust line can be measured reliably because the exhaust-gas flow of these lines is considered relatively stable.

Heat recovery in other heat exchangers can be calculated by using (15.9) and measurement from the temperatures and flow transmitters in input and output positions for each heat exchanger. Detailed design values of each heat exchanger, such as input and output temperatures, are out of this book's scope. If heat recoveries from each heat exchanger, PT, and ST are calculated, then a Sankey diagram can be drawn periodically.

As illustrated in Section 1.18, the only way to decrease the value registered by the EEOI is to reduce the quantity of used fuel, which means increasing heat efficiency in the overall heat exchange process.

As shown in Section 15.1.6.2, the MEPC required when the EEOI is used as a ship energy efficiency management tool, rolling average, or moving average should be calculated within the minimum period of time or number of voyages that is statistically relevant. But in the case of an EEOI monitoring system based on the methods mentioned above, the EEOI can be monitored every few minutes and can be calculated as a moving average with accumulated data. This moving average can be compared with calculated EEOI during the voyage. The greatest benefit of a near real-time EEOI monitoring system is that monitored EEOI data can help to find the optimum operation point of the whole ship's energy system. As mentioned in Section 15.2.2, and shown in Figure 15.9, information from the CAMS, PMS, RCS, SG/SM, safety system, and WHRS is provided to the M/E ECS and all are combined to find the optimum operating point of the main engine and used to

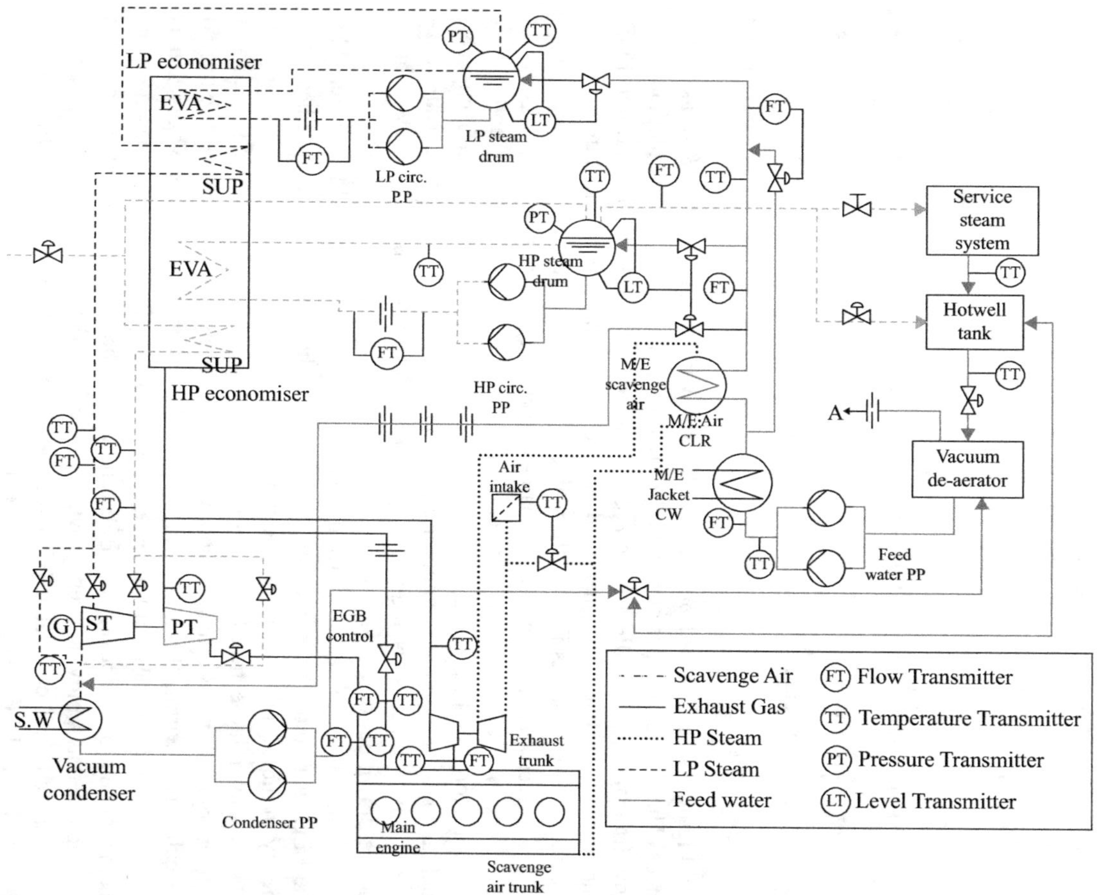

Figure 15.16 Diagram of waste heat recovery system based on the WHRSs of Wärtsilä and MAN B&W

manage the ship's energy. Even if the efficiency of the M/E decreases slightly, increasing heat recovery from exhaust gas by the WHRS may result in an improvement of the total heat efficiency.

Abbreviations

CAMS	Control, Alarm, and Monitoring System
EEDI	Energy Efficiency Design Index
EGB	Exhaust-Gas Bypass
EVA	Evaporator for steam
HP	High Pressure (steam system)
IMO	International Maritime Organization
LP	Low Pressure (steam system)
ME-ECS	Engine Control System for main engine
MEPC	Marine Environment Protection Committee
PMS	Power Management System
PTG	Power Turbine Generator unit
PTI	Power Take-In
PTO	Power Take-Out
RCS	Remote Control System for main engine
SG/SM	Shaft Generator/Shaft Motor
SFOC	Specific Fuel Oil Consumption
SMCR	Specified Maximum Continuous Rating
SPP	Steam Production Power (value for the available exhaust energy for steam production)
STG	Steam Turbine Generator unit
ST–PT	Steam Turbine – Power Turbine generator unit
SUP	Super Heater for steam
WHRS	Waste Heat Recovery System

References

[1] MEPC.1/Circ. 684,*'Guidelines for Voluntary Use of the Ship Energy Efficiency Operational Indicator (EEOI)'*, IMO Ref. T5/1.01, Aug 17, 2009

[2] Wartsila O-E-RT-WHR case study

[3] www.mandieselturbo.com/1017740/Press/Publications/Technical-Papers/Marine-Power/Low-Speed/Waste-Heat-Recovery-System.html, *'Waste Heat Recovery System (WHRS) for Reduction of Fuel Consumption, Emissions and EEDI'*, MAN Diesel & Turbo

[4] Jadhao J.S., Thombare D.G., 'Review on exhaust gas heat recovery for I.C. engine', *International Journal of Engineering and Innovative Technology (IJEIT)*, 2013, 2(12), pp. 93–100

Chapter 16

Regulation of ship-source pollution through international convention regimes

Proshanto K. Mukherjee[1] *and Jingjing Xu*[2]

16.1 Introduction

The menace of marine pollution has plagued the global maritime and environmental communities for several decades. Originally it was land-based sources, comprising effluents from factories, mills and other industrial complexes situated along the shores of rivers and lakes as well as sea coasts, that predominated. In terms of volume, quantity and deleterious effect, pollution of the seas from land-based sources still remains highest. With advancements in drilling technology coupled with numerous discoveries of offshore oil resources, pollution emanating from the seabed, exemplified by catastrophic blowouts such as the *Santa Barbara*, *Ixtoc-1* and *Deep Water Horizon* incidents, has become a matter of significant concern. Pollution of the marine environment through the atmosphere in the form of acid rain has also gained notoriety in recent times, with ships and aircraft identified as polluters, among other offenders. Aside from the sources mentioned above, ship-generated pollution stands out singularly as the one that attracts maritime interest the most.

It was not until the infamous *Torrey Canyon* disaster in 1967 that ship-source marine pollution drew public attention, if not worldwide, certainly among the traditional maritime states of Europe. Diplomatic initiatives that followed focused largely on the public international law and private law dimensions of oil spills. In 1969, the Intervention Convention[1] and the Civil Liability Convention[2] were adopted under the auspices of the International Maritime Organization (IMO). In subsequent years, several private law conventions were adopted to deal with the issues of liability and compensation. However, it often goes unnoticed that the first international convention on ship-source pollution was a regulatory convention which predated the *Torrey Canyon*. This was the International Convention for the Prevention of Pollution from Oil (OILPOL) 1954.[3] Since then numerous regulatory conventions and treaty instruments have been adopted, largely through the IMO, addressing virtually all varieties of pollutants emanating from ships.

[1]Dalian Maritime University, China
[2]Plymouth University, UK

This chapter dwells exclusively on regulatory conventions in the field of ship-source marine pollution. It is recognised that the aim of conventions and similar treaty instruments is to create and facilitate uniformity in international law, without which ships would perpetually be facing a 'patchwork quilt' international regime. It is also recognised that regulation, in the present context, focuses mainly on the health of the marine environment. Thus there are regulatory conventions that embrace ships' activities and their environmental impacts, including the impact created by the eventual demise of the ship in a ship-breaking yard. Virtually all of them are elaborations of their corresponding 'blueprint' prescriptions in the United Nations Convention on the Law of the Sea (UNCLOS), 1982.

There are at present altogether nine treaty instruments, including conventions and protocols. These are addressed in the chapter in as much detail as the context allows. Equal coverage is not given to each; some are discussed only in summary fashion. Some of the conventions are presented in synoptic form, covering their evolutionary development together with their basic principles and features; others are treated by making specific references to selected convention Articles and Regulations and providing summary explanations. Where appropriate, the method of comparative analysis is used to illustrate similarities and differences between and among conventions. Given its overwhelming functional importance, MARPOL[4] is examined in more detail than other instruments. It is also the most voluminous of all in size and content. Inevitably, therefore, the discussion starts with MARPOL, even though chronologically it is predated by the original London Convention on Dumping of Wastes.[5]

We start by looking at the substantive elements of MARPOL, which consists of six Annexes, each addressing a particular type of pollutant. They are successively oil, noxious liquid substances, packaged harmful substances, sewage, garbage, and air pollution. The Oil Pollution Preparedness and Response (OPRC) Convention of 1990[6] and its companion instrument the HNS Protocol of 2000 are then discussed. Apart from the London Convention on Dumping of Wastes mentioned above, the other related instrument is the Basel Convention of 1989,[7] which is an instrument of the United Nations Environment Programme (UNEP). This Convention in turn is associated with the recently adopted (not yet in force) Ship Recycling Convention (SRC) of 2009,[8] which is an IMO instrument. Aside from these, the Ballast Water Management (BWM) Convention of 2004[9] is of contemporary interest although it is not yet in force. Finally there is the Anti-Fouling Systems (AFS) Convention of 2002,[10] which is in force, and the Nairobi Convention on Removal of Wrecks.

The chapter concludes with an appraisal of the success of the regulatory system and what issues remain to be addressed or resolved in this important and challenging field of endeavour, given the dynamic strides being made in shipping technology.

16.2 Theoretical underpinnings

16.2.1 Legal framework: the marine pollution spectrum

As a prelude to the whole discussion it is submitted that in theoretical and philosophical terms, the overall legal framework of marine pollution lies across a spectrum stretching from public international law at one end to private law at the other; from preventive to mitigative to remedial measures in the ameliorative process. In the middle of the spectrum lies the regulatory segment, which is the subject of this chapter.

From the spectrum diagram depicted below it should be apparent that there are close interrelationships among convention instruments belonging to different segments of the spectrum, due to commonalities of subject matter. MARPOL is perhaps the most vivid example of this truism in terms of regulatory and private law interfaces. Annex I of MARPOL regulates discharges of oil and the CLC/Fund regime provides for liability and compensation in respect of damage suffered by victims of oil pollution. Annex II of MARPOL regulates discharges of noxious

Marine pollution spectrum convention interfaces

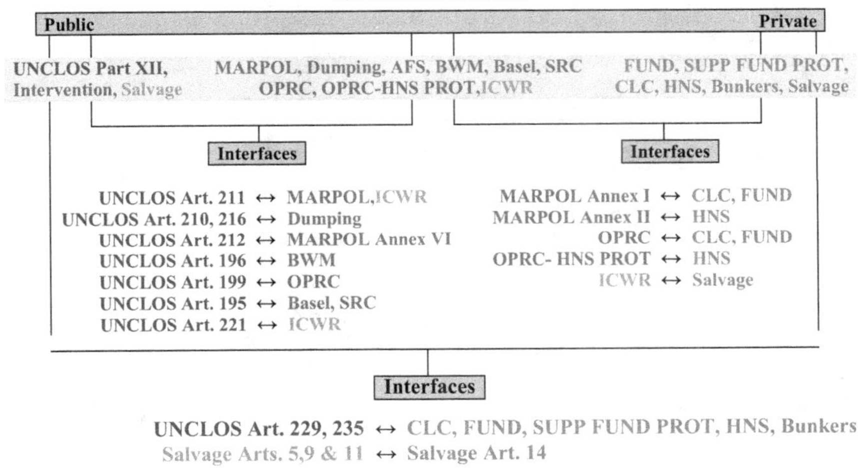

CONVENTIONS

United Nations Convention on the Law of the Sea (UNCLOS), 1982
International Convention Relating to Intervention on the High Seas in Cases of Oil Pollution Casualties (Intervention), 1969
International Convention for the Prevention of Pollution from Ships, 1973, as modified by the Protocol of 1978 relating thereto (MARPOL 73/78)
Convention on the Prevention of Marine Pollution by Dumping of Wastes and Other Matter (Dumping), 1972; Protocol of 1996
International Convention on the Control of Harmful Anti-fouling Systems on Ships (AFS), 2001
International Convention for the Control and Management of Ships' Ballast Water and Sediments (BWM), 2004
Basel Convention on the Control of Transboundary Movements of Hazardous Wastes and Their Disposal (Basel), 1989
International Convention on Oil Pollution Preparedness, Response and Co-operation (OPRC), 1990;HNS Protocol (OPRC-HNS Protocol), 2000
International Convention on Civil Liability for Oil Pollution Damage (CLC), 1992
International Convention on the Establishment of an International Fund for Compensation for Oil Pollution Damage (FUND), 1992; Supplementary Fund Protocol (SUPP FUND PROT), 2003
International Convention on Liability and Compensation for Damage in Connection with the Carriage of Hazardous and Noxious Substances by Sea (HNS), 1996
International Convention on Civil Liability for Bunker Oil Pollution Damage (Bunkers), 2001
International Convention on Salvage (Salvage), 1989
Nairobi International Convention on the Removal of Wrecks (ICRW), 2007
The Hong Kong International Convention for the Safe and Environmentally Sound Recycling of Ships (SRC), 2009

COLOR CODE

Preventive
Remedial
Preventive/Mitigative
Mitigative
Preventive/Remedial/Mitigative

liquid substances from ships, whereas the HNS Convention is the corresponding private law regime providing for liability and compensation in respect of victims of such pollution.

Ship-source pollution is of three kinds, namely, operational discharges, deliberate dumping, and accidental spills. The first and third are regulated through MARPOL and the second through the London Dumping Convention. The other conventions discussed do not neatly fit into this categorisation but are germane to a clear appreciation of the overall regulatory law framework of ship-source pollution and are addressed in their proper contexts.

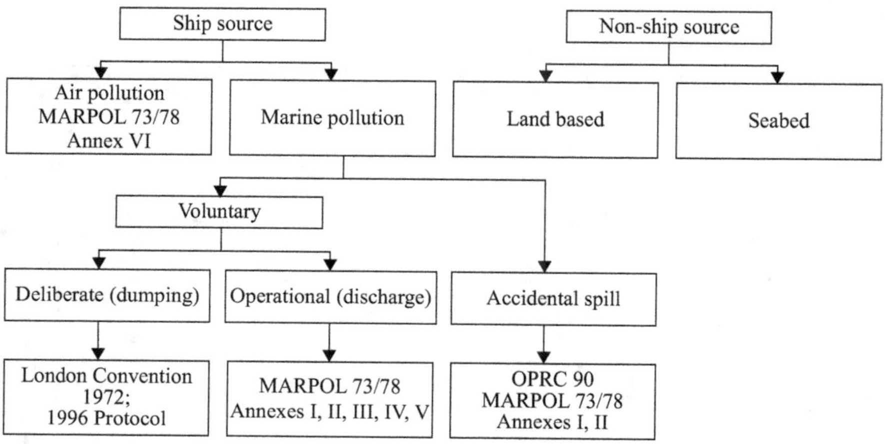

At this juncture, it is perhaps instructive to obtain an unequivocal appreciation of what is regulatory law in the first place.

16.2.2 Functional approach in regulatory law

It is submitted that legal theory, in terms of law-making in the field of ship-source oil pollution, can be depicted by utilising the application of functionalism, or the so-called functional approach. This approach, according to the present authors, means deviating from strict legal principles to attain certain policy objectives where no established legal principles are readily available for the purpose. The use of the functional approach requires a clear appreciation and articulation of the purpose of the law which is preceded by the envisaged policy.

The purpose in the present context is to prevent pollutants entering the seas from ships. The notion of prevention is *prima facie* a regulatory matter and is associated exclusively with public law, the object of which is to regulate public conduct for the welfare of the public. No such law existed prior to 1954 in the context of pollution from ships, at least in the international arena; there were no established legal principles which could provide the foundation for effectuating policy reflecting society's desire to have a clean marine environment through preventive measures imposed on ships. Under the auspices of the newly established International Maritime Organization (IMO), the world community came together and adopted for the first time an international instrument designed to prevent pollution of the seas by oil emanating from ships. As a ground-breaking effort it was an illustration of the functional approach to law-making.[11]

The functional approach is even more vividly illustrated by the creation of the International Oil Pollution Compensation Fund (IOPC Fund) to exact a measure of responsibility from owners of oil cargo for ship-source pollution where there was no existing legal principle under which they could be held liable in law, given that the cargo was in the custody of the shipowner.

16.2.3 Concept of regulatory law

As a preliminary observation it is noted that the concept of regulatory law, maritime or otherwise, is virtually alien to certain legal systems, both in substantive terms and with respect to incongruent use of terminology. The lack of a clear perception of what exactly is regulatory law within the scope of international treaty law can be a major hurdle to the understanding of ship-source pollution law, which is overwhelmingly regulatory in scope. In some jurisdictions which follow the civil law tradition, regulatory law is equated with administrative law, which is inconsistent with its perception in international treaty law and the way regulatory law is viewed typically in common-law jurisdictions. As noted above, the object of regulatory law, as the term implies, is to regulate public conduct; the public in the present context being entities engaged in shipping activities that pollute the marine environment. Regulatory law is promulgated through regulations, which can be substantive or procedural or both.[12]

In treaty law, regulatory regimes often have a non-mandatory element known as *para droit*, or soft law, which is absent in domestic legal regimes. Soft law is basically non-binding, but its application can be made binding at the international level by specific expression as exemplified by various IMO Codes which apply to both maritime safety and ship-source marine pollution issues.[13] At the domestic level, an IMO Code, even if it is non-mandatory, can be made binding through the vehicle of regulations. Guidelines are another species of soft law which are non-binding but can have a persuasive effect, which is useful.

In contrast to regulatory law, administrative law in common-law jurisdictions is a body of law which provides legal recourse to citizens or subjects of the state aggrieved by acts or omissions of public authorities which fail to carry out their statutory duties and mandates prescribed by law. It is based on legal principles such as *audi alterem partem*, or the right for both sides to be heard, the doctrine of fairness, discretion and the like, and is distinguished by its own peculiar remedies such as *certiorari*, *mandamus*, injunction and prohibition, which may be given in conjunction with damages as in civil liability actions.[14] There is no equivalent of this legal phenomenon in civil law jurisdictions, which is why regulatory law is sometimes, misleadingly for common lawyers, described as administrative law.

Notably, there is a concept of administrative law similar to that prevailing in common-law jurisdictions within international institutions such as the United Nations. While there is no parallel concept of administrative remedies in the arena of international treaty law, detention of a ship as a sanction is expressly provided for in conventions such as MARPOL, which can be effectuated administratively or judicially according to the domestic law and procedure of a relevant state party.

16.3 Regulatory ship-source pollution conventions[15]

16.3.1 MARPOL Convention

The MARPOL Convention was adopted in 1973. It was intended as a replacement for the earlier OILPOL of 1954, which addressed only oil as a ship-source pollutant. MARPOL, at its very inception, was far more comprehensive in scope than OILPOL. It covered five types of pollutants through its five original Annexes. In 1997, a Sixth Annex was added to include ship-source pollution entering the sea from the air and through the atmosphere. Thus, at present there are six Annexes to the Convention, that regulate six different types of ship-source pollutants. They are as follows:

> Annex I – Oil; Annex II – Noxious Liquid Substances; Annex III – Packaged Harmful Substances; Annex IV – Sewage; Annex V – Garbage; and Annex VI – Air Pollution.

The MARPOL Convention is the instrument that regulates ship-source pollution from operational discharges. It deals exclusively with preventive measures, as can be gleaned from its title 'International Convention for the Prevention of Pollution from Ships'. Although MARPOL is directed primarily to regulating operational discharges, there are several provisions that deal with design and construction of tankers and address such matters as damage control including intact, subdivision and damage stability, pertaining to accidental spills of oils and chemicals. These are, of course, strictly speaking, non-operational matters, but nevertheless they fall within the scope of the object and purpose of the convention, namely, to prevent ship-generated pollution.

As mentioned above, marine pollution emanates mainly from land-based sources such as industrial by-products, pesticides and herbicides and other effluents resulting largely from daily urban activities. Even so, considerable pollution is generated by ships, and in terms of quantity entering the oceans, oil undoubtedly remains the predominant pollutant, not only because of worldwide tanker traffic but also because of discharges of oil from bilges and machinery spaces in non-tankers.

It is also mentioned above that the 1954 OILPOL Convention dealt only with oil, as is evident from its title. But it is significant that much of it has been subsumed by Annex I of MARPOL. Of course, the regulatory regime is not static and amendments continue to be adopted as and when updating becomes necessary.

The MARPOL Convention originally adopted in 1973 went through a major revision through its 1978 Protocol. The adoption of the 1978 Protocol was the result of the deliberations of a diplomatic conference at the IMO on Tanker Safety and

Pollution Prevention (TSPP) which took place in the wake of the *Amoco Cadiz* disaster. The conference produced two sets of Protocols; one for MARPOL and the other for the International Convention for the Safety of Life at Sea (SOLAS). The SOLAS Protocol operates as an amendment requiring ratification or accession by SOLAS state parties who wish to take it on board. By contrast, the MARPOL Protocol is merged with the original Convention of 1973. By virtue of Article I paragraph 2, the 1973 Convention and its 1978 Protocol must be 'read and interpreted together as one single instrument'. The legal implication of this provision in terms of treaty law is that the state parties to MARPOL 73 must accept the 1978 Protocol. The Convention thenceforth came to be known as MARPOL 73/78. However, since the adoption of the 1997 Protocol to the Convention through which Annex VI on Air Pollution was added, and its subsequent entry into force, the Convention is now simply referred to as 'MARPOL', pursuant to a decision taken at the 56th session of the Marine Environment Protection Committee (MEPC) of the IMO.[16]

16.3.1.1 Articles of the Convention

The 1973 Convention consists of 20 Articles. Some reflect basic legal principles; others are contractual provisions in that they were negotiated by the parties. There are two Protocols constituting the main body of the 1973 Convention. Protocol I contains provisions concerning procedures to be adopted for reporting incidents involving harmful substances, pursuant to Article 8 of the Convention. Protocol II provides the arbitration procedures applicable to disputes between state parties which are to be settled in accordance with Article 10 of the Convention. The Protocols of 1978 and 1997 contain nine Articles each.

The substantive articles in the Convention proper are presented below in synoptic form.

Article 2 – Definitions

(2) 'Harmful Substances' are essentially those that cause harm to human health, living resources and marine life, damage amenities and interfere with other legitimate sea uses.

(3) 'Discharge' means release of effluent from ship howsoever caused, including escape, disposal, spilling, leaking, pumping, emitting or emptying.
It is noteworthy that 'discharge' does not include 'dumping' within the meaning of the London Convention, release of harmful substances from offshore exploration and legitimate marine scientific research for pollution abatement.

(4) 'Ship' means any vessel of any type operating in the marine environment and includes hydrofoil boats, air-cushion vehicles, submersibles, floating craft and fixed or floating platforms.

Article 4 – Violation:

Basically, flag State jurisdiction applies regardless of the place of violation; a coastal State has jurisdiction if a violation occurs in any place that is subject to its jurisdiction under international law. Needless to say, the above-noted provisions apply to state parties to MARPOL who are required to establish sanctions and impose penalties that are adequate in severity to discourage violations and must be equally severe regardless of the location of the violation.

Article 5 – Certification and inspection of ships

Certificates issued by a state party must be accepted by other state parties. Ships of state parties while in ports or offshore terminals of other state parties are subject to inspection, which in the first instance must be limited to verifying that a valid certificate is on board. If there are clear grounds for believing that the actual condition of a ship or its equipment is not substantially in conformity with the particulars in the certificate, a physical inspection may be carried out and the ship may be detained until it can proceed to sea without causing an unreasonable threat to the marine environment, except that it may be allowed to proceed to the nearest appropriate and available repair facilities.

Article 8 – Reports on incidents involving harmful substances

Ship-source pollution incidents must be reported without delay in accordance with Protocol I, mentioned above. Every party receiving such a report must notify the ship's flag state without delay and also the IMO.

Article 10 – Settlement of disputes

Any disputes between state parties relating to the application or interpretation of MARPOL must be settled by negotiation, or if that fails, by arbitration in accordance with Protocol II, mentioned above.

Article 12 – Casualties to ships

Where there is a major deleterious effect on the marine environment, a casualty investigation must be carried out.

Article 15 – Entry into force

For compulsory Annexes it is 12 months following the date on which 15 States, comprising 50% of world tonnage, became parties.

Article 16 – Amendments

Paragraph (2)(d) provides for adoption by 2/3 majority of parties present and voting. Paragraph (2)(f) provides for acceptance of amendments as follows:

- (i) For an Article, it is the date on which 2/3 of parties, comprising at least 50% of the gross tonnage of the world fleet, accept the amendment.
- (ii) For an Annex, either the '2/3 – 50%' rule applies or it is by 'tacit acceptance' on a date determined at the time of adoption, which must not be less than 10 months later, unless there is objection by not less than 1/3 of parties to the Convention, or by parties whose combined fleets comprise not less than 50% of the gross tonnage of the world fleet.
- (iii) For an Appendix to an Annex, only the tacit procedure applies.
- (iv) For Protocol I, the same procedure applies as for Annexes.
- (v) For Protocol II, the same procedure as for an Article applies.

Paragraph (2)(g) provides for the entry into force of an amendment six months following acceptance in each case, but this provision does not apply to parties who have expressly declined to accept or have declared that their express approval is required.

16.3.1.2 Regulations in the Annexes

The substantive regulatory law is contained essentially in the Annexes, which consist of Regulations. Annexes I and II are compulsory, which means for a state to be a party to MARPOL it must ratify or accede to those two Annexes. The remaining Annexes are optional but state parties are encouraged to subscribe to all of them for the sake of harmonious and globally consistent application of the Convention. The remaining Annexes also make the convention comprehensive and complete. These are added reasons for states to consider acceding to or accepting all the Annexes. Incidentally, all Annexes are now in force. The salient features of the MARPOL Annexes are summarised in the text below.

The IMO periodically publishes up-to-date Consolidated Editions of MARPOL. These are supplemented by amendments if any are adopted during the interim periods between these editions. The Consolidated Editions contain a unique feature, namely, the so-called Unified Interpretations which appear at the end of each of the respective texts of Annexes I, II, III and VI. These are not, in strict terms, part of the Convention, but they contain valuable explanatory elaborations of the highly technical regulations. If used selectively and judiciously, the detailed supplementary texts can be quite useful in the drafting of domestic legislation aimed at implementing the MARPOL Convention. They are of great practical utility for professional users of the convention such as surveyors, inspectors, maritime administrators and shipboard personnel. As well as this, the unified interpretations serve as an interpretive tool for tribunals charged with the task of applying relevant provisions of the Convention, whether in the context of judgements relating to a casualty or in the event of litigation over a dispute.

As mentioned earlier, the Articles represent the terms of the treaty proper negotiated by the delegations present and voting at a diplomatic conference. In that sense, the Articles have the status of principal legislation, whereas the Regulations contained in the Annexes are subordinate to the Articles. The hierarchy becomes relevant in the event of a conflict between an Article and a Regulation, in which case, in the opinion of the present authors, the Article will prevail. Instances of conflicts are few and far between if there are any at all; indeed, the correlation between certain provisions in Articles and in corresponding Regulations makes for the removal of uncertainties and fosters a sound depiction of the law for the benefit of users.

There are some common features among the Annexes which deserve to be highlighted. Among them is the notion of the 'special area'. It is a significant feature of the Convention in terms of the setting of discharge standards. Notably and incidentally, the concept of the 'special area' is also addressed in UNCLOS, but there are conceptual differences between the two conventions. Under UNCLOS, a coastal state may impose 'special mandatory measures' applicable to ships navigating in their exclusive economic zone (EEZ) where international rules and standards (presumably MARPOL) are 'inadequate to meet special circumstances'. The measures are to be communicated to the 'competent international organization' (presumably the IMO) and subject to determination by that organisation that the measures are warranted, the coastal state may adopt the special mandatory measures for 'special areas'. The criteria for the establishment of such a special area are:

> recognized technical reasons in relation to its oceanographical and ecological conditions, as well as its utilization or the protection of its resources and the particular character of its traffic.[17]

The UNCLOS criteria are partly similar to those of MARPOL set out below; however, whereas under UNCLOS, a special area is contemplated to be within the coastal state's EEZ, under MARPOL there is no such restriction. As a result, much larger bodies of water, even falling outside the maritime zones of a coastal state, can be designated as a special area under MARPOL, as exemplified below.

With regard to MARPOL, discharges are totally prohibited in 'special areas' under Annexes I, II and V. There is no mention of special areas in the other Annexes because of the particular characteristics of the pollutants and the ways in which they are regulated under those Annexes. The generic definition of 'special area' is

> … a sea area where for recognized technical reasons in relation to its oceanographic and ecological condition and to the particular character of its traffic the adoption of special mandatory methods for the prevention of sea pollution is required.[18]

The special areas under the convention are the Mediterranean Sea area, the Baltic Sea area, the Black Sea area, the Red Sea area, the Gulfs area, the Gulf of Aden area, the Antarctic area, the North Sea area, the North-West European waters and the Wider Caribbean Region. They are identified by reference to geographical coordinates or other characteristic features circumscribing their respective perimeters. Not all the above-mentioned areas are included in each Annex. With respect to Annex I, the North Sea area and the Wider Caribbean Region are not special areas. The only special areas under Annex II are the Baltic Sea, the Black Sea, and the Antarctic. Under Annex V all except the Gulf of Aden area and the North-West European waters are special areas. Annex VI contains something akin to a special area, known as 'emission control areas' for nitrogen oxide (NO_x), sulphur oxide (SO_x), and particulate matter. In respect of NO_x the only emission control area at present is the North American area as described in Appendix VII to Annex VI. With respect to SO_x and particulate matter, the present emission control areas are the Baltic Sea area as defined in regulation 1.11.2 of Annex I, the North Sea as defined in regulation 5.1(f) of Annex V, the North American area, and the United States Caribbean Sea area as described in Appendix VII to Annex VI. Other emission control areas may be designated by the IMO in accordance with criteria and procedures set out in Appendix III to Annex VI.

Since all discharges are prohibited in special areas, vessels are required to retain all their wastes on board and discharge them to shore-based reception facilities. To complement this requirement, state parties are required to provide adequate reception facilities at locations ashore which they must designate. It is recognised that providing waste reception facilities is a costly affair, especially so for developing countries. Over the years since MARPOL was adopted in 1973, this has been an issue for discussion within and outside the IMO, regarding how such facilities are to be financed. The financial implications of this important preventive prescription in the Convention can be considerably burdensome for some countries, but equally, without adequate reception facilities a major objective of the convention will remain unfulfilled.

Exceptions can be made to compliance with the discharge standards under all Annexes, irrespective of whether the location of the discharge is a special area or waters where limited discharges are allowed, if the non-compliance is deemed to be necessary for saving life at sea or for securing the safety of the ship.[19]

Annexes I, II and IV and VI contain another important preventive feature. These are the provisions requiring vessels to be properly surveyed and accordingly certificated. These provisions have now been harmonised with corresponding requirements under the SOLAS and LOADLINE Conventions. The relevant certificates are the International Oil Pollution Prevention (IOPP) Certificate under Annex I, the International Pollution Certificate for the Carriage of Noxious Liquid Substance in Bulk, otherwise referred to as the NLS Certificate, under Annex II, the International Sewage Pollution Certificate under Annex IV and the International Air Pollution Prevention (IAPP) Certificate under Annex VI. Record books are required to be maintained under Annexes I, II and V, known respectively as the Oil Record Book, the Cargo Record Book and the Garbage Record Book. Under Annex V there is also a requirement for each vessel to have a Garbage Management Plan. The Convention requires member states to treat violations of the Convention as offences under domestic law and provide for appropriate sanctions.

A significant feature of MARPOL is the 'tacit acceptance' procedure for amending various elements of the Convention set out in Article 16. This methodology provides that following the adoption of an amendment, it is deemed to be accepted if no objections from a specified number of member states are received by the IMO within a pre-established period of time allocated for the purpose. Once it is accepted, the amendment enters into force on a date fixed by the relevant IMO Committee. It cannot enter into force if it is rejected by one-third or more of the state parties whose combined merchant fleets represent at least 50% of global gross tonnage.

When a foreign ship is in a port or offshore terminal of a state party to the Convention, that state can assert port state jurisdiction over that ship and exercise port state control.[20] The general requirements are provided for in Article 5 of the Convention and the detailed control procedures in relation to operational requirement are set out in the respective Annexes. For Annex I the relevant provision is Regulation 8A, for Annex II it is Regulation 15, for Annexes III and V, the provision resides in the respective Regulation 8 of each of those Annexes, and in Annex VI it is Regulation 10. A detailed depiction of all the Annexes of the Convention, even in summary form, would be somewhat disproportionate, given the generality of the scope of this chapter. However, it is considered expedient and useful to provide some description of the salient features of Annex I, given that oil is the most common ship-generated pollutant of all.

Annex I

At present there are 43 Regulations in Annex I, which are grouped under seven chapters. The first chapter contains general provisions. Chapter 2 deals with procedures pertaining to surveys and certification. Chapter 3 prescribes the requirements for control of pollution from machinery spaces of all ships. Requirements for control of pollution from the cargo areas of oil tankers are contained in Chapter 4. Chapter 5 consists of only one Regulation prescribing requirements for the carriage on board of a shipboard oil pollution emergency plan (SOPEP). In Chapter 6 there are requirements for reception facilities, and in Chapter 7 there is a single Regulation prescribing special requirements for fixed or floating platforms. Chapter 8 contains Regulations that provide for pollution prevention during oil cargo transfers at sea between oil tankers, and Chapter 9, containing the last Regulation, deals with special requirements for the use and carriage of oils in the Antarctic area.

In the aftermath of the infamous *Erika* oil spill, decisions were made by the MEPC at its 45th and 46th sessions, to expedite the phasing out of single-hull tankers and bring the double-hull requirements into effect sooner than originally contemplated. Considerable pressure was exerted on the IMO by the member states of the EU to that effect. Eventually, the final phasing-out date for Category 1 tankers was brought forward from 2007 to 2005, and the final phasing-out date for Category 2 and 3 tankers, from 2015 to 2010. (The categorisation of tankers is contained in the Regulations.)

The incorporation of the double hull requirement in MARPOL generated extensive debate within the IMO. Its supporters pointed to the Oil Pollution Act, 1990 (OPA 90) of the United States to advance their position. It is notable in this context that *Exxon Valdez*, which ran aground in Alaskan waters in 1989 and caused a massive oil spill, was in fact a double-hulled tanker. It is also to be noted that other technologies are available that are equally, if not more, effective, such as the Colombi-Egg design. However, the predominant view appears to be that the double-hull alternative is the most feasible technologically as well as financially.

It is beyond the scope of the present chapter to delve into the details of each Regulation or even to identify them by name or title, given the fact that there are 43 of them altogether. However, it is important for the reader to carefully note the definitions set out in Regulation 1, the exemptions and waivers in Regulation 3, and the exceptions in Regulation 4. The provisions relating to exceptions in Regulation 4 have featured in controversies over the validity of Directive 2005/35/EC of the European Parliament and of the Council relating to ship-source pollution. Almost all the remaining Regulations are purely technical in nature.

It should be noted that there are three Appendices to Annex I, one containing a list of oils and the other two containing standard forms for the IOPP Certificate and Supplements, and the Oil Record Book. The text on Unified Interpretations to Annex I contains five Appendices.

Annex II

With regard to Annex II it must be noted that important changes have been made to it in recent years. These are set out in summary form below. Needless to say, to fully appreciate the import and significance of these changes, one must review the whole Annex in light of its previous version.

The first notable point is that there are no longer any special areas under this Annex. In effect all seas everywhere are 'special areas' for the purposes of NLS discharges. Chemical tankers are required to comply with not only the Annex II (environmental regulation and operation) provisions but also the IMO Construction and Equipment of Ships Carrying Dangerous Chemicals in Bulk (BCH) or International Bulk Chemical (IBC) Codes (design and other operational requirements).

Among the important Regulations, Regulation 6 provides for NLS categories X, Y and Z and 'other substances'. Regulation 11 deals with design, construction, equipment and operations. Notably, vegetable oil carriers are exempted from Regulation 11 requirements pursuant to Regulation 4.3. Regulation 13 provides for discharge restrictions and standards according to categories.

Annexes III, IV and V

As mentioned earlier, it is beyond the intended scope of this chapter to enter into a rendition of the details of all MARPOL Annexes; hence the selectivity. Even so, it is felt that some reference should be made to the very basic attributes of Annexes III, IV and V. In this vein, it is pointed out that Annex III, dealing with packaged harmful substances, is in a sense the dry counterpart of NLS, addressed in Annex II, but it is apparent from the brevity of the provisions that the drafters considered it somewhat less consequential to the MARPOL package as a whole. There are, of course, discharge restrictions and prescriptions for packaging of materials. The definition of 'harmful substance' mentioned above is obviously relevant in this regard.

Annex IV deals with discharges of sewage from ships. It was the last Annex to enter into force, which points to the lower consideration of its importance within the overall regulatory framework of MARPOL. Incidentally, sewage is a particularly troublesome pollutant when it comes to pleasure vessels of the smaller variety, which are often not fitted with the required equipment for containment of sewage or for its transfer to shore reception facilities. Such vessels often feel free to pump the sewage into the open sea or semi-enclosed waters such as in a harbour.

Annex V deals with garbage, an environmental menace, particularly with respect to cruise ships and the waters mostly frequented by such ships, such as the Mediterranean Sea and the Caribbean region, which is the only special area for this Annex. Plastic bags in which shipboard garbage is collected are non-biodegradable and are particularly harmful to cold water seas such as the Arctic, where the environment is ecologically fragile. There are requirements in the Annex for comminuters and incinerators to be installed on board certain types of ships.

Annex VI

Arguably, air pollution, addressed in Annex VI, is not exactly ship-source because the exhausts emanating from a ship do not directly enter the marine environment. In UNCLOS the reference is to 'pollution from and through the atmosphere', which is a more accurate description of the phenomenon. The main factor involved in this pollution type is that it is generated by exhaust emissions from ships which enter the atmosphere and beyond, generating what is loosely referred to as pollution of the air. What matters from a marine environmental perspective is that through an indirect route the seas are polluted. In relation to the scope of MARPOL, therefore, the pollution is originally ship-generated and is a by-product of the ship's operation, which makes it a part of 'operational discharges'.

Exhaust emissions from ships have long been viewed as a serious threat to the atmospheric environment and also to the marine environment through their entry into the oceans via the atmosphere. As mentioned earlier, this Annex VI was adopted through a Protocol in 1997. It consists of four chapters; Chapter 1 contains 'General', matters including a host of definitions, and the usual exceptions and exemptions in Regulations 1 and 3 respectively; Chapter 2 deals with surveys and certification and port state control through Regulations 5 to 10 and detection of violations and enforcement through Regulation 11. Chapter 3 is of utmost substantive importance as it regulates and controls emissions of ozone-depleting substances through Regulation 12, NO_x through Regulation 13, SO_x and particulate matter, otherwise known as 'black carbon', through Regulation 14, and volatile organic compounds (VOCs) through Regulation 15. Chapter 4 provides for Regulations concerning energy efficiency for ships and includes the concept of 'energy efficiency design index' (EEDI) addressed in Regulations 20 and 21, and requires ships to carry a 'ship energy efficiency management plan' (SEEMP) in Regulation 22. Regulation 23 deals with the promotion of technical cooperation and transfer of technology.

Several Regulations in Chapter 4 are designed to deal with the operational efficiency of ships aimed at reducing emissions of greenhouse gases (GHG) including carbon dioxide (CO_2), which is a major contributor to the phenomenon of global warming. It is apparent that efforts to regulate and reduce CO_2 emissions from ships continue to progress at the IMO. Discussions at the MEPC are ongoing but the exercise is yet to be completed.[21] Consensus has not been reached and there are some difficulties because of divided opinions among industrialised and developing states. At the 67th session of the MEPC, held in October 2014, the third IMO GHG study was approved, providing further updated estimates on GHG emissions from ships.[22] This shows that the subject continues to be open for discussion and the IMO has the resolve to continue pursuing the matter.

The first 18 Regulations are extant; regulations 19 to 23 are prospective. There are presently seven Appendices to Annex VI and one prospective Appendix VIII. They cover standard forms for certificates and a range of technical matters.

16.3.2 Oil Pollution Preparedness and Response Convention (OPRC)

The Oil Pollution Preparedness and Response Convention (OPRC) is another regulatory Convention in the realm of ship-source pollution. Its uniqueness lies in its object and purpose, reflected in the title. The notion of preparedness and response projects both a preventive and a remedial element through the mitigative stage stretching across the marine pollution spectrum. It also illustrates the continuum of preventive measures to be taken before an anticipated pollution event, the event itself and the remedial actions that need to be taken after the fact. The spectrum and the continuum taken together represent the strategic framework within which the regulatory law of the OPRC Convention operates.

As background, it is notable that the genesis of the Convention was the recognition by the IMO of the need to develop an international legal and operational framework through a convention instrument to address the issue of combating serious oil pollution incidents through co-operative measures. A number of incidents occurring in the late 1980s provoked the move. The MEPC, in response to a request by the IMO Assembly, embarked on the task of preparing a draft convention for consideration at a diplomatic conference convened in 1990. The Convention was designed to address combating measures in relation to major oil spills resulting from casualties such as collisions and groundings. The world maritime community was already acquiring experience in this field through the activities of regional organisations or centres established and operating under the joint efforts of the IMO and the United Nations Environment Programme (UNEP). The first such centre was the Regional Oil Combating Centre (ROCC), established in Malta under the Mediterranean Action Plan (MAP) of UNEP. The name of this centre was later changed to the Regional Marine Pollution Emergency Response Centre (REMPEC). MAP was established pursuant to the Barcelona Convention, a regional UNEP convention serving the Mediterranean region. Similar centres or organisations were set up in other regions of the world. One of these was REMPEITEC, located in Curacao, for the Caribbean and Central American region under the Cartagena Convention. While these organisations were essentially UNEP bodies, the IMO acted as the executing authority and provided directions for their policies, priorities and overall operations. In November 1990, at the end of the diplomatic conference, the OPRC Convention was adopted together with ten Conference Resolutions attached to the Final Act. The following are among the salient features of the convention:

In the Preamble, reference is made to 'the need to preserve the human environment in general and the marine environment in particular' and to 'serious threat by ... oil pollution incidents involving ships, offshore units, sea ports and oil handling facilities'. There is also express reference to the 'polluter pays' principle as a general principle of international environmental law alluding to the private law aspect of liability and compensation for pollution damage manifested through the Civil Liability Convention (CLC) and the Fund Convention.

The general provisions and definitions are contained in the first two Articles. Article 3 requires ships of state parties to carry on board SOPEPs. Similar plans are required to be carried on offshore units. Under Article 6, operators of such units must coordinate their plans with those of the relevant coastal state authorities' systems established for preparedness and response in cases of oil pollution incidents. There must be included in every national system of a coastal state a national contingency plan. Article 4 of the Convention sets out the procedures to be followed for reporting an event involving a discharge or probable discharge of oil, or the presence of oil at sea which has been observed. When such a report is received, certain actions need to be taken by the receiving states. These are set out in Article 5. Article 7 requires state parties to the Convention to co-operate among themselves in their efforts to respond to oil pollution incidents. Co-operation among state parties with regard to research and development projects in the context of preparedness and response is also necessary and is provided for in Article 8. Such co-operation contemplates promotion and exchange of results of research and development. These may include state-of-the-art technologies, surveillance techniques, methods for containment and recovery of spilled oil, dispersion and clean-up of oil pollution, mitigation of damage and restoration of the marine environment affected by the spill. In this connection, Article 9 contemplates technical co-operation among state parties for training and transfer of technology. Article 10 deals with promotion of bilateral and multilateral co-operation in preparedness and response. Under Article 12, the IMO is to undertake certain functions and activities. These include provision of information, technical services, technical assistance and promotion of training and education. The procedures for amending the Convention are found in Article 14, including amendments through tacit acceptance.

16.3.3 Dumping of wastes

The principal and common characteristic of MARPOL and the London Convention on Dumping of Wastes is that both are preventive in scope. However, MARPOL regulates operational discharges from ships as its main purpose, even though there are numerous other matters also pertaining specifically to tankers which are governed by that Convention. As distinguished from 'operational', deliberate dumping of wastes at sea is the theme of the London Convention and the object and purpose of it is to prevent or control that activity. Thus MARPOL deals with discharges and the London Convention with dumping; and the definitions of the two terms in the two respective conventions have been designed to be mutually exclusive through express provisions making the distinction clear.

A significant observation to be made in this regard is that a vessel transporting waste from land with the object of disposing of it at sea is at once under the dumping regime of the London Convention as well as under MARPOL with respect to its two different activities or functions. Since it is a ship producing pollutants incidental to its normal operations, it is governed by MARPOL with respect to its operational discharges. At the same time the *raison d'être* of the ship is to transport waste for disposal at sea and for that function it is subject to the rules of the London Convention.

The London Convention adopted in 1972, then known as the London Dumping Convention in ordinary parlance, was originally designed to reflect a policy under which dumping of waste was permissible at sea unless it was prohibited or controlled by the Convention. In other words, ships were free to dump so long as the activity was not barred or restricted by the Convention. In 1996, the thinking behind that policy was changed dramatically to reflect a far more stringent stance. A protocol was adopted pursuant to which all dumping was henceforth to be prohibited in the first instance, unless the Convention allowed the particular substance to be dumped under a regime of permits issued by the state in question.

16.3.4 Basel Convention

The Basel Convention deals primarily with the transboundary movement of hazardous wastes. It was adopted in 1989 under the auspices of UNEP with the basic objective of controlling and regulating the export and import of hazardous and other wastes and entered into force in 1992. Aside from addressing transboundary movements (TBM), which is its prime characteristic, it also provides for the environmentally sound management (ESM) of such wastes. If a comparison is to be made between the London and the Basel Conventions, both of which deal with wastes at sea, it must be appreciated that for wastes from land loaded on board a ship with the deliberate object of disposal at sea, the governing regime is the London Convention. By contrast, where wastes originating from land are loaded and transported on board a ship destined for disposal in another state, that is, where wastes are exported from one state for importation into another, the Basel Convention is the governing regime. In both cases MARPOL is applicable for the regulation of operational discharges from ships. To clearly stress this point, both the London and Basel Conventions contain similar clauses expressly stating that they do not apply to wastes generated on board as a result of normal shipboard operations.

An important aspect of the Basel Convention is that it also serves as the current international regime for the regulation of ship-breaking operations. The use of this Convention is explained by the proposition that a ship traversing the oceans on its last voyage heading towards its ultimate destination, where it is to be broken and dismantled forever, is in essence hazardous waste executing a transboundary movement and should be treated as such to justify the application of the Basel Convention. In this context it must be noted that a new IMO Convention has been adopted to deal specifically with the subject of ship recycling. It is the SRC, 2009, discussed below.

The Basel Convention is typically a regulatory, non-self-executing convention in terms of treaty law. There are virtually no provisions in the convention that directly impinge on ships or shipowners/operators. The obligations under the convention are mainly imposed on or directed to state parties.[23] This is a fundamental tenet of the Basel Convention. In contrast, MARPOL, which is also a regulatory convention, may be construed as partially self-executing, in that obligations, for example, those relating to discharge standards, are imposed directly on ships, their crew and their owners/operators. There are also non-self-executing provisions in MARPOL, such as the obligation to provide adequate reception facilities, which are directed to state parties.

The obligations under Basel, in so far as ships are concerned, are largely indirect. For example, a ship which is to be used to transport hazardous substances by sea across international boundaries is required to notify the authorities of the loading and the discharging states concerned and seek their permission to load and/or discharge the substances. However, it is only possible for a ship to do so under the national legislation of the relevant state, which state, if it is a party to the convention, is obliged to put into place the necessary statutory/legal framework to implement the convention.

The teleology of the Basel Convention can be gleaned from its extensive preamble, consisting of 24 clauses. In summary, the object and purpose of the Basel Convention is to protect human health and the environment against the adverse effects resulting from the generation and management of hazardous wastes where such wastes are transported across international boundaries. The convention purports to impose strict control over the movement and ultimate disposal of such wastes.

Under the Basel Convention, 'wastes' are 'substances or objects which are disposed of or are required to be disposed of by the provisions of national law'. In attempting to understand the definition of 'wastes' one needs to appreciate the definition of 'disposal' under article 2 paragraph 4 of the Basel Convention. In Basel, the term means any operations specified in Annex IV of the Convention. It is notable that Annex IV has the caption 'Disposal Operations' and consists of a list of operations divided into two groups. Under the first group, entitled Section A, 15 operations are listed which do not lead to the possibility of resource recovery, recycling, reclamation, direct reuse or alternative uses. Section B consists of 13 operations which may lead to resource recovery, recycling reclamation, direct reuse, or alternative uses.

The crux of the Basel regime is not simply the issue of disposal of wastes as in the London Convention, where it is characterised as 'dumping' but rather the focus is on disposal of 'hazardous' wastes, although certain wastes described as 'other wastes' are also subject to the regime. The other essential element of the regime, as mentioned above, is that it only applies when there is a 'transboundary' movement of such wastes, which is subject to regulation under the Convention. The significance of these two elements is reflected in the title of the Convention as well as in several of the preambular clauses. It is therefore pertinent, in the above context, to determine what constitutes 'hazardous waste' under the Convention.

The characteristics of the wastes controlled under the convention are identified in terms of their categories. The categories and characteristics are set out in Annexes I, II and III. Article 1 provides that the scope of the Convention extends to hazardous wastes and other wastes. Paragraph 1(a) refers to substances listed in Annex I which *prima facie* are hazardous wastes unless they do not possess any of the characteristics set out in Annex III. This aspect of paragraph 1 reflects an express objective characterisation of what is intended to be considered as hazardous waste.

By contrast, subparagraph (b) reflects a subjective determination, through the domestic legislation of a state party to the Convention, of what constitutes hazardous waste. For the purposes of the convention the term 'other wastes' applies to any wastes belonging to any category contained in Annex II that are subject to transboundary movement. These are household wastes and residues arising from their incineration.

A purposive approach to the Basel Convention would *prima facie* involve controlling the taking on board of hazardous wastes covered by the convention for transboundary transportation and their ultimate disposal. However, the convention purports to cover as well the management of such wastes from their generation through to their disposal, including the transportation phase.

16.3.5 Ship Recycling Convention

The Hong Kong International Convention for the Safe and Environmentally Sound Recycling of Ships (SRC) was adopted under the auspices of the IMO in 2009. It is not yet in force and will only happen when '15 states, of which the combined gross tonnage accounts for no less than 40 per cent of that of the world and the combined maximum annual ship recycling volume over the past 10 years is no less than 3 per cent of their combined gross tonnage' ratify or accede to it. The formula is unusual but suits the subject matter. The Convention provides a comprehensive so-called cradle to grave legal regime covering the entire life of a ship and addressing not only environmental but also human health concerns.

The work leading up to the Hong Kong Convention started within the MEPC in 1998. It was recognised then that the IMO had a role to play in the regulation of ship-breaking, which was given the new name of ship recycling, apart from and in addition to the efforts undertaken by the Basel Convention Secretariat and the ILO. This included the preparation of a ship before the start of recycling operations and acting in a coordinating role. The goal of the Convention is to specifically regulate ship recycling globally in terms of safety, environmental and human health issues efficiently and effectively. The regulatory framework addresses design, construction and operational requirements consistent with safety and environmental protection principles to be observed by ship recycling facilities at the same time allowing them to maintain efficiency. To put it another way, one aim of the Convention is to ensure compliance with the Convention by ship recycling facilities and to establish satisfactory control and enforcement measures.

The Convention comprises two components. The first consists of the Preamble and 21 Articles setting out the broad and basic principles and attendant obligations. The second is the Annex which contains the 'Regulations for Safe and Environmentally Sound Recycling of Ships'. In the Regulations there are four chapters and seven Appendices, which constitute the essence of the substantive law relating to the regulation of ship recycling activities. Chapter 1 contains definitions and other general provisions. Incidentally, the definition of 'ship' is broad enough to encapsulate any type of ship operating or having operated in the marine environment subject to the Convention. The term 'ship recycling facility' is defined as 'site, yard or facility used for the recycling of ships'. Chapter 2 sets out the requirements for the design, construction, operation, maintenance and survey and certification of ships. It also deals with the preparation of a ship for recycling. Chapter 3 provides the requirements for ship recycling facilities in detail, including their authorisation to operate and how they are to be controlled. This chapter also deals with requirements for developing ship recycling facility plans and ESM of hazardous substances on board a ship proceeding to a facility for recycling. Chapter 4 provides for reporting requirements to be complied with before and after a recycling operation. Appendices 1 and 2 contain lists of hazardous materials subject to control under the Convention and Appendices 3 to 7 contain the standard forms prescribed for use in the Convention. There are also a number of recommendatory guidelines, which are non-mandatory but useful in terms of assisting parties to comply with the provisions of the Convention, some of which are quite complex.

There are requirements for the qualification, control and operational management of facilities to ensure that they operate in a safe and eco-friendly manner. It is well known that ships destined for recycling often contain hazardous and toxic materials and ozone-depleting substances which pose serious environmental, safety and health risks. The Convention provides for control measures to address these issues, including the requirement for the development of an 'Inventory of Hazardous Materials' which must be carried on board a ship throughout its life, this being a manifestation of the 'cradle to grave' approach taken by the Convention. The unique control and enforcement mechanism under the Convention is a remarkable feature pursuant to which flag states, port states and recycling states are given different roles consistent with their duties and obligations regarding ships heading for recycling. In this context it is important to note that compliance with the Convention may not always be as satisfactory as may be expected by the global community, simply because such matters as monitoring of activities and authorisation and control of ship recycling facilities are essentially dependent on shore-based efforts and levels of stringency observed by recycling states. Incidentally, the major ship recycling states are developing countries, where national priorities are heavily based on economic policies and considerations at a national level.

The Hong Kong Convention has in many ways made remarkable strides in the field of regulation of ship recycling, but there is still room for improvement. Even so, as a regulatory convention governing this important aspect of ships and shipping, and a first attempt by the international community to grapple with this serious problem through convention law, due recognition should be given to its merits.

16.3.6 Ballast Water Management Convention (BWM)

The International Convention for the Control and Management of Ships' Ballast Water and Sediments, 2004, otherwise referred to as the Ballast Water Management Convention, deals with alien species carried in ships' ballast tanks that are environmentally harmful and sediments left in ships' ballast tanks after the ballast waters are discharged.

The phenomenon of alien species travelling in ballast tanks of ships has been recognised as an environmental problem since the advent of steel-hulled vessels over a century ago. While sea water used as ballast is essential for the safety, stability and efficiency of ships, ballasting also results in invasive species entering the ship in one marine environment and being discharged into the waters of another, causing serious ecological, economic and health hazards. To combat this problem, scientists, mainly in certain developed countries, have been engaged in research and development on a continuing basis. The traditional method of ballast water interchange has not been entirely successful in resolving the problem. In the absence of a universal regulatory framework to address this issue several states have unilaterally introduced their own legal regimes. Initially a proposal was made at the IMO to add a seventh Annex to MARPOL to introduce a regulatory regime that would apply globally, but after considerable debate it was decided that a new and separate convention was a better approach. Thus, the BWM was adopted in February 2004. It is not yet in force. The IMO initiated the 'Globallast' project to provide technical assistance to developing countries to prepare for the legal and practical implementation of the convention when it enters into force for the state concerned.

Article 1 consists of definitions of convention terms. Article 2 requires state parties to attempt to achieve ultimate elimination of the transfer of harmful alien species through BWM, refrain from causing greater harm to the marine environment, the property and resources of any state, and human health, and to follow IMO recommendations regarding threats and risks relating to BWM outside national jurisdictions. Article 3 contains a list of parties subject to the Convention. Article 4 requires ships of state parties to comply with the Convention, including its Annexes, and state parties to develop national policies for BWM. Article 5 requires state parties to have adequate reception facilities for sediments in ports and terminals where ballast tanks are cleaned or repaired. Article 6 encourages state parties to promote scientific and technological development on BWM, make related information available, and monitor its effects within their jurisdictions. Article 7 requires state parties to ensure that their ships are surveyed and certificated in accordance with the Regulations in the Annex. Article 8 states that violations under the Convention are prohibited, and if evidence of violations is submitted to flag or port state authorities that they must investigate and commence proceedings, where necessary under their national laws, and impose appropriate sanctions. Article 9 provides for port state inspections of ships subject to the Convention, which are to include examination of Ballast Water Record Books, verification that valid certificates are on board, and sampling of ballast water. Article 10 requires state parties to co-operate in the detection of Convention violations and enforcement of the Convention.

16.3.7 Anti-fouling Systems

The International Convention on the Control of Harmful Anti-fouling Systems on Ships (AFS) was adopted in 2001 and entered into force on 17 September 2008.

The Convention stems from the recognised need to eliminate marine growth accumulating on the steel hulls of ships through means that are less environmentally harmful. Marine growth consists of natural organic deposits which can potentially have serious implications for shipping. The economic impact of this natural phenomenon is manifested in significant reductions of ship speed, depending on the volume of the marine growth, resulting in corresponding reductions in the ship's earning capacity. In practical terms, lack of speed imposes constraints on a ship's ability to meet, for example, charter party requirements relating to arrival, loading, and discharging times for cargo.

The phenomenon was first discovered when steel hulls of ships came into vogue in the early twentieth century. When its commercial impact was realised, the shipping industry resorted to the use of certain varieties of anti-fouling paints which contained organotin compounds acting as biocides. These were found to be harmful to the flora and fauna and other organisms in the marine environment through which the ship traversed. The anti-fouling systems also produced the risk of chronically toxic substances entering the waters, causing ecological damage and having harmful effects on human health. This environmental detriment thus generated another serious implication for the shipping industry. The AFS Convention was designed to promote eradication of marine growth through the use of alternative, less harmful or totally harmless substances in anti-fouling systems replacing the ones that resulted in notably harmful consequences for both the marine environment and human health.

16.3.8 Nairobi Convention on Wreck Removal

The Nairobi International Convention on the Removal of Wrecks was adopted in May 2007 but is not in force as yet. That will happen, if it does, twelve months after ten states have ratified or acceded to the Convention.

The work of the IMO in arriving at this Convention took an inordinately long time, for various reasons. It is apparent that the subject was not considered to be of imminent importance in the years following its initiation. Expediency in its development was precluded by other pressing issues entering the IMO Agenda and especially the envisaged priorities of the Legal Committee. As a result, deliberations on its development were periodically suspended. The shipping industry doubted the need for such a convention, which in their view addressed a subject that could be adequately covered by existing insurance mechanisms in so far as a shipowner's liability for a wreck was concerned. However, coastal states expressed concerns over situations where a ship did not carry adequate insurance or carried no insurance at all. Except for the ship-source pollution conventions relating to liability and compensation for pollution damage, there was no global regime for compulsory third party liability insurance in respect of ships purporting to call at ports or offshore terminals or to enter the maritime zones of coastal states where such states were entitled to exercise jurisdiction over pollution matters. Individual coastal states had started to institute the requirement for compulsory third-party liability through national legislation but it was not a widespread practice.[24]

Under the usual hull and machinery (H&M) insurance cover, a shipowner as the assured can claim for a constructive total loss (CTL) in respect of a wreck. In such instance, under the doctrine of abandonment, the assured issues a notice of abandonment by virtue of which the owner releases the wreck and the associated legal interest in favour of the insurer and the insurer provides indemnification for a total loss.[25] However, as a result of this, the insurer in effect becomes the owner of the wreck and is subject to the liabilities associated with its removal. Insurers, therefore, are reluctant to acknowledge receipt of a notice of abandonment. This raises legal implications which are treated differently by common and civil law courts.

In the absence of any mandatory international regime, wreck removal costs were and continue to be borne by port authorities or other branches of coastal state governments out of necessity for navigational safety reasons and prevention or avoidance of pollution. Several coastal states therefore see merit in an international convention regime, but are reluctant to submit themselves to convention rules applying in their territorial seas and port approaches where they already have fairly comprehensive wreck removal legislation. On the other hand, the Convention being applicable up to the outer limits of the EEZ, or up to 200 nautical miles from the baselines, which is a notable attribute of the Convention, is attractive to them due to limitations on their jurisdiction to that extent under international law. The shipping industry felt that the reach of the Convention extending to the EEZ was unnecessary because wreck removal problems were concentrated on waters close to port and harbour approaches, especially in fairways and navigation channels. Eventually compromises were made in the Convention under which coastal states can apply Convention provisions in their territorial seas at their discretion; otherwise their domestic laws would apply. But the limits of liability and compulsory insurance provisions under the Convention would remain applicable in those waters.

The Nairobi Convention applies in situations where a physical or environmental hazard exists as a result of the existence of a wrecked ship or cargo. In terms of compulsory wreck removal certification, the Convention apples to ships of 300 gross tonnage and above, which translates into over 70,000 ships, which obviously places an onerous responsibility on state parties.[26] The Convention provides for reporting requirements and measures for locating, marking, and determining the hazard profile of a wreck. State parties must facilitate wreck removal by owners or a government authority in the event of non-compliance by an owner or where urgent action becomes necessary. The liability regime for shipowners is strict liability, except where there may be conflict with the obligations in other ship-source pollution conventions concerning liability and compensation. Under the Convention, shipowners are entitled to limit their liability according to any international regime such as the LLMC or any applicable national limitation regime. Claimants have direct action rights against insurers up to the amounts set out in the wreck removal certificate. Ships subject to certification under the Convention must carry insurance at least in accordance with LLMC 1996 levels, which must be evidenced by certification approved by a state.[27]

Whether or not the Nairobi Convention will eventually enter into force remains to be seen. At present it is not possible to speculate, but perhaps acceptance of the Convention by the maritime community at large would be forthcoming if the scope of application was limited to the territorial sea. The counter argument, of course, is that parties do have the discretion to limit the application of the Convention to their territorial seas.

16.4 Sanctions and the penal law dimension

It is well recognised that all regulatory instruments fall within the domain of penal law. In other words, non-compliance must be punishable by the imposition of sanctions. However, there are issues in this regard which need to be elaborated contextually. First, it is pertinent to note that regulatory conventions, including those dealing with ship-source pollution, such as the ones discussed in this chapter, provide for violations but do not prescribe sanctions except for detention of ships which are deficient in terms of safety or seaworthiness or pose marine environmental risks in respect of themselves and others. Violations typically found in regulatory conventions must be transformed into offences through domestic legislation and appropriate sanctions must be provided to make them effective in practice. This leads to the issue of how such an offence should be characterised, particularly in the light of the adage that the punishment must fit the crime.

The characterisation of an offence and its attendant sanction may be regulatory or criminal depending on its severity. In the common-law system, regulatory offences, being less serious, are distinguished from criminal offences. They are also referred to as public welfare offences. As the description implies, these are acts that offend or are detrimental to the welfare of the public. In respect of such offences no proof of *mens rea* is required; the prosecution need only prove the *actus reus* of the offence to succeed. But if such an offence is characterised simply as a strict or absolute liability offence, the sanction, which must be commensurate with the seriousness of the offence, often does not provide an adequate measure of deterrence and the perpetrator has little or no incentive to comply with the law.

Ship-source pollution violations are typically characterised as public welfare offences in the legislation of numerous common-law jurisdictions. Unless the offence leads to serious consequences, it cannot be characterised as a *mens rea* offence, which at any rate would be difficult to prove. A solution to this dilemma is found in the alternative characterisation known as the 'halfway house' used by courts in Canada and Australia, and in United Kingdom legislation giving effect to MARPOL violations.[28] Under this method of dealing with a ship-source pollution offence, in the first instance the offence is treated as a strict or absolute liability offence but the accused is then afforded a defence to prove on a balance of probabilities that he exercised due diligence to avoid committing the offence and that he did not act negligently. If he is able to discharge his reversed burden to the satisfaction of the court, he is exonerated; otherwise a more severe penal sanction can be imposed than he would have otherwise received, if it was a strict or absolute liability offence.[29]

16.5 Summary and conclusion

In this chapter an attempt has been made to provide the reader with an appreciation of the international regulatory regime of the law of ship-source pollution derived from the conventions in the marine pollution spectrum and by reference to the continuum of ship-source pollution serving as the basis for discussion. The theoretical underpinnings, including identifying the sources and categories of ship-source pollution, have been presented as preliminaries to the main body of the chapter. An explanation of the precise nature of regulatory law has been presented and the connection with Part XII of UNCLOS, which serves as the constitutional framework providing the 'blueprints' for specific conventions of the IMO and one of UNEP, has been discussed. The regulatory conventions in question have been reviewed in as much detail as is deemed to be relevant in view of the context. The MARPOL Convention has been discussed quite thoroughly, in particular the provisions of Annex I, viewed as one of primary importance, and Annex VI, which has come into the forefront of considerations at the IMO because of concerns with pollution generated by exhaust emissions from ships and the resulting knock-on impact on global warming. The other regulatory instruments appearing in the spectrum have been addressed, some in more detail than others, and the penal law implications of regulatory conventions have been touched on to bring closure to the substantive elements of the chapter.

It should be apparent to the reader that there is still a considerable amount of unfinished business associated with this subject matter. Ship-source pollution is a matter of ongoing concern periodically inflamed by incidents provoking public outrage and calls for accelerated action at the international level. Shipping activities and ship operations increasingly involve environmental implications. As is evident from the numerous amendments to and protocols for the regulatory conventions in the field of ship-source pollution which have been adopted over the years and the continuing work of the IMO in the MEPC, much remains to be achieved. This is particularly relevant to the continuing deliberations on Annex VI of MARPOL. The BWM Convention has come close but has not yet entered into force despite the concerted efforts expended at the IMO. The Hong Kong SRC still has a number of issues that need particular attention paid to them for it to attract the requisite international support to enter into force. Discussions between the IMO and the Basel Secretariat on the scope of application of MARPOL vis-à-vis the Basel Convention are continuing and the Nairobi Convention on Wreck Removal has not yet attracted sufficient support to allow it to enter into force.

It would be fair to conclude, in the light of the discussions in this chapter, that the regulatory law of ship-source pollution is still in a state of evolution in many respects, but the prognosis is positive, given the achievements to date of the international maritime community at the IMO and its continued interest and efforts.

Acknowledgements

The authors gratefully acknowledge the research assistance of Ms Huiru Liu of Dalian Maritime University in the preparation of this chapter.

Acronyms

BCH	Construction and Equipment of Ships Carrying Dangerous Chemicals in Bulk
BWM	Ballast Water Management
CTL	Constructive Total Loss
EEDI	Energy Efficiency Design Index
EEZ	Exclusive Economic Zone
ESM	Environmentally Sound Management
GHG	Green-House Gases
H&M	Hull and Machinery (in insurance)
IBC	International Bulk Chemical
ILO	International Labour Organization
IMO	International Maritime Organization
IOPC	International Oil Pollution Compensation fund
IOPP	International Oil Pollution Prevention
OILPOL	International Convention for the Prevention of Pollution from Oil
OPRC	Oil pollution Preparedness, Response, and Co-operation
MAP	Mediterranean Action Plan
MARPOL	International Convention for the Prevention of Pollution from Ships
MEPC	Marine Environment Protection Committee
NLS	Noxious Liquid Substance
SOLAS	international convention for the Safety Of Life At Sea
SOPEP	Shipboard Oil Pollution Emergency Plan
SRC	Ship Recycling Convention
TB	Trans-Boundary movement
UNCLOS	United Nations Convention on the Law Of the Sea

Endnotes and references

1. International Convention Relating to Intervention on the High Seas in Cases of Oil Pollution Casualties, 1969.
2. International Convention on Civil Liability for Oil Pollution Damage, 1969.
3. International Convention for Prevention of Pollution of the Seas by Oil, 1954.

4. International Convention on Prevention of Pollution from Ships, 1973 Modified by Its Protocols of 1978 and 1997 Together with Several Amendments.
5. Convention on the Prevention of Marine Pollution by Dumping of Wastes and Other Matter, 1972 Amended by Its Protocol of 1996.
6. International Convention on Oil Pollution Preparedness, Response and Co-operation, 1990 and OPRC-HNS Protocol, 2000.
7. Basel Convention on the Transboundary Movement of Hazardous Wastes and Their Disposal, 1989.
8. Hong Kong International Convention for the Safe and Environmentally Sound Recycling of Ships (SRC), 2009.
9. International Convention for the Control and Management of Ships' Ballast Water and Sediments, 2004.
10. International Convention on the Control of Harmful Anti-fouling Systems on Ships, 2001.
11. OILPOL 1954; *see supra*, note 3.
12. Notably, in the civil law system the terms 'regulation' and 'regulate' are used in respect of statute law in general. In European Union law, 'regulation' has yet another particular connotation, the consideration of which is beyond the scope of this discussion.
13. Examples are the International Bulk Chemical (IBC) Code, the International Gas Carrier Code (IGC) Code and the International Code for the Construction and Equipment of Ships Carrying Liquefied Gases in Bulk.
14. See Peter Leyland and Gordon Anthony, *Textbook on Administrative Law*, Seventh Edition, Oxford University Press, 2012 for further explanations.
15. For a synoptic overview of MARPOL, OPRC, London, Basel, AFS, and the BWM Conventions, see Proshanto K. Mukherjee and Mark Brownrigg, *Farthing on International Shipping*, 4th edn, Springer Publications, 2012, pp. 274–94.
16. See Introduction to MARPOL Consolidated Edition, 2011, published by the IMO.
17. See UNCLOS Article 211 (6), in particular subparagraph (a).
18. See Annex I Regulation 1 (11) and Annex V Regulation 1 (3).
19. See Regulation 4 of Annex I, which cross-refers to Regulations 15 and 34. These two Regulations explicitly prescribe the discharge standards.
20. The ground rules regarding port state jurisdiction are set out in Article 218 of UNCLOS. One important feature of this provision is that port state jurisdiction can be asserted only if the ship in question enters a port or offshore terminal of the coastal state voluntarily. This implies that such entry under *force majeure* conditions will not attract port state jurisdiction and port state control cannot be exercised for blacklisting or 'greylisting' but that does not prevent a ship from being detained if it has become unsafe or is a threat to the marine environment, until the relevant deficiencies are rectified. If it is not possible to fix the deficiencies in that port, the ship may be allowed to proceed to the nearest repair yard. See UNCLOS Article 226, paragraph 1(c).

21. See Proshanto K. Mukherjee and Mark Brownrigg, *Farthing on International Shipping*, 4th edn, Springer Publications, 2012, p. 285, and pp. 290–3.
22. Maritime Reporter, 23 October 2014.
23. See Article 4, captioned 'General Obligations of the Basel Convention'.
24. Several Caribbean states, in revising their national merchant shipping legislation following the IMO Model Maritime Code, included compulsory third party liability insurance provisions as a condition for foreign ships to enter their ports or offshore terminals.
25. Rubina Khurram, 'Total Loss and Abandonment in the Law of Marine Insurance' (1994) JMLC, Vol. 25, No. 1, p. 95.
26. Proshanto K. Mukherjee and Mark Brownrigg, *Farthing on International Shipping*, 4th edn, Springer Publications, 2012, pp. 327–8.
27. Proshanto K. Mukherjee and Mark Brownrigg, *Farthing on International Shipping*, 4th edn, Springer Publications, 2012, pp. 327–8.
28. R. v. City of Sault Saint Marie, [1978] 40 Can. Crim. Cas. (2nd) 353 (S.C.C.).
29. For a detailed discussion, see Proshanto K. Mukherjee, 'The Penal Law of Ship-Source Marine Pollution: Selected Issues in Perspective' in *Liber Amicorum Judge Thomas A. Mensah: Law of the Sea, Protection of the Marine Environment and Settlement of Disputes*, September 2007, Max Planck Institute for Comparative Public Law and International Law, pp. 463–96.

Chapter 17

Foresight application for transport sector

Joanna Ejdys, Joanicjusz Nazarko*, Łukasz Nazarko* and Katarzyna Halicka**

17.1 Introduction

The unpredictability of the conditions in which organisations operate – resulting from the dynamics and complexity of the environment – is the central issue of strategic management. Uncertainty about the future of enterprises, industries and sectors forces managers to look for new instruments and methods, on the one hand, to determine future states and, on the other hand, to create that future. In the case of a variable and unpredictable environment, a manager can choose one of three paths: (1) adopting a passive attitude – thus losing market position, (2) using past practices, often not adapted to current conditions, (3) shaping their future in a systematic, structured way, while taking advantage of upcoming opportunities in the environment [1]. If one chooses the third path, the application of foresight as a valuable instrument of the 'creation' of the future should be considered. The primary objective of foresight is to anticipate and shape the future strategies of organisations, regions or branches. It strengthens the unit's ability to anticipate emerging opportunities and threats in the environment, to identify dependence paths (networks), and key success factors (drivers).

The experience gained from the utilisation of foresight studies accumulated up to date indicates a huge potential for its further application. Foresight has been used successfully in the process of building coherent and functional images of the future at national and regional levels as well as those of industries and individual organisations. The scope of foresight research includes not only the future of technology and scientific research but also the future of various other dimensions of the socio-economic reality. Transport systems and technological solutions used in the sector have been subjects of interest for numerous foresight studies.

The aim of this chapter is to present good practices and potential benefits of using foresight studies in the process of creating the future of the broadly understood transport sector. To begin with, the authors define the concept of foresight and the characteristics of foresight studies. Next, they point to potential areas of

*Białystok University of Technology, Faculty of Management, Poland

application, leading to the typology of foresight studies. Finally, they review the existing applications of foresight research in the transport sector. The chapter is concluded with the indication of the four groups of potential benefits of foresight studies at the sectoral level.

17.2 The essence of foresight

In times characterised by high dynamism of the environment, inference about the shaping of phenomena in future is associated with the need to process and interpret information from multiple sources. In the long term, traditional methods of forecasting, owing to their assumptions concerning the maintenance of the status quo of the factors influencing a predicted phenomenon, may be unreliable. Foresight is the method which allows for predicting the future, taking into account the variability of the environment. Foresight is a systematic process, shaping the future of technology, science, economy, environment and society. Its main objectives are to identify the emerging trends, phenomena and technology, and to strengthen the areas of strategic research offering significant potential for economic and social benefits for a dozen, or several dozen, years.

The literature on the subject contains many definitions of foresight. However, this term was used first in 1985 by J.F. Coates, who defined 'foresight' as 'the process by which the forces shaping the long-term future can be understood, and which should be taken into account in policy-making, planning and decision-making'. Table 17.1 shows selected definitions of the term.

According to the authors the accurate definition of this concept is presented in the Polish literature. According to this definition, foresight is understood as 'the process of creating the society's culture of thinking about the future, in which stakeholders, that is scientists, engineers, representatives of the industry and public administration employees are involved in setting the strategic directions for research and technological development' [8].

Taking into account all the above definitions, it can be noted that the significant attributes of foresight include (Figure 17.1):

- *Anticipating*: coming down to predicting social, economic and technological needs, and predicting and designing the directions of their development.
- *Inclusivity*: being the participation of a wide range of diverse participants (i.e. the stakeholders) in the process of formulating a socially agreed vision of the future.
- *Social networking*: focusing on the formation of relationships between stakeholders of the research aimed at supporting the process of collecting and processing of the information coming from the environment.
- *Strategic vision*: being the desired future, based on the rational prerequisites of the present.
- *Activities*: consisting in mobilisation to particular steps towards the desired future.

Table 17.1 The diversity of the definitions of 'foresight' – selected explanations of the term

Foresight is . . .	
the process by which the forces shaping the future can be understood, relevant to policy, planning and decision-making. Both qualitative and quantitative methods are involved in this process in order to monitor the signals on emerging trends that are the implications for the policy. In the face of the passage of time and changes in conditions implementation of this policy becomes, therefore, more appropriate, flexible and more reliable. Foresight is not planning, but only its element [2]	J.F. Coates
the process, which aims to predict the long-term future of science, technology, economy and society by identifying strategic areas of science and technology to ensure the maximum economic and social benefits [3]	B. Martin
a systematic manner of assessing of the development of science and technology, which can have a big impact on the competitiveness of enterprises, creating wealth in the economic sense, and the quality of life [3]	L. Georghiou
the ability to create and maintain the high quality of coherent vision, and to use it in a manner, which is to provide measurable benefits to organisations by detecting adverse environmental conditions, support of the development of the organisation's policy, and formation of strategy in order to examine new markets, products and services [2]	R.A. Slaughter
a deliberately structured process that combines the expectations of different actors towards technologies in order to formulate the strategy of future, which takes into account the broader economic and social development [4]	A. Webster
a systematic, participatory process of building a medium- and long-term vision, aimed at today's decisions and mobilisation of joint actions [5]	I. Miles
active predicting of the future, and therefore not only identifying the possible future events, but also planning appropriate actions that will allow for making best use of what is going to happen and at the same time minimising the potential risks [6]	M. Dietrich
intense, recurring periods of open reflection, creation of networks, consultation and discussion, leading to a common development of the vision of the future and a sense of partnership built around this vision, in order to use long-term opportunities created by the development of science, technology and innovation. It is discovering a common space of open thinking about the future and incubation of strategic approach [7]	J. Cassingena Harper

Source: Own work.

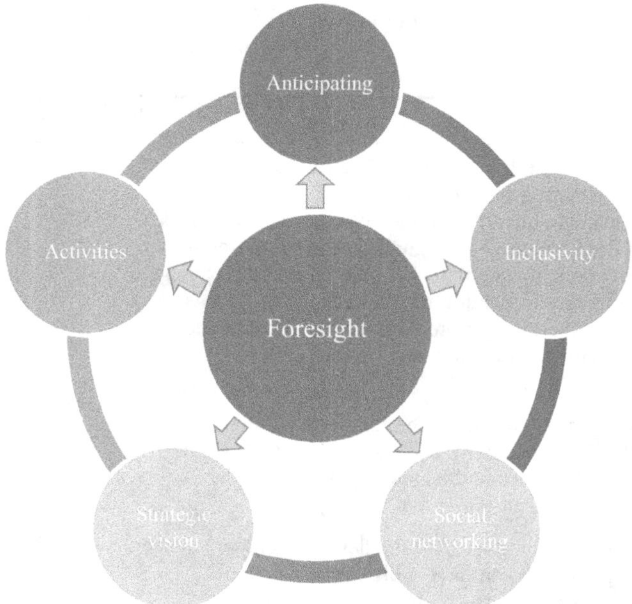

Figure 17.1 Features of foresight
[Source: own elaboration on the basis of Reference 5.]

Foresight is often wrongly identified with forecasting. However, the aim of foresight is not prediction, but 'the determination of certain (desirable) visions of development and shaping the future in the social process' [9]. A forecast is only a single image of reality, the point in the area of uncertainty, and therefore can safely be applied to phenomena that are less sensitive to disturbing events [10]. On the other hand, foresight, through the use of scenarios, allows the study of the behaviour and interaction of trends in a large area of uncertainty. It allows a broader look at the phenomenon and its consideration from a greater distance (usually 5–30 years) than forecasts. In contrast to the forecast, it takes into account the discontinuity of events through the identification of unprecedented events, the so-called wild cards and weak signals [11]. This way it is possible to study the future more comprehensively. Foresight is not a substitute for forecasting or strategic planning [12]. Each of these plays a separate role and often complements the others.

In order to examine the frequency of the use of foresight a database of scientific publications, Web of Science was analysed. The choice of the database was dictated by its availability. Basing on the search of databases using keywords such as *foresight*, in the last ten years, 1,956 articles were identified (Figure 17.2).

The growing interest in foresight studies is also confirmed by the number of uses of the term 'foresight' in bibliographic databases of scientific journals (Figure 17.3).

After a thorough analysis of the literature, it can be concluded that foresight is a combination of intuition, research methods, anticipative analysis and examination

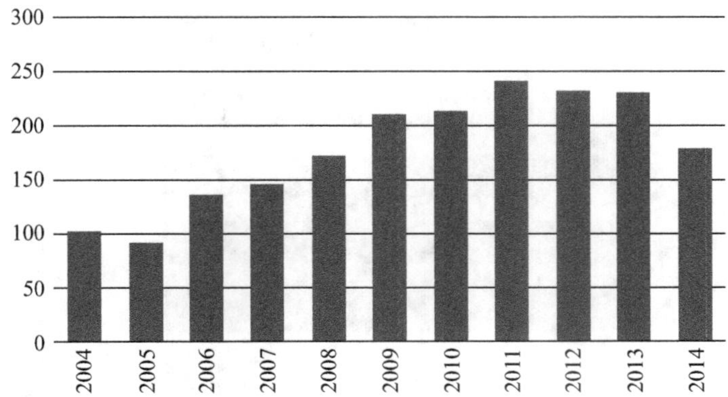

Figure 17.2 *Number of publications in the database and Web of Science on foresight*
[Source: own elaboration.]

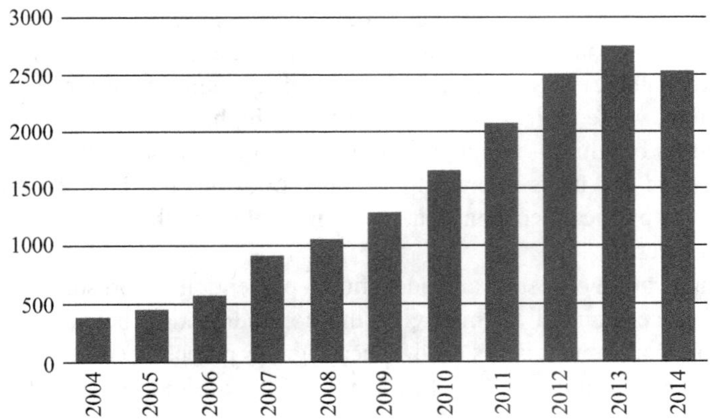

Figure 17.3 *The number of uses of the term 'foresight' in the Web of Science in the years 2004–2014*
[Source: own elaboration.]

of the development of trends. It should be an ongoing process and is usually carried out in the form of research projects [13]. The aim of implementation of the foresight approach is, among other things:

- Determining the direction of scientific policy – drafting policy guidelines and determining the plan for the use of funds for research and development.
- Determining the priorities aiming at the identification and selection of the most promising options from the entire spectrum of possibilities – provision of rational prerequisites determining the allocation of funds for research, identification of needs for scientific apparatus, or the assessment of future requirements for qualified scientific personnel.

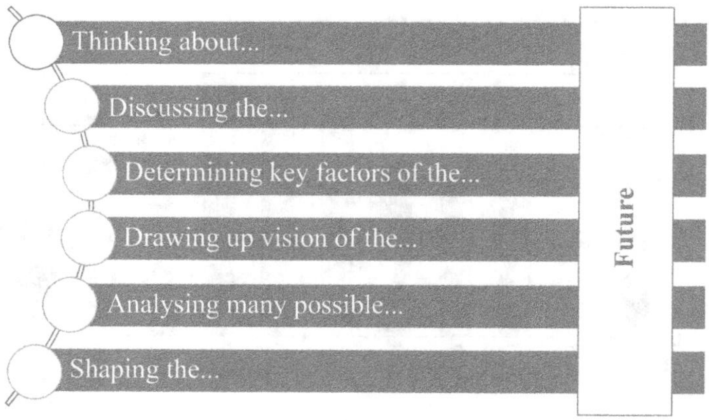

Figure 17.4 The purpose of foresight studies
[Source: own elaboration.]

- Anticipating trends in science and technology – providing information on the opportunities and risks of future scientific research, and identification of unprecedented events, that is, events of which the probability of occurrence is negligible, while their impact would be very high.
- Consensus building – stimulating the involvement of scientists, various institutions, and end users in order to establish priorities for the R & D sphere.
- Promoting political decisions that are compatible with the preferences of R & D stakeholders.
- Communication of results and education – perception of foresight research as a social process aimed at improving the communication between participants and potential beneficiaries, and educating the potential stakeholders and all those who have influence on the creation of the desired vision of the future of science and technology.

The purpose of foresight studies is primarily (Figure 17.4): (1) thinking about the future (analysis of the situation and development trends); (2) discussing the future (taking into account the views of different environments); (3) determining the key factors (e.g. political, economic, social, cultural, environmental, technical); (4) drawing up visions of the future (scenario building); (5) consulting – analysing many possible 'futures'; (6) shaping the future (recommendations).

It should be emphasised that foresight research should be carried out in compliance with the triangulation principle: theoretical, of methods, researchers and data sources [13]. Theoretical triangulation is the use of different theories for interpretation of the phenomenon. Methodological triangulation involves using multiple test methods to evaluate the same phenomenon [14]. The triangulation of researchers is based on the use of multiple stakeholders, representing various scientific communities, for the interpretation of the same phenomenon.

Data triangulation is based on the use of data from various sources for the analysed phenomenon [15].

Foresight is a future-oriented study, aimed in particular at informing, among others, policy makers and other opinion-forming environments about the desired direction of long-term growth and change, thus facilitating the formulation of innovation policies and strategies [16]. The foresight projects' results help those involved in agreeing upon scenarios of future development and harmonising the activities of social partners (government, local government, academia and business, small and large companies, sectors of the economy), and provide assistance in establishing the criteria for funding science and technology [17].

Owing to the lack of a commonly accepted equivalent of the word 'foresight' in Polish and in many other languages, the English term is used in the literature. The most accurate possible meaning of the term 'foresight' may be its translation as 'active building of an image of the future'.

17.3 Types of foresight initiatives

Taking into account the primary objective of the foresight studies, which is to create the future in a systemic and participatory manner, the scope of their applications, and thus the typology of research, are very extensive. Despite the fact that in the world the vast majority of foresight projects were and still are initiated and financed by the state [18], foresight is being used increasingly both at organisational and at governmental level. A classification of foresight studies can be made taking into account the institutional, territorial and conceptual dimensions of foresight. The institutional perspective reflects the type of entities interested in developing and implementing the foresight research results (companies, local authorities). The territorial perspective reflects the spatial object (country, region), which is of interest to the foresight research. The conceptual dimension of foresight studies relates to the subject of the research, that is the search for answers to questions about future states of a reality (technology, development of the industry, the transport system, the development of the health system, education, etc.). Table 17.2 presents examples of foresight projects, taking into account their typology.

Many of the identified foresight projects concern the transport sector. The foresight research categories most commonly used in the literature and practice include regional foresight, corporate foresight (strategic or organisational), technological foresight and regional economic foresight (business information foresight).

In 'A Practical Guide to Regional Foresight', developed by FOREN, *regional foresight* is defined as the implementation of anticipation, participation, network structure, vision and actions on a limited territorial scale, where proximity is a determining factor [19]. Regional Foresight is a specific process of creating a vision and undertaking action within a limited territory characterised by a specific concentration of growth factors [20]. The main objectives of foresight activities carried out at the regional level are analysed in two areas: facilitating the

Table 17.2 Foresight research typology with examples

Foresight dimension	Criteria	Examples of foresight initiatives
Territorial	Supranational	Creating prospective value chains for renewable road transport energy sources in Nordic countries Sectoral Innovation Foresight: (sector: automotive, food and drink, knowledge-intensive services, aerospace, and wholesale and retail) Transportation & Logistics 2030
	National	Competitiveness Monitor: an Integrated Foresight Platform for the German Leading-edge Cluster in Logistics Future of Super Intelligent Transport Systems for the Netherlands Forecasting of long-term innovation development in Russian economic sectors: results, lessons and policy conclusions (sector: the aircraft industry, commercial shipbuilding)
	Regional	Future Fuel Technology for APEC Regions
	City/metropolis/district	Research, Technology and Innovation Policy in Vienna Technology for Industry Foresight – Kocaeli 2012 An Innovation System for the Manchester City Region Imagine Dublin 2020
Institutional	International organisation	FreightVision UE
	Private sector organisation (strategic, corporation foresight)	Shell energy scenarios to 2050
	Government (central, local)	FreightVision Austria 2050
	Non-government organisation	Foresight on Advanced Technologies in Poland Corporate Foresight in Small and Medium-Sized Enterprises
Conceptual	Technology/sectoral/branch/ industry	Démarche Prospective Transport 2050. For a Better French Transport Policy
	R & D	Research and Innovation Futures RIF VERA – Forward Visions on the European Research Area
	Socio-economic development	The Future of Learning: New Ways to Learn New Skills for Future Jobs (A Foresight Study)
	Selected part of reality	Quality and Leadership for Romanian Higher Education (QLHE) Foresighting Food, Rural and Agrifutures in Europe Future Strategies for Ageing Management in the Working World of Salzburg Province
	Business information foresight	Regional business information foresight. Scenarios of the innovation development of the enterprises of Mazovia

Source: Authors' own elaboration based on database http://www.foresight-platform.eu.

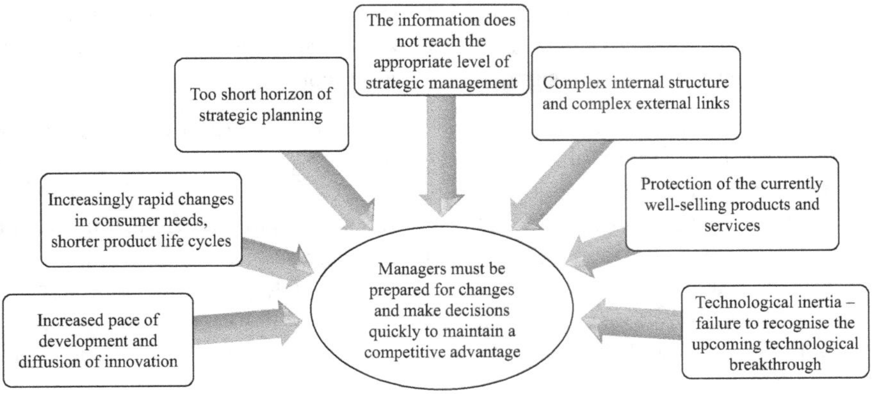

Figure 17.5 Why do companies need foresight?
[Source: Reference 22.]

implementation of the policy of development and the mobilisation of regional stakeholders [13].

Corporate foresight is a systematic, participatory process taking place in an enterprise, which aims to support strategic decisions through anticipation, analysis, and interpretation of long-term directions of social, economic and technological development, as well as actively shape alternative visions and ideas about the future [21].

Branch (Sectoral) foresight is a nationwide initiative (less often regional), analysing and anticipating possible future scenarios and the development of technology in a given sector of the economy, sponsored by the state, and sometimes by research institutions, industry associations and individual companies. An example of industry foresight could be foresight in just the transport sector. The main reason for companies to employ foresight is the fact that managers must be prepared for changes and make decisions quickly to maintain a competitive advantage (Figure 17.5).

Technology foresight is a systematic means of assessing those aspects of scientific and technological development which could have a strong impact on industrial competitiveness, wealth creation and quality of life [23]. Technology foresight involves long-term views of technology-related developments, aimed at informing ongoing decision-making, and using a wide range of formal techniques, an interdisciplinary approach, the involvement of wise pools of experts but also stakeholders, and creating networks that promote a shared view of the developments discussed, along with joint commitment to the approaches or actions emerging from these exercises [24].

Business information foresight is aimed at setting strategic directions and defining areas of economic development, affecting the development of the region. The concept was introduced by J. Nazarko, who pointed out that business information foresight can be identified by determining its informational function as relating mainly to economic information. The determinant of the regional nature of foresight in this case is the spatial dimension referring to the processes occurring or having an impact on the development of a regional system [13].

Regardless of the application field, the foresight research tool is associated with the system of innovation, both at regional and at organisational level. In this context, the objectives of foresight studies mainly concern:

- Future research, in terms of determination of the RDI policy priorities (research-development-innovation policy).
- Reorientation of the innovation strategy.
- Indicating the technological capabilities necessary to implement the innovation strategy which is a source of competitive advantage.
- Stakeholder involvement in the creation of the policy of innovation and the implementation of innovation strategies.
- Building a network of connections within the innovation system (the triad: science-business-government).

The chapter presents the extent to which foresight studies were used in the process of creating the broadly understood transport industry. Authors have indicated good practices in this area, and the main results of the foresight research.

17.4 Examples and good practice of foresight application in the transport sector

Issues of mobility, transport and logistics constitute an attractive research area for various kinds of forward-looking activities. The future of moving people and goods from one place to another may be seen as a variable that is heavily dependent on a wide array of heterogeneous factors. This provides a space for an interdisciplinary future-oriented reflection where various economic, social, technological, environmental, political, legal etc. trends and phenomena are taken into consideration. At the same time, the future of mobility and transport is a variable that strongly influences the development trajectories of human civilisation as a whole. It is not surprising, therefore, that companies, governments and other organisations have been conducting foresight studies in the field of broadly understood mobility.

The aims and scope of foresight initiatives depend mainly on the actors performing (sponsoring) such studies. This general observation also applies to future-oriented studies in the field of transport, logistics and mobility. Companies that maintain any sort of a foresight activity (from simple systems based on indicators and extrapolations, through causality analysis and weak signals identification, to integrated foresight systems that convert future insights into strategic actions at the managerial level) see it as a means of achieving competitive advantage and increasing shareholder value in the long term. Public entities (government agencies, EU bodies), in turn, usually see foresight as an effective policy intelligence tool helping to design regulations, incentives and procurement programmes that will push the mobility sector in the direction of a desirable future. In this context, Figure 17.6 complements the foresight typology presented in Table 17.2 (specifically, foresight's institutional dimension).

Figure 17.6 Aims of foresight as a derivative of a foresight sponsor
[Source: authors' own elaboration.]

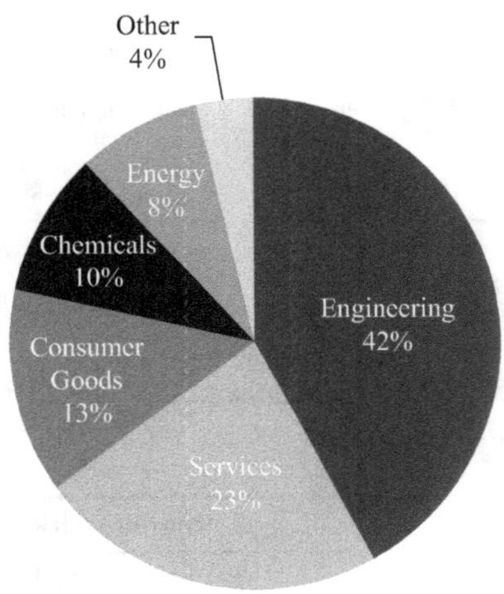

Figure 17.7 Multinational companies participating in the corporate foresight benchmarking study
[Source: Reference 29.]

17.4.1 Corporate foresight in mobility, transport and logistics

It is quite understandable that companies do not eagerly disclose all their foresight work, as this may deprive them of the advantage in the highly competitive and fast-changing environment. Thus it is not possible to provide a precise answer to the question about the scope and intensity of foresight application by the companies operating in the mobility, transport, and logistics sectors. General studies of corporate foresight show, however, that the transport industry, transport services and logistics companies are among the leaders in terms of corporate foresight use. A benchmarking report on strategic foresight in multinational corporations includes such companies in the two top categories: 'Engineering' and 'Services' (Figure 17.7). Other studies also feature enterprises related to the notion of mobility [25–28].

Some enterprises, however, share the information about their internal foresight processes, presumably in order to build the image of a modern and innovative company and to benefit from the insights and reviews of their work by external specialists. Volkswagen may be seen as a company speaking openly about their foresight methodology (not so much about the results, however). The company's underlying foresight concept is the reconciliation of the two somewhat opposing (but not mutually exclusive) notions: strategy (implying focus and clear orientation) and innovation (implying openness and diffusion). The process was divided into four stages, presented in Figure 17.8.

In the course of the foresight project implementation some good practices and recommendations were developed. They are presented in Table 17.3.

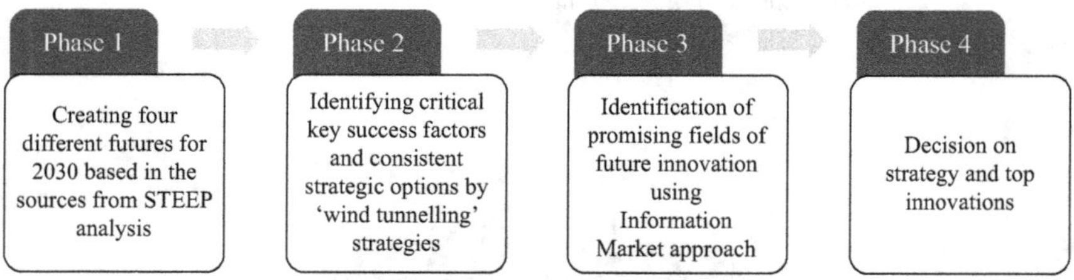

Figure 17.8 Foresight project phases at Volkswagen
[Source: Reference 30.]

Table 17.3 Good practices and recommendations for corporate foresight developed at Volkswagen

Heading	**Principle**	**Recommendations**
Organisational enablers	Futures thinking must be owned and adequately supported.	Supported by top management Allocated sufficient resources and capacity Cross-functional teams Embedded in business units
Process enablers	The process must be pluralistic, engaging and focused.	Clear objectives Sufficiently broad to generate new thinking Accepting of multiple futures Consultative: internally and externally
Implementation enablers	Implementation must be strategic and sufficiently embedded.	Make the process transparent Outputs relevant to corporate strategy Iterative Integrated into planning process Established audience

Source: Reference 30.

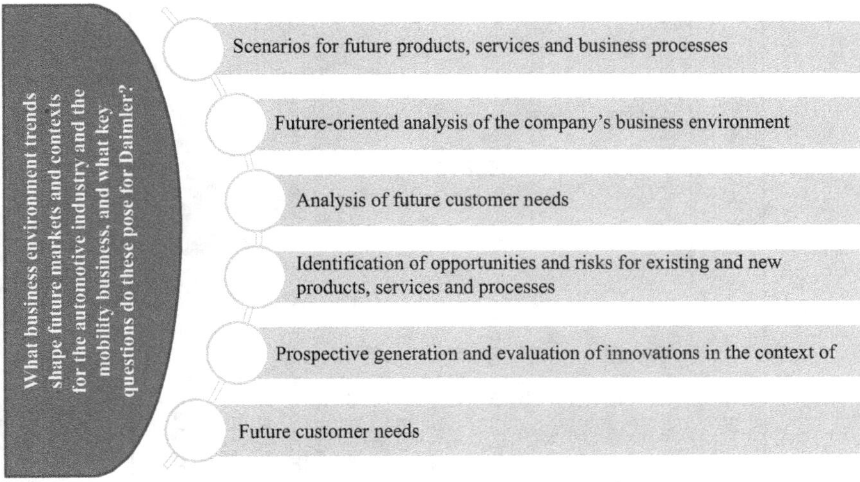

Figure 17.9 Main types of activity at Daimler's Society and Technology Research Group
[Source: Reference 32.]

Another notable example of foresight application in corporate activity is the work of the Daimler Chrysler's Society and Technology Research Group (STRG), which since 1979 has been systematically reflecting on the conceivable long-term trends influencing the automotive market. This makes Daimler one of the earliest founders of a company-based futures research group with a portfolio of more than 300 foresight projects [31]. The STRG underlying philosophy reflects the core foresight approach to the notion of the future, namely that that it is not possible to reliably predict the long-term future. However, it is possible (and necessary) to prepare for an uncertain future by considering possible alternative developments and analysing the forces influencing them. In this light, the STGR quest is to find answers to the question: 'What business environment trends shape future markets and contexts for the automotive industry and the mobility business, and what key questions do these pose for Daimler?' [32]. To accomplish this mission STRG performs five types of activity, presented in Figure 17.9.

The types of activity presented above translate into specific research fields. Examples of such research fields are laid out in Table 17.4.

Similarly to Volkswagen, Daimler's vast experience in corporate foresight has resulted in some lessons learned (collected by F. Ruff) applicable to foresight activities in the mobility sector and beyond (Table 17.5).

These recommendations help face the classic challenge in automobile industry (and, for a matter of fact, not only in that industry), the fact that the companies are forced to take decisions today that will have consequences for the next few decades, owing to the vehicles' life cycles (from design studios to obsolescence) [31].

17.4.2 Sectoral foresight in mobility, transport and logistics

Unlike the corporate sector, where most foresight knowledge is concealed from the general public, government agencies and EU bodies provide a considerable amount

Table 17.4 Foresight research fields at Daimler

Foresight research field	Premise	Examples of research questions	Project example
Mobile societies of the future	Success or failure of the products and strategies of an automotive company depend on their long-term fit with the societal business environments.	• Which societal factors will drive the mobility patterns of the future? • Which relevance and which forms will the mobility of persons and goods have in the future? • Where are fundamental changes and escalation risks recognisable?	Future perspectives of mobility and transport in Asia
Markets and customers of the future	The increasing complexity of markets is driven by changes in target groups and customer needs.	• Where and how will the future (private) customer live? • Who will transport which goods how and where (commercial customer)? • Which requirements will customer groups in changing business environments have?	Needs of new customers regarding a new product segment
Enterprises and value creation of the future	Corporate value creation and business models are subject to continuous change, which is strongly driven by changes in the external business environment of companies.	• Which market developments in other branches could have an impact on the automotive industry? • Which technologies affecting future developments are relevant for future value creation? • How can the collaboration in innovation management be improved?	Informed Employee Workspheres

Source: Reference 32.

Table 17.5 Good practices and recommendations for corporate foresight developed at Daimler

Heading	Principle	Recommendations
Business model	Corporate foresight embedded in and closely connected to core processes of the company	• Distinguishable impact on corporate decisions has to be evident and appreciated by top management. • Foresight agenda and the transfer of results have to be closely connected with core processes of the company. • Funding model which mixes elements of (indirect) corporate research funding with contract-based (direct) assignments by internal customers results in a healthy balance of 'thinking forward' and 'being relevant'.
Diversified customer portfolio	Direct, strongly interactive relationship with a diversified portfolio of internal customers	• Keeping in mind that the demand for foresight is neither continuous nor always growing. • Proceeding to unravel new relevant topics and accessing new customers with these topics. • Corporate foresight has to anticipate the life cycle of relevant topics at the boundaries of the company.
Corporate foresight and science	Corporate foresight practitioners have to keep close connections to the scientific communities	• Foresight in a company is not about doing science and basic research but about creating and brokering specific knowledge types (content) and by designing and conducting structured interaction about business and corporate issues (process).
Methods	Corporate foresight has to invest strongly in the development and validation of methods	• Methodological pluralism and activity is essential but an end in itself. • Corporate customers generally do not want methods but solutions.
Make or buy corporate foresight	Reasonable choice between own corporate foresight unit and external foresight consulting	• For small and medium-sized companies, that have a limited number of external risks and a limited need to capture long-term uncertainty, project-based and externally assigned foresight is usually a reasonable option. • Large and multinational companies with an internal foresight unit can leverage a competitive edge over external services (generation of exclusive knowledge inaccessible to competition, deeper understanding of company's business logic, no 'start-up' costs at every foresight initiative).

Source: Reference 32.

of information on foresight projects they have conducted and sponsored. As shown in Table 17.6, specialised sectoral foresight research focused on mobility, transport and logistics has been conducted for at least a decade (this applies to the specific understanding of foresight presented at the beginning of this chapter; one may be certain that a general future-oriented reflection on mobility and transport – not necessarily called explicitly foresight – has a much longer history).

An interesting recent example of a study exploring the future of logistics is 'Delivering Tomorrow Logistics 2050 – A Scenario Study'. It was conducted by a specialised German consulting company and commissioned by Deutsche Post AG. Five alternative scenarios of the global logistics environment were developed as a result of the project. They are presented in a synthetic form in Table 17.7.

Satisfying the more inquisitive reader, the applied scenario building methodology was diligently described in the report. It was based on the environmental scanning and key factor analysis. The factors influencing the studied subject (i.e. logistics in 2050) were identified, systematised and classified. Eventually, 27 factors (out of 60) were distilled. They were further evaluated by the experts from the point of view of their impact and uncertainty. As a result, 14 key factors that formed the backbones of the five scenarios were determined. Next, the factors were set in a morphological box to project different possible states of each factor in order to create qualitatively different scenario narratives. Additionally, this allowed the validation and enrichment of the scenarios as carried out. Finally, the crucial stage of identifying the implications for the logistics industry was conducted.

17.5 Benefits from using foresight approach

The use of foresight studies in the long term may be the source of many benefits for the transport organisation. The experience of the authors concerning the implementation of foresight studies allows the identification of four basic groups of benefits, which include:

- early warning system
- foresight as a tool of innovation creation
- creation of the current and future image of the organisation
- engagement and participation tool (Figure 17.10)

Foresight as an early warning system allows organisations to:

- prepare for unexpected changes in the environment, both positive and negative
- prepare for the use of future market opportunities
- track changes in the society: consumers and competitors
- overcome current problems by proposing unconventional organisational, economic, business, and technological solutions in the long term, developed as a result of social consensus
- minimise the negative effects of the variability and unpredictability of the environment

Table 17.6 Sectoral foresight and technological foresight exercises related to mobility, transport and logistics

Project	**Authors/leading institutions**	**Years**
Foresight for Transport: A Foresight Exercise to Help Forward Thinking in Transport and Sectoral Integration	ADELPHI Research, University of Cardiff, NESTEAR, ALAMO Online	2004
FreightVision – Freight Transport Foresight 2050	AustriaTech	2008–2012
FreightVision Austria 2050	Federal Ministry of Transport, Innovation and Technology of Austria	2012
Towards a Sustainable Transport System – Supporting Economic Growth in a Low Carbon World	Secretary of State for Transport, United Kingdom	2007
Analysis of factors influencing the development of transport infrastructure until the year 2030 – A Delphi-based scenario study	S.W. Schuckmann *et al.*, 'Technological Forecasting & Social Change', 2012, No. 79, pp. 1373–87	2012
Scenarios of technology development in road construction materials	General Directorate for National Roads and Motorways, Warsaw, Poland	2011–2013
Innovation for our future mobility	European Commission	
Foresight for transport	The Interdisciplinary Centre for Comparative Research in the Social Sciences – ICCR, ADELPHI Research, University of Cardiff, NESTEAR, Grupo ALAMO	2002–2004
Security of Road Transport Networks		2009–2012
STAR-TRANS 'Strategic Risk Assessment and Contingency Planning in Interconnected Transportation Networks'	Project coordinator: INTRASOFT International SA	2009–2012
'Delivering Tomorrow' Logistics 2050 – A Scenario Study	Z-punkt GmbH	2012
Europe INNOVA (transport)	European Commission	

(Continues)

Table 17.6 (Continued)

Project	Authors/leading institutions	Years
Competitiveness Monitor: an integrated Foresight Platform for the German Leading-edge Cluster in Logistics	European Foresight Platform	
Future of Super Intelligent Transport Systems	European Foresight Platform	2011
Transportation & Logistics 2030	European Foresight Platform	2010
Démarche Prospective Transport 2050 – For a Better French Transport Policy	European Foresight Monitoring Network	2007
AGORA 2020 – Transport, Housing, Urbanism and Risk	European Foresight Monitoring Network	2005
Transport and Mobility in an Enlarged Europe – 2020	European Foresight Monitoring Network	2004
Foresight Vehicle Technology Roadmap 2020. Technology and Research Directions for Future Road Vehicles	European Foresight Monitoring Network	2004

Source: Own elaboration.

Table 17.7 Five alternative scenarios of the global logistics environment developed in the framework of the 'Delivering Tomorrow' project

Scenario name	Core idea	Key factors
Scenario 1: untamed economy – impending collapse	When the world is driven by materialism and faces frequent natural disasters.	Materialism, resource exploitation, rising incomes, pollution, climate change, free trade, Arctic mining, Asia, natural disasters, urban congestion, unsustainable growth, consumption.
Scenario 2: mega-efficiency in megacities	When megacities become epicentres of green growth.	Urbanisation, megacities, automation, sustainability, wealth, green growth, collaboration, virtual reality, interconnectedness, efficiency, city logistics, global hubs, urban-rural divide, robotics, technology, rent-and-use, supergrid.
Scenario 3: customised lifestyles	When individualisation becomes pervasive and 3D printing dominates manufacturing and households.	3D printing, FabShops, dematerialisation, global hubs, affluence, creativity, consumption, regional trade, progress, unique lifestyles, decentralisation, prosumption, Internet security, customisation, individualisation, digital piracy, home fabbing.
Scenario 4: paralysing protectionism	When globalisation is reversed and protectionist barriers are raised.	Resource scarcity, contraction, nationalism, productivity decline, ageing societies, customs delays, international conflict, regional blocs, fragmentation, stagnation, security, protectionism, nationalised logistics.
Scenario 5: global resilience – local adaptation	When frequent catastrophes lead to a paradigm shift away from efficiency maximisation to vulnerability mitigation and resilience.	Vulnerability, climate change, supply disruption, regional trade, international collaboration, energy security, backup infrastructure, resilience, decentralised solutions, redundant system, supply security, disaster response, flexible technology, disasters, energy logistics, automation.

Source: Reference 33.

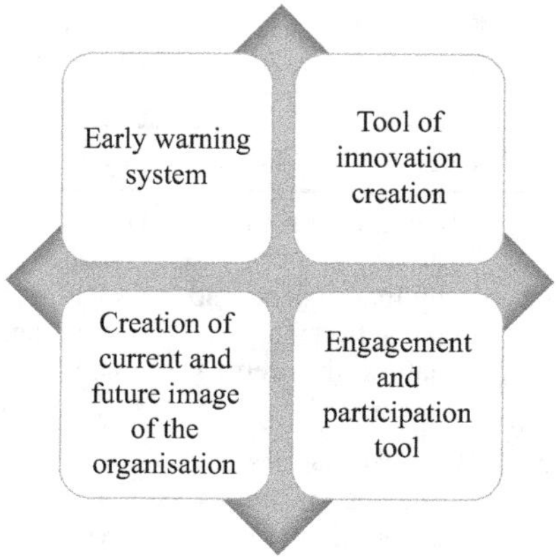

Figure 17.10 Foresight – key benefits for organisations
[Source: authors' own elaboration.]

Foresight as a tool for creating innovation allows organisations to:

- redefine and create the future-portfolio of a company
- identify future niches
- identify the changing needs of consumers
- interact with key players in the business and business-related processes, such as customers, suppliers, legislators – creating new technologies, products and services
- provide strategic information to decision-makers by identifying the prioritised directions for the development of technologies, products and services
- concentrate the funds on key future areas

Foresight as an instrument creating the current and future image of the organisation allows:

- internal and external stakeholders to identify with the organisation
- explore the unique features of the organisations distinguishing it from others

Foresight also means involvement and participation, expressed through:

- the participation of all stakeholders in the process of building a shared vision of the future
- stakeholder involvement in the creation of the future
- a sense of responsibility for the present and the future of the organisation

17.6 Summary

The rapid development of technology and increasing consumer demand, with the simultaneous pressure to adapt operations to the principles of sustainable development, present new challenges for transportation systems. Questions about the future are still valid: traveller information, traffic management, vehicles, public transport, emergency, electronic payment and safety issues.

The answers to questions about the future of transport systems can be sought on international, national, regional and local levels. Foresight is the tool enabling the creation of the desired vision of the future with the intrinsic involvement of all stakeholders. So far, the practical application of the foresight studies in the transport sector has confirmed its applicability and effectiveness.

The results of foresight projects can contribute to the strengthening of research and development, as well as to the growth of the innovation and competitiveness of the entire transport sector. The participatory nature of foresight research allows for mutual co-operation and the diffusion of knowledge in the process of solving problems and shaping the future. This type of co-operation attracts the interest of entrepreneurs, representatives of science, and the wider public. Foresight studies may also attract the interest of the representatives of international organisations (e.g. International Road Transport Union, International Union of Railways, SMDG), ministries at the national level responsible for the development of transport, transport associations (e.g. Freight Transport Association, International Air Transport Association), companies in the transport sector, academic institutions dealing with the issues related to the development of transportation systems (e.g. Transport Research & Innovation Portal – TRIP).

The methodology of foresight research, based on the use of both quantitative and qualitative methods, fosters social consensus while ensuring the implementation of the most innovative solutions consistent with the principles of sustainable development.

Abbreviations

STEEP	(foresight term) Social, Technological, Economical, Environmental and Political
STRG	Society and Technology Research Group (Daimler Chrysler)
TRIP	Transport Research & Innovation Portal

References

[1] Ejdys J. 'Overcoming problems associated with uncertainty of the environment by using foresight approach'. *Economics and Management*. 2013; 18: 331–339

[2] Miles I., Cassingea Harper J., Georghiou L., Keenan M., Popper R. 'The many faces of foresight' in: Georghiou L., Cassingea Harper J., Keenan M., Miles I., Popper R. *The Handbook of Technology Foresight. Concepts and Practice*, Northampton: Prime Series on Research and Innovation Policy, Edward Elgar Publishing; 2008

[3] United Nations Industrial Development Organization. *UNIDO Technology Foresight Manual. Organization and Methods*; 2005

[4] Salmenkaita JP., Salo A. 'Emergent foresight processes: industrial activities in wireless communications'. *Technological Forecasting and Social Change*. 2004; 71(9): 897–912

[5] Keenan M. Miles I. *A Practical Guide to Regional Foressight*. Seville: Institute for Prospective Technological Studies, FOREN Network; 2001

[6] Detrich, M. 'Foresight to nie prognoza'. *Sprawy Nauki. Biuletyn Ministra Nauki i Szkolnictwa Wyższego*. 2004: Available from http://www.sprawynauki.waw.pl/?section=article&art_id=1169 [Accessed 30 Dec 2014]

[7] Cassingena Harper J. *eForesee Report*. Malta Council for Science and Technology; 2003

[8] Jakuszewicz I., Kononiuk A., Magruk A., Nazarko J. 'Inicjatywy foresight w Polsce i na świecie' in Kiełtyka L. (ed.), *Multimedia w organizacjach gospodarczych i edukacji*. Warszawa: Wyd. Difin; 2006

[9] Halicka K. 'Priorytetowe technologie polskiej energetyki w perspektywie 2030 roku'. *Przegląd Elektrotechniczny*. 2014; 9: 105–108

[10] Halicka K. 'Przyszłość polskiej energetyki w świetle badań foresight'. *Rynek Energii*. 2013; 2: 10–17

[11] Nazarko J., Glińska U., Kononiuk A., Nazarko Ł. 'Sectoral foresight in Poland: thematic and methodological analysis'. *International Journal of Foresight and Innovation Policy*. 2013; 9(1): 13–38

[12] Ejdys J., Nazarko Ł. 'Foresight gospodarczy – instrumentem orientacji na przyszłość'. *Prace Naukowe Uniwersytetu Ekonomicznego we Wrocławiu*. 2014; 340: 651–664

[13] Nazarko J. *Regionalny foresight gospodarczy. Metodologia i instrumentarium badawcze*. Warszawa: ZPWiM; 2013, p. 122

[14] Denzin N.K., Lincoln Y.S. 'Scenario planning: a collage construction Approach'approach'. *Foresight*. 2009; 11(2): 19–28

[15] Denzin N.K. 'Using triangulation in nursing research'. *Journal of Advanced Nursing*. 1996; 24: 123

[16] Okoń-Horodyńska E. 'Foresight – czyli jak określać priorytety rozwoju innowacji' in Szabłowski J. (ed.). *Zarządzanie innowacjami – teoria i praktyka*. Białystok: WSFiZ; 2006, vol. 6, p. 31

[17] Nazarko J. 'Kształtowanie polityki proinnowacyjnej regionu np. Foresightu Technologicznego "NT for Podlaskie 2020"'. *Optimum – Studia Ekonomiczne*. 2011; 2(50): 241–251

[18] Popper R., Keenan M., Miles I., Butter M., Fuenta G.S. *Global Foresight Outlook 2007, The European Foresight Monitoring Network* [online]. 2008. Available from http://www.foresight-network.eu/files/reports/efmn_mapping_2007.pdf [Accessed 30 Dec 2014]

[19] JRC-IPTS, PREST, CM International. *A Practical Guide for Regional Foresight, FOREN – Foresight for Regional Development Network* [online]. 2001. Available from http://foresight.jrc.ec.europa.eu/documents/eur20128en.pdf [Accessed 30 Dec 2014]

[20] Klepka M. 'Foresight' in Matusiak K.B. (ed.). *Innowacje i transfer technologii. Słownik pojęć*. Warszawa: PARP; 2011, p. 83

[21] Müller A.W. *Strategic Foresight – Prozesse strategischer Trend- und Zukunftsforschung in Unternehmen*, Dis. Switzerland: Universität Zürich; 2008

[22] Rohrbeck R. *Corporate Foresight: Towards a Maturity Model for the Future Orientation of a Firm*. Heidelberg and New York: Physica-Verlag, Springer; 2010

[23] Georghiou L. 'The UK technology foresight programme'. *Futures*. 1996; 28(4): 359–377

[24] Georghiou L., Harper J.C., Keenan M., Miles I., Popper R. (eds.). *The Handbook of Technology Foresight*. Northampton: Edward Elgar, Publishing, Inc. Northampton; 2008

[25] Becker P. *Corporate Foresight in Europe: A First Overview. Working Paper*. European Commission 2002. Available from ftp://ftp.cordis.europa.eu/pub/foresight/docs/st_corporate_foresight_040109.pdf [Accessed 30 Dec 2014]

[26] Daheim C., Uerz G. 'Corporate Foresight in Europe: Ready for the Next Step?', Second International Seville Seminar on Future-Oriented Technology Analysis: Impact of FTA Approaches on Policy and Decision-Making, Seville; 28–29 September 2006

[27] Hammoud M.S., Nash D.P., 'What corporations do with foresight'. *European Journal of Futures Research*. 2014; 2: 42

[28] Rohrbeck R., Schwarz J.O. 'The value contribution of strategic foresight: Insights insights from an empirical study of large European companies'. *Technological Forecasting & Social Change*. 2013; 80: 1593–1606

[29] Rohrbeck R., Mahdjour S., Knab S., Frese T. *Benchmarking Report: Strategic Foresight in Multinational Companies*, European Corporate Foresight Group; 2009

[30] Rudzinski C.V., Uerz G. 'Volkswagen: Open Foresight at the Front End of Research Innovation' in Gassmann O. and Schweitzer F. (eds.), *Management of the Fuzzy Front End of Innovation*. Switzerland: International Publishing Switzerland; 2014. pp. 295–299

[31] 'Future Gazing', *Mercedes-Benz next. Magazin für Innovation, Technologie, Mobilität*, Daimler AG 2010, http://next.mercedes-benz.com/en/future-gazing/, [Accessed on 12 Jan 2014]

[32] Ruff F., *Society and Technology Foresight in the Context of a Multinational Company*, EU-US Seminar: New Technology Foresight, Forecasting and Assessment Methods, Seville; 13–14 May 2004

[33] Deutsche Post AG (Germany). *Delivering Tomorrow. Logistics 2050 – A Scenario Study*; 2012

[34] Barré R., Keenan M. '*Revisiting foresight rationales: what lessons from the social sciences and humanities?*' in Cagnin C., Keenan M., Johnston R., Scapolo F., Barré R. (eds.). *Future-Oriented Technology Analysis — Strategic Intelligence for an Innovative Economy*. Berlin: Springer; 2008

[35] Ejdys J., Ustinovicius L., Stankeviciene J. 'Innovative application of contemporary management methods in a knowledge-based economy –

interdisciplinarity in science'. *Journal of Business Economics and Management*. 2015; 16(1): 261–274 doi:10.3846/16111699.2014.986192

[36] Kopyciński P., Mamica Ł. *Operacjonalizacja metodologii badań foresight*, Kraków: Małopolska Szkoła Administracji Publicznej Akademii Ekonomicznej w Krakowie; 2006

[37] Singleton R.A., Straits B.C. *Approaches to Social Research*, New York: Oxford University Press; 2009

Chapter 18

Aeronautical air-ground data communications: current and future trends

Fabien Garcia, Alain Pirovano*, Mickaël Royer* and Quentin Vey**

The aeronautical industry is evolving toward the Intelligent Transport System paradigm. Those changes are led in part by framework projects such as SESAR (Single European Sky Air traffic management Research) directed by Eurocontrol and NextGen (Next Generation Air Transportation System) directed by the Federal Aviation Administration (FAA). Those two projects aim at integrating 4D trajectories, that is space-time trajectories, in flight planning and to develop System Wide Information Management (SWIM) services. Those two concepts will allow for better planning of flights and hence reductions in fuel usage, flight costs and delays for users.

These innovations will require communication systems allowing efficient information sharing and transmission between aircraft and ground systems. Until recently air-ground communications consisted in analog voice communications. More recently new cellular and satellite systems were deployed that allow digital data communication between the different stakeholders of the air traffic activities (i.e., air traffic control (ATC), airlines and their pilots).

This chapter presents the current state of aeronautical communications in Section 18.1 and then turns to recent solutions that are proposed in research projects in Section 18.2.

18.1 Aeronautical air-ground data communications

18.1.1 Overview

This section gives an overview of the aeronautical air-ground communication domain and its context, properties and organization. It should be noted that solutions presented here are mainly dedicated to data communications between airlines and their crew or between pilots and air traffic controllers.

18.1.1.1 From analog voice to data link

The earliest communication with aircraft involved mainly ground staff and pilots. They were based on visual signaling using for instance colored paddles or hand signs.

*ENAC (École Nationale de l'Aviation Civile), France

The first aeronautical radio link for air-ground communications has been proposed at the beginning of the twentieth century by AT&T allowing the first voice communications between ground personnel and pilots. After World War I new radio communication systems offering greater range and better performances were developed. It was only in 1935 that airborne radios were considered reliable and efficient enough to be widely deployed on aircraft. These air-ground communication means were proposed in order to increase air safety. From that moment, the Very High Frequency (VHF) band was mainly used for radiotelephony services. Those used technologies have evolved but the main principle is still the same today: the VHF reserved bandwidth (currently from 117.975 MHz to 137 MHz) is split in several channels with a spacing to ensure efficient sharing of resources without interferences between channels. Today, the airband is split in 2,280 channels with a spacing of 8.3 kHz. A voice communication use either VHF or High Frequency (HF) (from 3 MHz to 30 MHz) voice radios which has been further augmented with SATCOM (Satellite Communication) since the early 1990s. Hence, voice communications are possible even in oceanic areas. Nevertheless, considering the increasing number of aircraft, the lack of resources made it necessary to foresee new solutions in order to avoid congestion. A solution, known as data link or digital data link, is based on new way to exchange between end users. Data link offers the ability to transmit short and relatively simple digital messages between aircraft and ground stations using communication systems that are mainly based on VHF or SATCOM. This way to communicate was introduced in 1978 by ARINC (Aeronautical Radio Incorporated) as ACARS (Aircraft Communications Addressing and Reporting System) with the aim to reduce crew workload and improve data integrity. This system, still in use in some airspaces, uses VHF channels initially dedicated to voice communication.

In the early 1980s, ICAO (International Civil Aviation Organization) decided to create a special committee known as FANS (Future Air Navigation System) with the mission: to identify and assess new concepts and technologies, which may have future benefits for the development of international civil aviation. Then, during the 1990s, Boeing then Airbus have developed their FANS products that were then joined as FANS-1/A. In order to be FANS-1/A compliant an aircraft has to be outfitted with a set of technologies ranging from specific communication hardware to applications such as CPDLC (Controller Pilot Data Link Communications). All these successive steps give today an heterogeneous and relatively complex aeronautical world where air-ground data communications may be based on ACARS, FANS 1/A or even FANS 2/B with the last improvements.

18.1.1.2 Properties and traffic classes

Regarding the standards and recommendations known as SARPS (Standards And Recommended PracticeS) and provided by ICAO in their documents, four categories of communications are identified with different requirements. Air Traffic Service Communications (ATSC) and the Aeronautical Operational Control (AOC) Communications that are considered as safety related, and the Airline Administrative

Control (AAC) and the Aeronautical Passenger Communications (APC) that group non-safety related applications:

- ***ATSC**, Air Traffic Service Communications (critical).* This class regroups communication between pilot and ATC to ensure the safety, speed and efficiency of the flight. For instance, these services may be related with meteorological information or route information during the flight.
- ***AOC**, Aeronautical Operational Control (critical).* This class regroups communication required for the exercise of authority over the initiation, continuation, diversion or termination of flight for safety, regularity and efficiency reasons. For instance, it includes airline companies communication with their aircraft such as maintenance messages, fuel levels, exact departure and arrival time.
- ***AAC**, Airline Administrative Control (noncritical).* This class regroups communications for the exchange of aeronautical administrative messages. Examples of AAC are information regarding passengers (list of passengers, connections), or special cleaning requests.
- ***APC**, Aeronautical Passenger Communications (noncritical).* This class regroups communication relating to the non-safety voice and data services to passengers and crew members for personal communication.

It has to be noted that critical communications (ATSC and AOC) follow very specific international rules defined by ICAO and are based on dedicated physical systems with very stringent quality of service requirements. In this chapter, we mainly focus on communication systems dedicated to these critical services.

18.1.1.3 Actors and organizations

Many actors and organizations are involved in services design, standardization, deployment and maintenance for communication systems dedicated to data link. We propose to consider four classes of actors and organizations.

The first class includes *aviation authorities*. Their main objectives focus on the definition of the principles and techniques for international air navigation. They promote the planning and development of international air transport in order to ensure safe and orderly growth. The ICAO is the most important organization of this class. The ICAO Council adopts standards and recommended practices concerning air navigation known as SARPS. Eurocontrol is the European Organization for the Safety of Air Navigation working for seamless European air traffic management. It coordinates and plans ATC for all of Europe.

The second class relates to *air transport industry*. It includes main plane makers, like Airbus or Boeing, and communication equipment suppliers. These different actors are involved in communication systems dedicated to data link in the sense that they have to design, produce and install specific equipment that are in compliance with the aviation authorities regulations and guidelines.

The third class relates to *Data link Service Providers* (DSP). These organizations are responsible for the reliability of the transmission media and the integrity of the messages. They are expected to create and manage the different means for data links between the aircraft and the ground. They operates a network of ground stations deployed to provide VHF, HF, and SATCOM coverage in continental and oceanic airspaces. The two primary DSP are ARINC and SITA.

The last class lists the *users*. ANSP (Air Navigation Service Providers) belong to this class. They are government departments, state-owned companies or sometimes private organizations. As example we can cite DGAC (Direction Generale de l'Aviation Civile), particularly the DSNA (Direction des Services de la navigation arienne) entity, in France or FAA in the United States. ANSPs belong to this class because they are responsible for the ATC. The users class includes also the airlines, for example, a company that provides air transport services for traveling passengers and freight.

18.1.2 Current communication systems

18.1.2.1 Overall topology

As underlined in Section 18.1.1.1, successive steps in the history of aeronautical communications give today an heterogeneous and relatively complex domain where air-ground data communications may be based on ACARS, FANS 1/A or even FANS 2/B with the last improvements. Considering the objectives of this chapter, it mostly has to be underlined that an aircraft, particularly long-range ones, is equipped with several communication means in order to ensure that it will always find a way to communicate during its flight. Figure 18.1 gives a typical overview of

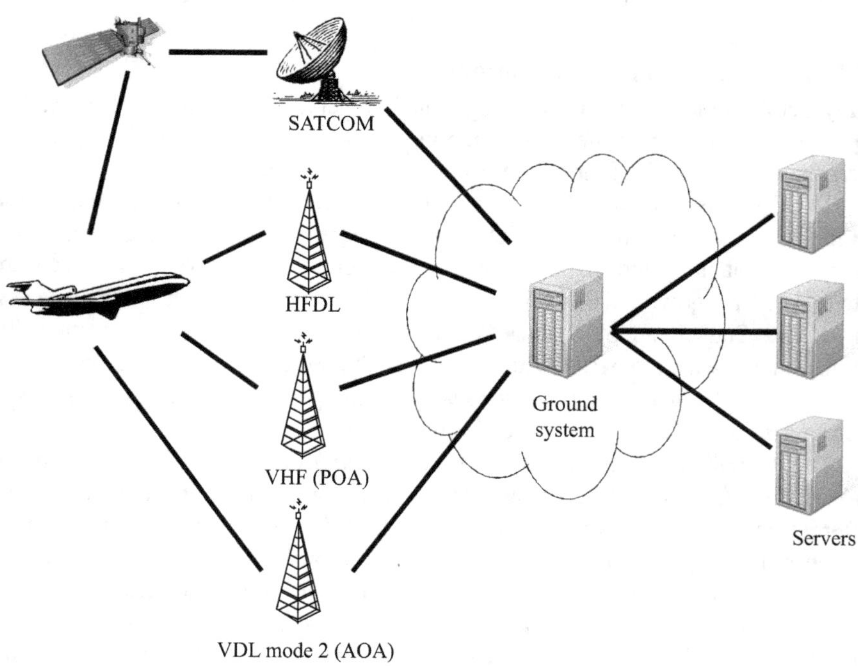

Figure 18.1 Air-ground communication systems

main existing means for air-ground data communications. As shown, the existing links may be based on:

- VHF with Plain Old ACARS for direct air-ground links in continental airspaces
- VHF Data Link mode 2, a more recent mean for direct air-ground links in continental airspaces especially in Europe
- High Frequency Data Link (HFDL) with ACARS for mostly direct air-ground links in oceanic airspaces or
- SATCOM links on dedicated frequency (L-band) for mostly direct air-ground links in oceanic airspaces

18.1.2.2 Subnetworks for air-ground communications

The first air-ground communication based on digital messages, known as ACARS, was introduced in 1978 by ARINC. It was aimed at improving the integrity of data exchanged with the aircraft crew by replacing voice by digital text messages. The term ACARS refers to the complete air-ground system and thus is used to designate different elements: airborne systems, the air-ground subnetwork, ground systems, the network services and even the applications using the network. The first air-ground link used by ACARS was VHF based. It offered 2,400 bps throughput. Due to their limited range, the ground stations deployed for VHF ACARS ground station cover mainly continental areas. This first version of ACARS is sometimes named POA, for Plain Old ACARS, in order to differentiate it with further enhanced versions. During the 1990s, DSP proposed ACARS via geostationary satellite links to overcome the limitations of continental coverage. More recently, ACARS has even been provided on existing aeronautical HF systems in order to provide HFDL.

Regarding SATCOM, Standards for Aeronautical Mobile Satellite Service (AMSS) have been developed by the ICAO and described in SARPS. AMSS defines a mobile-satellite service in which mobile earth stations are located on board aircraft and it has been designed to be a subnetwork of the Aeronautical Telecommunication Network (ATN) defined by ICAO as a digital data communications network for the aeronautical community. The AMSS may provide data and voice communications between aircraft and ground systems. Satellite systems for aeronautical safety communications operate in the mobile satellite service radio frequency bands included in the L-band (1–2 GHz) mainly because of the global coverage provided. Besides, ICAO authorizes only some satellite systems, for instance Aero-H/H+/I/L proposed by Inmarsat or Iridium which provide services in L-band. Inmarsat operates four geo-stationary satellites that cover about 97% of the earth's surface. The Iridium system is based on a constellation of 66 cross-linked satellites (plus seven spares) that create its network of global coverage.

Still in deployment phase, VDL mode 2 is the main system proposed in Europe as communication mean for air-ground data communications. It delivers data at 31.5 kbps using Differential 8 Phase Shift Keying (D8PSK) modulation at VHF frequencies. This is over 13 times faster than the VHF ACARS 2.4 kbps rate. The obtained spectral efficiency is the highest possible that can be supported by a 25 kHz channel while providing a range of 200 nautical miles. Aircraft and ground stations exchange their user data using data frames. Several frames can be sent before an acknowledgment is required (window management). VDL Mode 2 will result in faster transmission than ACARS of multi-block messages such as wind databases, flight plans, and so on. When tuned to the Common Signaling channel (CSC) for VDL Mode 2 (136.975 MHz), the aircraft listens for broadcast Ground Station Identification Frames (GSIFs), identifying the service provider and services offered by that ground station. After hearing one or more GSIFs, the aircraft may decide which ground station is the best (based on service provider preference, offered services or signal strength for instance) and then makes a logical connection to the selected ground station. As the aircraft transits through the airspace, it keeps track of all ground stations that it can hear. When the signal quality of the currently connected ground station degrades due to range effects, the aircraft then selects the best candidate ground station from the list and makes a logical connection to this ground station (hand-off). Hence the logical air/ground connections make a seamless transition as the aircraft flies through the airspace.

To avoid collisions with transmissions from other ground stations and/or aircraft, frames are only sent when a receiver detects that the VHF channel is free. This system is called Carrier Sense Multiple Access (CSMA) and takes steps to reduce the probability that aircraft transmit simultaneously. This in turn results in reduced message delay and high success rates under heavy loading conditions. A major benefit of CSMA is the small overhead since each ground station and aircraft is independent and does not require synchronization or timing signals from a central source.

The AVLC (Aviation VHF Link Control) protocol is used to establish a logical link between aircraft and the ground station. AVLC is derived from the well-known HDLC protocol (High-Level Data Link Control) that has been widely used in ground networks.

Since aircraft flying in VDL mode 2 domains but still equipped with pure ACARS exist, an adaptation layer has been developed to allow them sending ACARS messages. This protocol, named AOA (ACARS Over AVLC), is used in this case over the VDL mode 2 protocol stack.

Finally, it has to be underlined that some emerging communication systems, which aim to overcome the limitations of the existing ones, are currently studied in some project. For instance, we can cite LDACS (L-band Digital Aeronautical Communications System) as new cellular continental system proposed in the context of the SESAR project or the future aeronautical satellite system designed by ESA in IRIS program.

18.2 Future trends

In the face of increasing air traffic, current systems would not be able to cope with the increase in communication needs. Furthermore, new services such as in-flight real-time weather advisory would be hard or costly to deploy on those systems.

These considerations have led several stakeholders and research laboratories in the aeronautical domain to start research projects in order to define new aeronautical communications systems able to make the transition toward the future of air traffic.

SESAR and NextGen, funded respectively by Eurocontrol and the FAA aim at improving the aeronautical systems by pushing forward SWIM services. Such services need new communication systems but up to now, no radically new architecture was proposed. The two main direction of research in those two projects are based on cellular and satellite based solutions (such as the already named L-DACS and IRIS).

Other projects nevertheless try to go beyond the current class of systems and to promote new paradigms in aeronautical communications.

One of the first propositions on this way is to use inter aircraft links in order to create a Mobile Ad hoc Network (MANET). This kind of ad hoc network is dubbed Aeronautical Ad hoc Network (AANET) and is described in detail in Section 18.2.1. These networks are composed of wireless inter aircraft links and use aircraft on the way to the destination in order to forward data from other aircraft. This kind of network allow communications over wide areas without a fixed infrastructure or satellites and could therefore be used over oceans where no infrastructure is present for cellular networks, and to reduce the cost of deployment and operation associated to satellites.

Based on AANETs, the aeronautical communication paradigm can be pushed further. One can go from sending data to a specific address associated to one or more aircraft or ground system to sending data to all systems in an area. This is the domain of geographical routing, also called geocasting, which received a lot of attention in the general context of ad hoc networks. Sending data to a geographical zone could allow new services such as localized situation reports (e.g., weather advisory, dust plumes reports).

Finally, if sending to an area can lead to new services, an even greater potential arise from sending data to entities interested in receiving informations based on the content of the information and the context of the entity. One example would be an aircraft registering to receive all information about flight conditions on its trajectory. This kind of context aware transmission allows for new services while reducing the volume of data being sent or the amount of entities receiving messages they do not use. Furthermore it allows total decoupling between sender and receiver as no common address need to be known, only the type of information and the context of the receiver (e.g., the location of an aircraft, its type) are required. Those solutions are presented in detail in Section 18.2.2.

18.2.1 Aeronautical ad hoc network

18.2.1.1 Introduction to aeronautical ad hoc networks

An ad hoc network is a network in which the lower layers (up to the network layer) are directly handled by every node, that is, each node can act as a source, a relay or a destination for data, without a central managing entity or any infrastructure. They are sometimes referred to as mesh or multihop networks.

They can be opposed to cellular networks, in which a central entity manages the communication and relays data for every node within range. In these latter, nodes out of range of the relay can't communicate at all, even if they are in range of other nodes.

MANETs [11] are a class of ad hoc networks in which the nodes are mobile, and their study is an active research area. The mobility introduces several difficulties, such as routing in a dynamic topology. MANETs are also subclassed into several categories, matching particular needs and operational conditions.

Vehicular Ad hoc Networks (VANETs) and AANETs are some of these subclasses. In VANETs the nodes are generally ground vehicles like cars and trucks, whereas in AANETs they are civil aircraft (from commercial or general aviation). The similarities and differences between these two groups are regrouped in Figure 18.2. VANETs have been more intensely studied than AANETs, and the similarities between them allow us to consider approaches inspired from works on VANET on some sides of the AANET (e.g., routing along an airway may be similar to routing along a highway).

Expected drawbacks and advantages of AANET

Drawbacks

- *Connectivity*: Since the forwarding in an ad hoc networks is done directly by the nodes of the network, the connectivity (i.e. the percentage of aircraft connected to the ground with a multihop path) of AANET relies heavily on the aircraft density. Indeed, if there are no other aircraft in the whereabout of a particular data sender or at some point on the packet path, the data will not be delivered. Nevertheless, it has to be noted that, because of the very high availability requirements for critical communications, there must be several independent communication systems aboard civil aviation aircraft. Thus AANET should be considered as a complementary mean of communication.

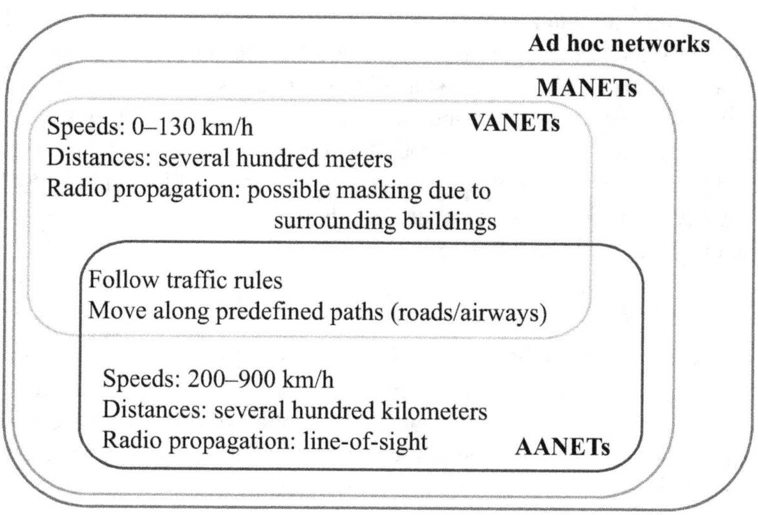

Figure 18.2 Properties diagram of AANET and VANET

- *Gateway Bottlenecks*: Even if only a few ground gateway are needed to ensure connectivity, their limited number may become problematic for the end-to-end data exchange speed. They may become the bottleneck of the network. In consequence, the capacity available for all the aircraft may be limited by the number of ground station.

Advantages

- *Cost*: As a result of their light ground infrastructure (only gateways to other networks are required), the deployment cost of an AANET can be very low. Currently investigated architectures use omnidirectional antennas and cheap on-board equipment, which will also lower the cost to equip the aircraft. In addition, there is no need of a third-party communication infrastructure like a satellite link, so the operational cost should also be low.
- *Robustness*: The decentralized architecture of an AANET requires it to be self-configuring and self-healing. As a consequence, the resulting network will be able to self-(re)configure in case of disruption, which increases the robustness of this communication system. For example, in Figure 18.3, if any relay on the path in bold arrows fails, the path drawn in dotted arrows can be used.
- *Coverage*: The coverage area of an AANET is virtually unlimited, as long as there are enough aircraft between one given node and the ground to create a multihop path. For example, some studies have shown that AANET are able to create a network all over the Northern Atlantic Tracks for commercial aviation with a few ground stations.

18.2.1.2 Known results

Connectivity

The connectivity of AANET has been studied for both continental traffic and oceanic traffic, using real aircraft position data (based on radar detection, or by

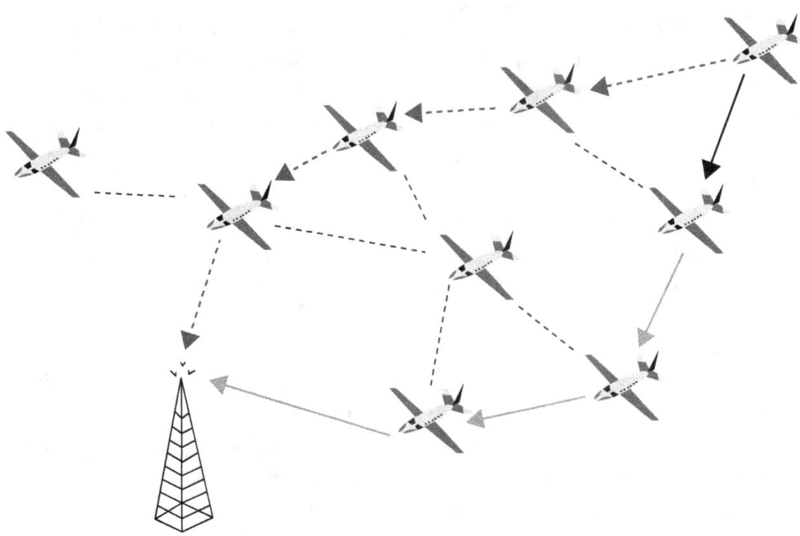

Figure 18.3 Illustration of AANET robustness: two possible routes for data

position report from the crew) [2]. In this study, the continental traffic considered was recorded over France for one whole day, and the oceanic traffic was focused the North Atlantic Tracks (NAT). The NAT are airways defined daily by ATC and used by aircraft to travel between Europe and North America.

In the continental study, most of the flights happened between 6 am and 9 pm. The conclusion was that with a radio range of 200 km is enough to ensure a 97% average connectivity over the French sky, and a range of 100 km ensures a connectivity of 96% between 6 and 21. The amount and location of ground stations had very little influence on connectivity during the day, it only enhanced connectivity during the night, when there was fewer aircraft in flight.

The oceanic study showed that a 350 km range ensured an average connectivity of 90%. In this study, the 11 ground stations were located where it was possible along the tracks and on islands.

Throughput assessment

The throughput has been estimated using the same real position data as in the connectivity. The TCP (Transmission Control Protocol) and greedy sources are used because TCP fairly adapts the sending rate to the available capacity, and this represents a worst-case scenario (saturation of the capacity). Using 1 Mbps links and a shortest geographic path routing, the mean aircraft throughput was measured at 38.3 kbps for the continental case, and at 68.2 kbps for the oceanic case. These figures for the worst-case scenario may seem low, but they have to be compared to the capacity offered by current aeronautical communication systems, as presented in Section 18.1.2.

18.2.1.3 Technical challenges, proposed solutions

Technical challenges

The specificities of AANET identified in Section 18.2.1.1 imply constraints on the design of protocols well-suited to AANET.

- *Number of nodes*: The high number of nodes (up to 1,600 simultaneous flight over the whole Europe, 400 over the north Atlantic) requires highly scalable protocols, particularly for the network layer, which handle the packet route definition and the transmission. Given the considered radio ranges, the amount of nodes in range of each other can be high, so the access layer protocols will also have to be scalable.
- *Speed of the nodes*: The typical cruise speed of an airliner is around 900 km/h. It has a direct impact on the choice of the physical communication channel, and an indirect impact on the routing algorithm because of the dynamicity of the topology.
- *Geographic distances*: The geographic distances between the nodes must be taken into account in the design of the physical and link layer. The differences in the distances between nodes, from 300 m (1,000 ft, the minimum vertical separation) to 400 km (order of magnitude of ranges considered in previous studies), will indeed lead to frequent "near-far" situations. It will so require a trade-off between connectivity and interferences.

Proposed network architecture

Physical layer: Direct Sequence Spread Spectrum (DSSS) is the technology foreseen for the physical communications between aircraft. It is a mature technology that allows collision recovery and simultaneous transmissions (under certain conditions). It is also fairly easy to use an adaptive spreading factor to adapt transmissions to interference conditions (i.e., reduce raw bit rate and increase robustness).

Automatic power control also has shown its usefulness in AANET. Since it adapts the radio power to the exact level required for the transmission between a given sender and receiver pair, thus reducing the interferences for further nodes. This results in an overall performance improvements.

The I and Q channels of an I/Q modulator can also be used as independent channels, for example, to segregate control traffic and data traffic, with a lower bit rate but a higher robustness for the control channel.

MAC layer: Medium Access Control (MAC) layers manage the concurrent transmission from different nodes by using algorithms and physical properties of the signal to avoid or recover packet collisions. Traditional MAC rely on Time Division Multiple Access (TDMA, which allocate a particular time slot to the sender), Frequency Division Multiple Access (FDMA, which allocates a particular frequency to the sender) or a mix of both. Although easy to implement for cellular networks, FDMA and TDMA require complex distributed resource allocation or time synchronization algorithms when used in ad hoc networks, which often impair the MAC layer efficiency.

Code Division Multiple Access (CDMA), brought by DSSS, has several advantages for *ad hoc* networks. It is not only more efficient than FDMA or TDMA [3], but it can also be easier to implement.

For example, Asynchronous CDMA (A-CDMA [4], based on SSA [5]) is based on the ability of spread spectrum transmissions to recover a certain amount of packet collisions with a common spreading code, provided that they are received with a small time difference (often less than the duration of a data bit).

Another example is the Cooperative CDMA-based Multi-channel Mac (CCM-MAC [6]), which is based on the ability of DSSS to fully discriminate two packets sent with different spreading code. It could also be improved with the independent I and Q channels and with an adaptive spreading factor.

Network layer and routing: Routing is a widely studied problem in ad hoc networks, but no universal solution has been found. As a result of the similarities between VANET and AANET and the amount of study on the former, VANET (and even MANET) routing algorithms should a good starting point for the design of an aeronautical routing algorithm. Several design philosophy are proposed:

- The proactive approach, in which the routes are computed beforehand, even if there is no message to send. Proactive algorithm are usually faster to start the first communication, but they consume also more resource because they require to send periodically signalization messages to compute or update the routes, even if the latters are not used.

- The reactive approach, in which the routes are only computed on request (i.e., when there is data to send to a particular destination). Reactive algorithms are usually slower to perform the first end-to-end transmission, but consume less resources than the proactive ones because they compute routes only on request, that is, when they are really needed.
- The hybrid approach, which combines both proactive and reactive strategies, in order to have the best of the two previous approaches.

Aside from the design philosophy the routing algorithm can be characterized with its routing metric, which will determine the chosen path. Of course, a routing algorithm can take several metrics into account to match the application scenario. Some interesting example (and some routing algorithm implementing them) are the following:

- *Number of hop*: Minimizing the number of relays is a way to reduce processing delay. Examples: AODV, ARPAM.
- *Load balancing*: Another way to reduce processing delay is to spread the load all over the network, thus reducing the length of the queues. Example: GLSR.
- *Geographic position*: The geographic position can help to route packets to the destination, for example, by forwarding the packet to the neighbor node closest to the destination. Thus it can reduce end-to-end delay. Examples: GPSR, GLSR, GRAA, LAR.
- *Link stability*: By choosing the most stable links one can increase route stability and reduce the amount of signaling required to re-compute a new route. This link stability can be measured through Doppler shift (a small Doppler shift means that the aircraft are following trajectories that are similar to each other) or with the knowledge of the velocity of the surrounding aircraft. Examples: MUDOR, ZBR.

Finally, the logical network topology produced by the routing can be "flat" that is, every node has the same role, or it can be hierarchic, that is, some nodes have different roles in order to optimize the resource allocation. An example of hierarchic routing algorithm are TOPO and ZBR, which create an overlay network with a subset of nodes, which serve as a backbone for the whole network.

18.2.2 Shifting the communication paradigm in AANET

The previous section illustrates the interest and the feasibility of AANET as an innovative mean of communication between aircrafts or between aircraft and the ground. This new network solution offers a higher throughput than traditional means like VDL2 and opens the door to new services in order to transmit more data from the ground to the aircraft such as real-time weather report.

In order to use this new solution as efficiently as possible, this section presents a new communication paradigm: *content-based routing*. After the definition of content-based routing, the Publish/Subscribe model is detailed since it is the most

Figure 18.4 Presentation of Publish/Subscribe paradigm

used model to provide content-based routing in a network. This section finishes with a state of the art about Publish/Subscribe systems adapted for AANET.

18.2.2.1 Definition of content-based routing

Content-based routing is a communication paradigm which differs in many ways from classical communication schemes where addresses are used to identify the recipients. These addresses can target a single receiver (unicast communication) or multiple receiver (multicast communication). In content-based routing, no addresses are used anymore. Instead, the recipient's selection is based on the content of the message correlated to the node's interests. This paradigm allows *space decoupling* between senders and receivers, meaning that a node does not need to know recipient's identities to send an message. The network is responsible to identify recipients and send messages to them, based on the content of these messages.

18.2.2.2 Publish/Subscribe solution

Publish/Subscribe systems are the most widespread solutions to provide content-based routing in a network. In this model, *events* are sent by *publishers* (event producer) and disseminated to *subscribers* (event consumer) by the network (Publish/Subscribe system). Figure 18.4 illustrates the API offered to applications by a Publish/Subscribe system.

The link between an event and subscribers interested by it can be based on different solutions.

- *Topic based*: In this model, clients subscribe to topics characterized by a name. Events are then published on topics and the Publish/Subscribe system is responsible for the forwarding of events to clients which subscribe to this topic. The topic-based subscription is quite simple but it's also limited since we can filter events only from a single attribute, the topic name.
- *Content based*: In content-based subscription model, clients base their subscriptions on the content of events. For example, the filter syntax "attribute – operator – pattern" can be used to define this subscription [7]. Thus, when an event is published, the Publish/Subscribe system has to forward it to clients whom subscriptions match with the event content.

The second approach is clearly the most appropriated to provide content-based routing and in the rest of this section, we will use the term "Publish/Subscribe" in place of *Content-Based Publish/Subscribe systems*.

18.2.2.3 State of the art

AANET studies are recent and concentrate on the feasibility of the approach, the MAC (Media Access Control) layer and routing protocols, as in Reference 2. No detailed studies have been conducted on the definition and validation of Publish/Subscribe systems adapted to AANET. Nevertheless, as explained in the previous section (Figure 18.2), AANET has many similarities with VANET, so it will be probable that Publish/Subscribe systems suited for VANET are also relevant for AANET. Such solutions can be regrouped in three categories.

The first one is *geographic routing-based solutions* [8 or 9]. Based on the assumption that each node know its position (thanks to a GPS receiver, for example) and each event/subscription is linked to a geographical region. In a VANET, announces of traffic jam or accident are easily related to a geographical region. Solutions of this category use a geographic routing protocol to dispatch events from publisher to subscribers. Each event is transmitted to nodes located in the linked area. The same process is applied for subscriptions. As illustrated in Figure 18.5, a node located in the intersection of the event and subscription regions will act as a proxy to forward events to interested subscribers.

The second category consists in proximity routing-based solutions [10,11] which can be seen as a specific case of geographic solutions. The main difference concerns the geographical area associated to events and subscriptions. For this kind of solutions, these regions are always centered on the publisher or subscriber (as shown in Figure 18.6). The routing protocol is then simpler but becomes inefficient

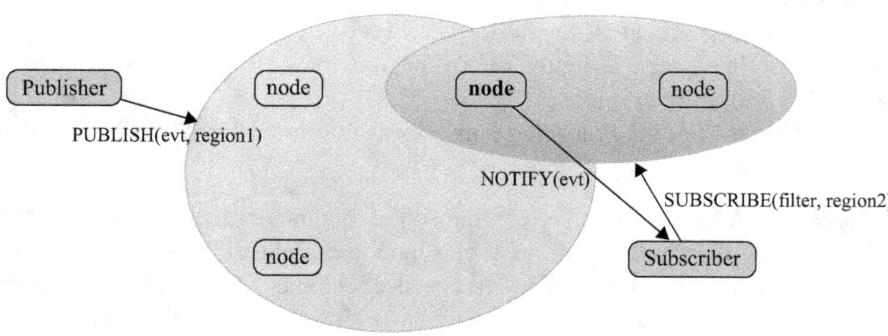

Figure 18.5 Geographic routing-based solutions

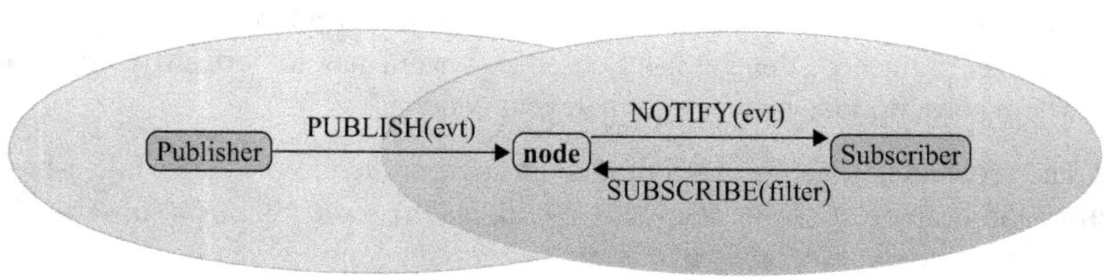

Figure 18.6 Proximity routing-based solutions

when the target region of an event is far from the publisher of the event (equivalent to a broadcast solution).

Finally, the last category of solutions is the overlay based solutions. A new component is introduced in these solutions, *brokers* which are responsible for dispatching events from publishers to subscribers. They are interconnected and thus form an overlay network. Most Publish/Subscribe middleware for network with infrastructure are overlay based solutions [12,13]. Thus, some Publish/Subscribe systems try to adapt this concept to VANET. For instance, Reference 14 proposes a protocol adapted from MAODV [15] to create an adaptive tree with brokers. This tree is then used to disseminate efficiently subscriptions to all brokers. Thus, when an event is published and arrives at a broker, it knows which brokers are interested and transmit the associated message to them.

18.2.2.4 Applications of the new paradigm

This section gives three examples of application for which Publish/Subscribe model is relevant and allows an efficient use of AANET resources.

The first one is the notification of weather events. Some weather situations such as storm or dust plumes are dangerous for aircraft. To avoid routes which are too close to these situations, aircraft must be aware of these weather events which can be detected by other aircraft or weather stations. The use of the Publish/Subscribe paradigm over AANET composed by commercial aircraft and ground station is appropriate to respond to this scenario. An entity which detects the storm acts as publisher of the event. The message contains the localization and the type of weather event and finally all aircraft send subscriptions to be aware of all weather events located near their trajectory. In this way, thanks to the Publish/Subscribe system, they will be notified as soon as a storm appears.

Another essential weather information for aircraft is wind. This phenomena has a big influence on the trajectory of an aircraft. A precise knowledge of this trajectory becomes primordial especially with the research that have been conducted recently concerning the use of the 4D trajectory to optimize ATC. Nowadays, wind prediction maps are built by weather center and given to aircraft to calculate those 4D trajectories. However, these predictions are not perfect and lead to difference between the calculated trajectory and the true one. To mitigate this issue, a new approach has been explored recently, based on the idea that an aircraft in flight knows the true wind close to its position and can send this value to aircraft which will go into this area. If each aircraft does the same thing, we create *a wind network* [16]. A Publish/Subscribe system will be an efficient solution to this scenario since an aircraft by default is not able to know which aircraft in the sky will be interested by this wind data.

The last example of application is related to communication between ground facilities of airline and their aircraft. As seen earlier in this chapter, data link communications have been used for years to exchange information between airlines and their aircraft, using networks such as ACARS. Nevertheless, these communications are always unicast even if sometimes point to multipoint communications are needed. Indeed, we can easily imagine an airline which needs to send a message to all its aircraft going to or coming from a specific airport. In this case, the use of a Publish/Subscribe system will be more appropriate thanks to the filtering capabilities offered by this approach that will be more efficient than many unicast communications.

18.2.2.5 Outcome and further work

First studies about the use of Publish/Subscribe system on an AANET shows that geographic routing-based solutions and proximity routing-based solutions are not adapted to the scenarios explained in the previous section. The main issue concerns efficiency of these approaches in term of resource usage. Most subscriptions or events are linked to large geographical regions and lead to large diffusion in the network. Consequently, an overlay-based solution seems more adequate. However, the number and the high mobility of nodes in AANETs are not compatible with the exiting approaches which build trees with all the nodes in the network. That is why a promising approach is the use of a clustering algorithm in the network to build an overlay where cluster heads will be brokers. The different forwarding strategies can be applied inside and outside cluster groups. For example, we can imagine a solution where subscriptions are broadcasted inside a cluster group and events are sent to all cluster heads.

References

[1] Scott Corson and Joseph. Mobile ad hoc networking (MANET): routing protocol performance issues and evaluation considerations. RFC 2501 (Informational), January 1999.

[2] Frederic Besse. *RESEAUX AD HOC AERONAUTIQUES*. PhD thesis, ISAE, Toulouse, February 2013.

[3] Zaihan Jiang and Mengchu Zhou. "Spread spectrum MAC protocol with dynamic rate and collision avoidance for mobile ad hoc network." *IEEE Transactions on Vehicular Technology*, 56(5):3149–3158, 2007.

[4] Indra. ANTARES Communication Standard Design Definition File. Technical note IRIS-AN-CP-TNO-610-ESA-C1, ESA, September 2013. DRL Nr: D020

[5] Norman Abramson. Spread aloha cdma data communications, 1995. WO Patent App. PCT/US1995/007,228.

[6] Yuhan Moon and Violet R. Syrotiuk. "A cooperative CDMA-based multichannel MAC protocol for mobile ad hoc networks." *Computer Communications*, 32(17):1810–1819, 2009.

[7] Marcos K. Aguilera, Robert E. Strom, Daniel C. Sturman, Mark Astley and Tushar D. Chandra. "Matching events in a content-based subscription system." In *Proceedings of the 18th Annual ACM Symposium on Principles of Distributed Computing*, PODC '99, pp. 53–61, New York, NY, USA, 1999. ACM.

[8] Davide Frey and Gruia-Catalin Roman. "Context-aware publish subscribe in mobile ad hoc networks." In *Proceedings of the 9th International Conference*

on Coordination Models and Languages, COORDINATION'07," pp. 37–55. Berlin, Heidelberg, 2007. Springer-Verlag.

[9] Ilias Leontiadis, Paolo Costa and Cecilia Mascolo. "Persistent content-based information dissemination in hybrid vehicular networks." In *Pervasive Computing and Communications, 2009. PerCom 2009. IEEE International Conference on*, pp. 1–10, March 2009.

[10] R. Meier and V. Cahill. "STEAM: event-based middleware for wireless ad hoc networks." In *Distributed Computing Systems Workshops, 2002. Proceedings 22nd International Conference on*, pp. 639–644, 2002.

[11] P. Eugster, B. Garbinato and A. Holzer. "Design and implementation of the Pervaho middleware for mobile context-aware applications." In *e-Technologies, 2008 International MCETECH Conference on*, pp. 125–135, January 2008.

[12] Antony I.T. Rowstron, Anne-Marie Kermarrec, Miguel Castro and Peter Druschel. "SCRIBE: the design of a large-scale event notification infrastructure." In *Proceedings of the Third International COST264 Workshop on Networked Group Communication, NGC '01*, pp. 30–43, London, UK, 2001. Springer-Verlag.

[13] P.R. Pietzuch and J.M. Bacon. "Hermes: a distributed event-based middleware architecture." In *Distributed Computing Systems Workshops, 2002. Proceedings 22nd International Conference on*, pp. 611–618, 2002.

[14] G. Cugola, A. Margara and M. Migliavacca. "Context-aware publish-subscribe: model, implementation, and evaluation." In *Computers and Communications, 2009. ISCC 2009. IEEE Symposium on*, pp. 875–881, July 2009.

[15] Elizabeth Royer and Charles Perkins. Multicast ad hoc on-demand distance vector (MAODV) Routing, 2000.

[16] Olga Rodionova, Daniel Delahaye, Mohammed Sbihi and Marcel Mongeau. "Aircraft trajectory prediction in North Atlantic Oceanic Airspace by Wind Networking, October 2014." *DASC 2014, 33rd Digital Avionics Systems Conference*, Colorado Springs, October 5–9.

Contributor biographies

Heather Eggins is a Visiting Professor at two UK universities, Sussex and Staffordshire, and is a senior member of Lucy Cavendish College, University of Cambridge. She is an editor of a number of books examining aspects of international higher education policy and was a Director of a learned society which is an NGO of UNESCO.

Anastasiya Yurchyshyna is a Researcher at the Institute of Service Science of the University of Geneva since 2009. Previously, she obtained three master degrees: the first in 2003 in Social Informatics from the National Taras Shevchenko University (Kiev, Ukraine), the second in 2005 in Risk-Quality-Decision from the University of Aix-Marseille III (France) and the third in 2006 in Economics and Management of Information and Knowledge from the University of Nice-Sophia Antipolis (France). Then, she carriedout her PhD research at the University of Nice-Sophia Antipolis (France) in the Department of ED STIC, and she obtained the PhD degree from this university in Computer Science in 2009 for her thesis: 'An ontology-based approach for modelling the process of conformity checking in construction'. Since 2009, she develops research in the domain of Service Science, notably centred in methods to develop cross-disciplinary services.

Michel Leonard is a Professor at the University of Geneva. His research domain is Information System and Service Science. His continuous research interests are the mix between informatics and the human activities in various sectors, in order to develop smart trans-disciplinary and trans-institutional services. Currently, he conducts researches on situational methods for services, which take into account organisational, informational and engineering aspects. He was the promoter of a Bachelor and a Master of Service Science at the University of Geneva. He is the promoter and cofounder of the Annual International Conferences on Exploring Service Science (IESS) since 2010.

Steve Kearns has worked for Transport for London for the past 15 years, including 7 years on the Central London Congestion Charging project and 5 years in Technology Delivery Group, developing a range of ITS projects. Prior to working for Transport for London, he spent 15 years as a transport planner in local authorities and community organisations in London where major projects included promoting the new Crossrail link, preparing the transport and business case for international rail station at Stratford, East London, which was used extensively during 2012 Olympic Games, and setting up transport schemes for elderly people and people with disabilities. He has a Master of Science degree in Transport Planning from University of Westminster. He is a member of the Chartered Institute of Highways and Transportation and was also a member of the European Union's ITS Urban Experts Group.

Marion Berbineau was born on September 18, 1962, in Toulouse, France. She received the Engineering degree in Electronics, Automatic and Metrology from Polytech'Lille (France) and the PhD degree in Electronics from the University of Lille, respectively, in 1986 and 1989. She is currently a full-time Research Director at IFSTTAR and a Deputy Director of the Components and SYStems Department that consists of 260 persons within 12 laboratories. Dr Berbineau is an expert in the field of radio wave propagation, electromagnetic field, and wireless systems for telecommunications and localization (particularly in the rail and ITS domains). She is involved in several EU and national projects as coordinator or participant. She is active as an expert for the GSM-R and future systems like LTE-R or 5G. She is the author and coauthor of several publications and patents. She is an IEEE member affiliated to Vehicular Technology Society and Communication Society.

Mohamed Kassab received the Engineering degree in Computer Sciences from ENSI (Tunisia), the Master's degree in Computer Networks from ENSI (Tunisia), and the PhD degree in Computer Sciences from Telecom Bretagne (France), respectively, in 2003, 2004 and 2008. From 2009 to 2013, he was a Postdoctoral Fellow at IFSTTAR (France). Since January 2014, he joined the University of Monastir as an Assistant Professor. His research interests include wireless network architecture, QoS management and mobility management in wireless networks, M2M communication, and software-defined networks. He is particularly interested in the study of these issues in transportation contexts.

Christophe Gransart is a Senior Researcher at IFSTTAR. He has had 15 years of experience in coordinating industrial and academic research projects dealing with distributed systems and middleware for transportation systems, V2V and V2I communications, and adaptive middleware. He was involved in various national and European projects. His main competencies are computer science, distributed architecture design, middleware expertise, safety and security, and ontologies.

Martine Wahl received the Engineering degree in Electronics in 1991 from the Polytech'Paris Sud (IFIPS) and the PhD degree from the Grenoble Institute of Technology (France) in 1997. She was an ATER teacher at the ENSIMAG Engineering School from 1997 to 1998. Since September 1998, she joined INRETS (IFSTTAR) as a full-time permanent research scientist. Her research interests include lidar sensors, fieldbusses, and wireless sensor networks for onboard and V2V systems.

Juliette Marais received the Engineering degree from ISEN Lille (France) and the PhD degree from University of Lille 1, respectively, in 1998 and 2002. Since 2002, she is a full-time researcher at IFSTTAR. Her research interests concern GNSS performance analysis and enhancement in transport environment, in particular focused on multipath effects mitigation and safety assessment. She is involved in national and European projects on GNSS-based ITS. She is the author and coauthor of several journals and conference papers.

Divitha Seetharamdoo was born in Mauritius Island, on September 16, 1978. She received the MSc degree in Telecommunication Engineering and PhD from Paris-East University and the University of Rennes I, respectively. From 2006 to 2007, she was a Postdoctoral Fellow at the Ecole Polytechnique Fédérale de Lausanne. Since January 2008, she joined INRETS (IFSTTAR) as a full-time permanent research scientist. Her research interests include antenna design, metamaterials, and their application in electromagnetic compatibility. She is the author and coauthor of several journals and conference papers.

Corinne Blanquart is a Senior Researcher at IFSTTAR (French Institute of Science and Technology for Transport, Development and Networks). She is heading the SPLOTT research unit (productive systems, logistics, transport organisation and work), and she is conducting researches on freight transport and logistics. More specifically, her areas of research are retailers' logistics in France and Germany. She received a PhD degree in Economics from Lille University in 1998 and her habilitation (professonial thesis) in 2009. She teaches courses in transport policies, rail transport and analysis of freight demand. She is responsible of the module 'actors and institutions' in the Post-Master professional certificate 'rail and transport systems'. She was a visiting scholar at the DLR in Berlin in 2011–2012.

Thomas Zéroual is an Economist and a Research Professor working in an International Business School in Paris. He teaches supply chain management and sustainable development in multiple institutions such as Institute of Political Sciences (IEP) and USTL (French University). From 2005 to 2008, he worked at the French Transport Institute (IFSTTAR). His main research areas are service, innovation and sustainability measurement in the retail sector. His most up-to-date interest is on sustainable transport policy and its impacts on three particular areas such as local, national and European.

Ousmane Diallo is a Lecturer in the Department of Informatics at the University of Assane Seck of Ziguinchor, Senegal. He received his PhD degree in Computer Science and Engineering from the University of Beira Interior. He is also a researcher, a member at the Instituto de Telecomunicações, Portugal, and a member of the NetGNA Research Group (http://netgna.it.ubi.pt). His current research interests are distributed databases for wireless sensor networks, real-time databases, wireless sensor networks, performance evaluation and Petri Nets.

Joel J.P.C. Rodrigues is a Professor in the Department of Informatics at the University of Beira Interior, Covilhá, Portugal, and a Senior Researcher at the Instituto de Telecomunicações, Portugal. He is the leader of NetGNA Research Group (http://netgna.it.ubi.pt), the Chair of the IEEE ComSoc Technical

Committee on eHealth. He has been general chair and TPC Chair of many international conferences, including IEEE ICC and GLOBECOM. He has authored or coauthored over 400 papers in refereed international journals and conferences, a book, and three patents. He had been awarded several Outstanding Leadership and Outstanding Service Awards by IEEE Communications Society and several best papers awards. He is a licensed professional engineer (as senior member), a member of the Internet Society, an IARIA fellow, and a senior member of ACM and IEEE.

Mbaye Sene received the PhD degree in Computer Science from the University of Paris Dauphine, France, in 2002, and a Master's degree of Management of Organisations in the same university in 2003. Currently, he is an Associate Professor at the University Cheikh Anta Diop, Senegal. His main research interests include performance evaluation of distributed systems, mobile communication networks and web service systems.

Chung-Ming Huang is a Distinguished Professor in the Department of Computer Science and Information Engineering, National Cheng Kung University, Taiwan. His research interests include wireless and mobile network protocol design and analysis, green computing and communication, media processing and streaming, and innovative network applications and services. More information about him and his lab can be found at http://www.mmnetlab.csie.ncku.edu.tw.

Shih-Yang Lin is an Outstanding Researcher. He got his PhD degree in Computer Science and Information Engineering, National Cheng Kung University, Taiwan, ROC. His research interests include wireless communication, vehicular network and vehicular applications. He has published eight journal papers more than eighteen conference papers and three patents.

Alexey Vinel is a Professor of Data Communications at the School of Information Technology, Halmstad University, Sweden. His research interests include wireless networking, vehicular ad-hoc networks and intelligent transportation systems. More information about him can be found at https://scholar.google.com/citations?user=H4WCOr8AAAAJ.

Magnus Jonsson is a Professor at Halmstad University, Sweden, and he is also Vice Dean and Director of Research at the School of Information Technology. He has published more than hundred scientific papers and book chapters in the areas of real-time communication, real-time and embedded computer systems, industrial communication, wireless and vehicular communication, and optical interconnection networks.

Mike Hayward is a Consulting Engineer based in the UK specialising in Intelligent Transport Systems. He is retained by several leading organisations including the World Bank, URS Corporation and TRL as a senior technical adviser on Electronic Toll Collection systems. He has over 30 years of experience of deploying intelligent transport systems, and during his career he has held senior positions in several engineering consultancies and completed projects for many well-known clients in the Highways, Automotive, Airports and Transit sectors. He is a chartered member of the Institution of Engineering and Technology, past chairman of the IET Automotive and Road Transport profession network, and a fellow of the Institute of Highways and Transportation.

Chi-Sheng Lin received BE and ME degrees from the Department of Electrical Engineering at NCNU in 2004 and 2006, respectively. He was in the obligatory military service from 2006 to 2007. In 2013, he received his PhD degree from NCU. His research interest is broadly in ITS, wireless communication, mobility management and resource allocation. Now, he is a Senior Engineer in Institute for Information Industry, Taiwan.

Jia-Chin Lin is a Distinguished Professor in the Department of Communication Engineering, National Central University, Taiwan and a visiting fellow in the Department of Electrical Engineering, Princeton University, USA. His research interests include wireless transmission technologies, signal processing for communications and signal synchronization techniques.

Mikael Lind is with the University of Borås, Viktoria Institute, and Linköping University, Sweden. He is the Director of the Informatics Department and the Founder of the InnovationLab at the School of Business and Informatics in Borås. He is also the Cofounder of the Swedish GSI (Graduate School of Informatics), and he is active in different international communities such as Language/action and Pragmatic Web.

Anders Brödje is a Senior Coordinator of Research and Innovation at the Swedish Maritime Administration, where he is actively involved in several research and development projects, among others the pan-European Accseas, MonaLisa and MonaLisa 2.0, where he will also serve as Co-Project Leader. Parallel to this, he is a Part-time Researcher at Chalmers University of Technology, Sweden. His main area of research is Human Factors – aspects of VTS, focusing primarily on training of VTS operators.

Sandra Haraldson is a Senior Researcher in Sustainable Business at Viktoria Swedish ICT HQ, Göteborg, Sweden. Her projects include 'Future Airports' and 'Mona Lisa 2.0 – Sea Traffic Management'.

Mikael Hägg is a Head of Division of Maritime Operations at Chalmers University of Technology, Gothenburg, Sweden.

Richard Watson is a Regents Professor and the J. Rex Fuqua Distinguished Chair for Internet Strategy in the Department of MIS at the University of Georgia's Terry College of Business. He is also the Research Director for the Advanced Practices Council of the Society for Information Management. He is a Visiting Researcher at Viktoria Swedish ICT in Göteborg and Co-international Coordinator for the Addis Ababa University. He received his PhD degree in IS. He is a Fellow of the Association for Information Systems and of the Royal Society of Arts. He has published over 160 articles and written or edited more than 10 books in ecological sustainability, energy informatics and information systems leadership.

Cristian De Mitri is the CEO of Eggup.net. He is a Biomedical Engineer and he was an Ex-Navy Cadet at Navy School 'Francesco Morosini' of Venice. Before he started his company, he spent six years in a multinational company as System Integrator involved in the design of Combat Management System for the Navy.

Raffaele Loscialpo received specialisation in Microprocessors (Electronic Engineering) through a digital project about a robotic video tracking system. Since then, he spent all the 18 years of experience in multinational companies (all of them starting with A: Alstom, Altran, Abbott), first as Project Designer, then Project Manager, Product Manager, Business Development Manager, Business Unit Director and Operations Director. He started a company (innovation consulting start-up in Bologna) across industries (automotive, automation, biomedical, electronics) that became the most profitable in Italy. He merges long experience in people management and coaching with business acumen and entrepreneurial spirit. He recently founded Qhaosing®, a cloud community inspiring on Change, taken from the homonymous book 'Qhaosing®'.

Domenico D. Bloisi is an Assistant Professor at Sapienza University of Rome, Italy. He received his PhD degree in February 2010 from 'La Sapienza'. From April 2009 to March 2010, he worked as an Associate Researcher with the VIPS lab, University of Verona, Italy. His main research interests are related to intelligent surveillance (including object detection, visual tracking and multiple sensor data fusion).

Luca Iocchi is an Associate Professor at Sapienza University of Rome, Italy. His main research interests include cognitive robotics, action planning, multi-robot coordination, robot perception, robot learning and sensor data fusion. He is author of more than hundred refereed papers in journals and conferences in artificial intelligence and robotics, member of the program committee of several related conferences (IJCAI, AAAI, AAMAS, ICRA and IROS), guest editor for journal special issues and reviewer for many journals in the field.

Daniele Nardi is a Full Professor at Sapienza University of Rome, Italy. His current research interests are in the field of artificial intelligence and cognitive robotics. He is a Trustee of RoboCup and past president of the RoboCup Federation. From 2005 to 2008, he also served as the Cochair of IEEE Technical Committee on Safety, Security and Rescue Robotics. He is active in the context of autonomous rescue robots and successfully coordinated a team participating on several editions of RoboCup Rescue since the year 2002.

Michele Fiorini is a Chairman of the IET Italy Network, member of the e-Navigation technical working group of the International Association of Marine Aids to Navigation and Lighthouse Authorities, and Project Engineering Manager for the 'Automatic National System of Radar Control for Maritime Areas of Poland' project realised by Selex ES. He served as Expert Evaluator for the Baltic Sea research program BONUS (science for a better future of the Baltic Sea region), and was Session Chairman for Clean Mobility and Intelligent Transport Systems at the 2013 Euro-Asia Economic Forum (EAEF-2013), Xi'an, China.

Pier Francesco Sammartino is a Scientific Project Officer with the Maritime Affairs Unit of Joint Research Centre of the European Commission, in the field of maritime awareness, surveillance and anomaly detection. Prior to this position, he was an External Contractor at Selex-ES in the field of UAVs and airborne platforms (2012–2014), a Post-Doc with the JRC in the field of radar antenna arrays (2009–2012) and a Radar Research Analyst with CEA technologies in Canberra, Australia (2008–2009). He holds a 5-year degree in Telecommunication Engineering from the University of Rome, 'La Sapienza' (2004) and a PhD degree in Electronic and Electrical Engineering from the University College London, UK (2009).

Michele Vespe is a Scientific Officer and Project Leader at the European Commission Joint Research Centre (JRC), where he works on maritime surveillance sensors, data fusion and maritime situational awareness. Between 2011 and 2013, he was a Senior Scientist at the NATO Centre for Maritime Research and Experimentation (CMRE), working on traffic knowledge discovery and anomaly detection. Until 2011, he was a Post-Doc Researcher at the JRC developing Synthetic Aperture Radar based maritime applications. Prior to this, he spent two years in the private sector as project engineer in the fields of remote sensing and data fusion. He holds a Engineering degree in Telecommunications from University of Florence (2003) and a PhD degree in Radar Signal Processing from University College London (2006).

Dario Tarchi is with the Maritime Affairs Unit at JRC where he is responsible for scientific activities concerning the development of innovative sensors and technologies for maritime surveillance. His main research interests concern the application of interferometric radar techniques and the development and testing of novel sensors and platforms for maritime surveillance, such as MIMO and passive radar systems, UAV platforms and radiolocation techniques.

Franco Oliveri received the 'Laurea degree in Electronic Engineering (system design and control system design)' in 1983 from the University of Genoa. Until 1997, he worked for the military industry, dealing with design of computers and communication systems. Subsequently, he was hired as a R&D Director in Malaysia by an important supplier of telecommunication systems. Eventually, he was a Vice President–Sales for Selex Communications. In 2006, he joined the JRC, in Ispra, where he worked on Secure Communication Systems, Software Defined Radio and Cognitive Radio Systems as well as on MIMO Radar. For the past three years, he has been the JRC Project Leader for the development of the Common Information Sharing Environment (CISE) for the EU maritime domain.

Francesco Papi received the Laurea degree in Control Engineering in 2007 and the PhD degree in Computer Science and Control Engineering in 2011, both from the University of Firenze, Italy. In January 2011, he joined the Sensors Department at Thales Nederland B.V., Hengelo, the Netherlands. From January 2013 to April 2014, he was a Research Fellow with the JRC, Ispra, Italy. In May 2014, he joined the School of Electrical Engineering and Computing at Curtin University, Perth, Australia. His research interests include linear and nonlinear estimation, multi-target multi-sensor tracking and data fusion.

Francesco Borghese received his Master's degree in Electronic Engineering from the University of Rome 'La Sapienza' in 1999. Since then, he has always worked in the communications industry, designing digital radio devices and systems to be operated in the maritime environment. He currently works for Elman Srl as responsible for all the AIS-related projects.

Giuseppe Aulicino is a Head of Radio-communications and Electronics Office at the Italian Coast Guard (www.guardiacostiera.it).

Antonio Vollero is a Lieutenant junior grade employed in the Research and Development Department of ITCG Headquarters since 2006. He performs his duty for the management of Italian AIS network and in the development of user requirements for the MARES project, the Mediterranean AIS Regional Server. He graduated in 2004 in Telecommunications Engineering and has a PhD awarded by Faculty of Engineering of the Roma 3 University as a result of a thesis with the title 'Distributed architecture for command and control platform based on Automatic Identification System.'

Nicole Schill is a marketing communications professional with over five years of experience in both technical writing and content marketing strategy. She is the Marketing Communications Manager at exactEarth adept at the strategic development and tactical execution of marketing campaigns and generating brand awareness globally. Nicole is a published freelance journalist and holds a BA from the University of Toronto.

Margaret Browning is an internationally recognized expert in AIS and ship-borne voyage data recoding technologies and has over 15 years of experience in maritime technology. She is also deeply involved in defining and evolving communications standards through various regulatory and technical bodies such as IMO, IALA, CIRM, RTCM and IEC. She holds a BS degree in Computer Engineering from Auburn University.

Jan-Hendrik Oltmann received a Dipl.-Ing. degree in Electrical Engineering from Darmstadt Technical University in 1992. In 1993, he joined the German Federal Waterways and Shipping Administration. Since then, he continuously contributed in a leading position to innovative international projects such as the Automatic Identification System (AIS), the technical architecture for e-Navigation, a terrestrial component of maritime radio navigation (Ranging Mode or R-Mode), the application of graph theory to the maritime domain (Route Topology Modelling) and to the domain of Vessel Traffic Services (VTS) at large. Besides these professional interests, he closely monitored the international 'green' movement since the 1980s and thereby encountered the topical question of his paper in this book: How 'green' is e-Navigation?

Yung-Ho Yu graduated from the Korea Maritime and Ocean University in 1974, after that he worked as a chief engineer on a commercial ship and a technical surveyor of KR (Korean Register of Ship Classification Society). After receiving the PhD degree in Control Engineering at the same university in 1990, he started working as a Professor at the university. Recently, he is interested in international standards of ship network and interfacing between standards, e-Navigation and e-Navigation compliance equipment and system.

Proshanto K. Mukherjee LLB (Dalhousie), PhD (Wales), Juris Dr (LLD) Honoris Causa (Lund University), Barrister and Solicitor, Notary Public (Ontario, Canada), FNI, AFRIN, Master Mariner. He is a Professor of Maritime Law, Dalian Maritime University; former Professor of Maritime Law and Director of Postgraduate Maritime Law, Lund University; former Vice President (Research), and now Professor Emeritus, World Maritime University; former Senior Deputy Director and Professor of Maritime Law, International Maritime Law Institute. He is a Honorary Research Fellow at Swansea University, Visiting Professor at Chung Ang University (Seoul, South Korea) and National University of Juridical Sciences (Kolkata, India), and has taught at the Hague Academy of International Law. He served 16 years at sea. He is author of Maritime Legislation and lead author of Farthing on International Shipping, Fifth Edition, and has published extensively on virtually all subjects in maritime law.

Jingjing Xu is a Professor of Maritime Law and Economics and also the Associate Dean for Research at the Faculty of Business, Plymouth University. She has published extensively on issues relating to the law, policy and management of maritime transport and played a lead role, or acted as an Expert Advisor, in a number of international maritime research projects. She is a Fellow of the Royal Institute of Navigation, and in addition to her role at Plymouth University, she is a Visiting Professor at World Maritime University, Lund University in Sweden, Shanghai Maritime University in China, and an Adjunct Professor at Australia Maritime College.

Joanna Ejdys is currently an Associate Professor at Faculty of Management, Bialystok University of Technology. She is a Vice Dean for Research and Development and an Editor of *Economics and Management Journal*. Her research interests include quality management, foresight studies, strategic management and technology management. She has also published more than 30 papers on foresight studies concerning regional and national science and technology innovation policy.

Joanicjusz Nazarko is a Full Professor at the Bialystok University of Technology, Poland. He serves as a Dean of the Faculty of Management and the Head of the Department of Business Informatics and Logistics. He is an experienced research leader and a recognised expert in the fields of foresight, forecasting, technology assessment, logistics and benchmarking. He has published over 200 publications and a number of expert assessments, projects and technical and economic elaborations.

Łukasz Nazarko is a Research and Teaching Assistant at the Faculty of Management, Bialystok University of Technology, Poland. In his academic activities, he focuses on foresight, innovation policy and technology assessment. He has taken part in a number of foresight initiatives on national, regional and sectoral levels. He has conducted research on foresight and innovation in Poland, Hungary, the UK and Japan.

Katarzyna Halicka, since 2008, is an Assistant Professor at the Bialystok University of Technology, at the Faculty of Management in the Department of Business Informatics and Logistics. She is also the Deputy Head of the Department of Business Informatics and Logistics at the Bialystok University of Technology, at the Faculty of Management. Her recent research interests are focused on foresight studies, future studies, forecasting and technology management.

Fabien Garcia is an Associate Professor at ENAC (Ecole Nationale de l'Aviation Civile – French Civil Aviation University). He received his PhD degree in 2003 from INSA (Institut National des Sciences Appliquèes), Toulouse. He then joined ENAC. He is teaching networking techniques at ENAC. His research activity is focused mainly on aeronautical networks, UAV communications and Network Quality of Service.

Alain Pirovano is a Professor at ENAC (Ecole Nationale de l'Aviation Civile – French Civil Aviation University) and the Head of the Communication Networks Research Group. He received his Engineering degree in Computer Sciences in 1994 and his PhD degree in 2001 from Telecom Paristech. He then joined ENAC. He is teaching networking techniques and advanced architectures. His research activity is focused mainly on aeronautical networks, satellite broadband systems and TCP/IP networks. In this domain, ENAC is involved in several cooperative projects within Aerospace Valley, a regional organization gathering academic and industry research departments, and SESAR (Single European Sky ATM Research) program.

Mickaël Royer is a PhD student at ENAC (Ecole Nationale de l'Aviation Civile – French Civil Aviation University). He is teaching networking techniques at ENAC. His PhD is focused mainly on aeronautical networks, especially on content-based routing in Aeronautical Ad-hoc networks. In this domain, ENAC is involved in several cooperative projects within Aerospace Valley, a regional organization gathering academic and industry research departments, and SESAR (Single European Sky ATM Research) program.

Quentin Vey is a PhD student at ENAC (École Nationale de l'Aviation Civile – French Civil Aviation University). His work is centred on Aeronautical Ad-hoc Networks as communication system for critical air–ground communication. He focuses more specifically on the challenges of medium access control and routing in such networks. He received a Master's degree in Computer Science and Telecommunication from the ENAC in 2013.

Index

www.ingramcontent.com/pod-product-compliance
Lightning Source LLC
LaVergne TN
LVHW011901200926
848484LV00002B/501